Handbook of Photoenergy

Volume I

Handbook of Photoenergy
Volume I

Edited by **Anderson Sheilds**

NY RESEARCH PRESS
New York

Published by NY Research Press,
23 West, 55th Street, Suite 816,
New York, NY 10019, USA
www.nyresearchpress.com

Handbook of Photoenergy: Volume I
Edited by Anderson Sheilds

International Standard Book Number: 978-1-63238-263-4 (Hardback)

Printed in China.

Contents

Preface

The element of light is one of the most fundamental ones in the universe. Even darkness is defined as the absence of light. Since it is found in such abundance, rather, since it prevails over every aspect of the universe it would be prudent to utilize it for our various purposes, be it scientific or not. Photoenergy can then be understood as the utilization of light as energy or fuel. Photoenergy is such a general term that one can say it covers a vast variety of related fields that concern itself with the manipulation of light. These can be photochemistry and photobiology, both of which study the effect of light on the various foundational elements of their field. One of the major applications of photoenergy can be seen in the functioning of solar cells and solar energy. Solar cells convert photoenergy or solar energy into electricity. One of the major advantages of such a source of energy is that it is a renewable source and for the foreseeable far future doesn't have any indication or chance of running out. This places it at the topmost tier when looking at alternative sources of energy to replace fossil fuels. Innovations in making better and more efficient technology to utilize photoenergy are thus the demand of the market in today's world.

This book is an attempt to compile and collate all available research on photoenergy. I am grateful to those who put their hard work, effort and expertise into these research projects as well as those who were supportive in this endeavour. I especially wish to thank my friends and family who have supported me at every step.

Editor

Environmental Life-Cycle Analysis of Hybrid Solar Photovoltaic/Thermal Systems for Use in Hong Kong

Tin-Tai Chow[1] and Jie Ji[2]

[1] *BEETRU, Division of Building Science and Technology, City University of Hong Kong, Kowloon, Hong Kong*
[2] *Department of Thermal Science and Energy Engineering, University of Science and Technology of China, Hefei 230026, China*

Correspondence should be addressed to Tin-Tai Chow, bsttchow@cityu.edu.hk

Academic Editor: Christophe Menezo

While sheet-and-tube absorber is generally recommended for flat-plate photovoltaic/thermal (PV/T) collector design because of the simplicity and promising performance, the use of rectangular-channel absorber is also tested to be a good alternative. Before a new energy technology, like PV/T, is fully implemented, its environmental superiority over the competing options should be assessed, for instance, by evaluating its consumption levels throughout its production and service life. Although there have been a plenty of environmental life-cycle assessments on the domestic solar hot water systems and PV systems, the related works on hybrid solar PV/T systems have been very few. So far there is no reported work on the assessment of PV/T collector with channel-type absorber design. This paper reports an evaluation of the energy payback time and the greenhouse gas payback time of free-standing and building-integrated PV/T systems in Hong Kong. This is based on two case studies of PV/T collectors with modular channel-type aluminium absorbers. The results confirm the long-term environmental benefits of PV/T applications.

1. Introduction

A photovoltaic/thermal (PV/T) system is a combination of photovoltaic (PV) and solar thermal devices that generate both electricity and heat energy from one integrated system. With solar cells as (part of) the thermal absorber, the hybrid design is able to maximize the energy output from an allocated space reserved for solar application. Air and/or water can be used as the heat removal fluid(s) to lower the solar cell working temperature and to improve the electricity conversion efficiency. Comparatively, the water-type product design provides more effective cooling than the air-type counterpart because of the favorable thermal properties. Those with flat plate collectors meet well the low temperature water heating system requirements. They are also ideal for preheating purposes when hot water at higher temperature is required.

While sheet-and-tube absorber is one common feature in flat-plate collectors, the use of rectangular-channel absorbers also has been examined extensively [1–3]. An aluminum water-in-channel-type PV/T collector design is recommended by the authors, with the prototypes well-tested under both free-standing and building-integrated manners [4, 5]. Through the adoption of the channel absorber design, the potential problem of low fin efficiency can be readily improved. Based on the thermosyphon working principle, the collector performance is found to have geographical dependence and working well at the warmer climate zones. In the Asia Pacific region, most large cities are dominated by air-conditioned buildings where space cooling demands are high. In these buildings, the exposed facades provide very good opportunity for accommodating the building integrated systems, hence, the BiPV/T. When a part of the solar radiation that falls on the building façade is directly converted to useful thermal and electric power, the portion of solar energy transmitted through the external facade is reduced. Hence, the space cooling load is reduced. Through dynamic simulation with the use of experimentally validated system models and the typical meteorological year (TMY) data of Hong Kong, the cost payback time (CPBT) of free-standing and building-integrated PV/T systems were found 12.1 and 13.8 years, respectively [6, 7]. The assessments were taken, respectively, at their best tilted and vertical collector positions for maximizing their system outputs. It is expected that these CPBT will be gradually shortened as the PV technology is in progressive advancement. In this paper, the environmental life-cycle analysis (LCA) of such hybrid solar systems as applied in Hong Kong is reported.

2. Environmental Life-Cycle Analysis

LCA is a technique for assessing various aspects associated with development of a product and its potential impact throughout a product's life [8]. Before a new energy technology is fully implemented, the environmental superiority over competing options can be asserted by evaluating its consumption levels (such as cost investments, energy uses, and GHG emissions) throughout its entire production and service life. In terms of economic analysis, a simplified approach is to ignore the time element so the cost payback time (CPBT) can be used. This is by adding together the cash inflows from successive years until the cumulative cash inflow is the same as the required investment. In analogy to the economical evaluation, two environmental cost-benefit parameters, the energy payback time (EPBT) and greenhouse gas payback time (GPBT), can be used to evaluate the time period after which the real environmental benefit starts [9]. EPBT is the period that a system has to be in operation in order to save the amount of primary energy that has been spent for production, operation, and maintenance of the system. It is the ratio of embodied energy to annual net energy output. In a BiPV/T system, for example,

$$\text{EPBT} = \frac{\sum_{\text{pvt}} + \sum_{\text{bos}} - \sum_{\text{mtl}}}{E_{\text{pv}} + E_t + E_{\text{ac}} - E_{\text{om}}}, \quad (1)$$

where $\sum_{\text{pvt}}$, $\sum_{\text{bos}}$ and $\sum_{\text{mtl}}$, are, respectively, the embodied energy of the PV/T collectors, of the balance of system (BOS), and of the replaced building materials; E_{pv} is the annual useful electricity output, E_t the annual useful heat gain (equivalent), E_{ac} the annual electricity saving of the HVAC system due to the space thermal load reduction, and E_{om} is the annual electricity consumed in system operation and maintenance activities. $\sum_{\text{mtl}}$ and E_{ac} can be omitted in free-stand PV/T system evaluation. Hence,

$$\text{EPBT} = \frac{\sum_{\text{pvt}} + \sum_{\text{bos}}}{E_{\text{pv}} + E_t - E_{\text{om}}}. \quad (2)$$

Similarly, in terms of greenhouse gas (GHG) emission, for BiPV/T

$$\text{GPBT} = \frac{\Omega_{\text{pvt}} + \Omega_{\text{bos}} - \Omega_{\text{mtl}}}{Z_{\text{pv}} + Z_t + Z_{\text{ac}}}, \quad (3)$$

where Ω stands for the embodied GHG (or carbon dioxide equivalent) emission and Z the reduction of annual GHG emission from the local power plant owing to the BiPV/T operation. And for the free-stand system,

$$\text{GPBT} = \frac{\Omega_{\text{pvt}} + \Omega_{\text{bos}}}{Z_{\text{pv}} + Z_t}. \quad (4)$$

Thus EPBT and GPBT are functions of the related energy system performance and their environmental impacts, like those of the power utilities, the building systems, local and overseas manufacturing, and transportation and on-site handling of PV/T collector system as a whole.

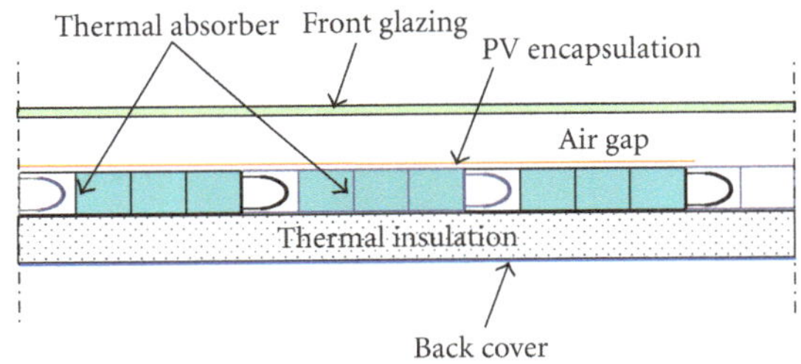

FIGURE 1: Cross-sectional view of the PV/T collector showing several absorber modules in integration (N.T.S).

FIGURE 2: Front view of free-stand PV/T collector system.

3. Aluminum Rectangular-Channel PV/T Systems

The sectional view of an aluminum rectangular-channel PV/T collector developed by the authors is shown in Figure 1. It is composed of the following layers: (i) front low-iron glass cover, (ii) crystalline silicon (c-Si) PV encapsulation, (iii) metallic thermal absorber constructed from extruded aluminum, (iv) thermal insulation layer with glass wool, and (v) back-cover steel sheet. The PV encapsulation includes TPT (tedlar-polyester-tedlar) and EVA (ethylene-vinyl acetate) layers at both sides of the solar cells. The rectangular-channel design strengthens the heat transfer and structural durability.

In a free-stand thermosyphon system, the PV/T collector carries a water tank with the natural water circulation via inter-connecting pipes. Figure 2 shows the external view. Water enters the collector at the lower header and leaves via the upper header. Table 1 lists the technical data of this PV/T collector for free-stand applications.

FIGURE 3: Front view of BiPV/T system with water tank at top of wall.

A BiPV/T system, on the other hand, is composed of an array of PV/T collectors that are integrated to the external wall of an air-conditioned building. See Figure 3 for reference. The water tank is located at the roof-top and the water circulation is again by means of thermosyphon. Table 2 lists the technical data of the BiPV/T wall system in our study.

4. Review of Previous Works on Flat Plate Collector Systems

4.1. Solar Hot Water Systems. The LCA works on domestic solar hot water (DSHW) systems in majority were from EU countries [10–13]. Streicher et al. [10] evaluated the EPBT of solar thermal systems by dividing the system into components. The cumulative energy demand was obtained by multiplying the weight of the main components with their respective cumulative energy demand values. They estimated that in Germany the DSHW systems have EPBT from 1.3 to 2.3 years. In their study, construction credit was given to the collector system in integrated roof-mounting mode. This is for the savings in building materials, transportation, and construction works. The collector itself accounts for 89% and 85% of the total embodied energy in the roof-integrated and open-stand systems, respectively. Tsilingiridis et al. [11] found that in Greece the materials used, including steel and copper, have the major contribution to the environmental impacts. Ardente et al. [12] found that in Italy the indirect emissions (related to production of raw materials) are about 80–90% of the overall GHG releases. Kalogirou [13] worked on a thermosyphon DSHW system in Cyprus. The system thermal performance was evaluated by dynamic simulation program. The LCA determined that 77% of the embodied energy goes to the collector panels, 15% goes to the steel frame, 5% goes to piping, and the remaining accounts for less than 3% of the total. Considerable amounts of GHG can be saved. The EPBT was estimated around 1.1 year.

Outside Europe, the study of Crawford et al. [14] in Australia showed that although the CPBT of DSHW systems can be 10 years or more, the corresponding GPBT can be only around 2.5–5 years. In their study, a conversion factor of 60 kg CO_2 eq/GJ was used to determine the GHG emission from the cumulative energy of the entire system. Arif [15] evaluated the environmental performance of DSHW systems in India. Based on the 100 litre-per-day and steady year-round usage, the EPBT was estimated 1.6–2.6 years, all depending on the local climates and also the collector materials in use. In the LCA work of Hang et al. [16] on a range of solar hot water systems in USA; dynamic thermal simulation was again applied.

TABLE 1: Collector and technical design data of free-stand PV/T system.

Design parameters	Data
Glazing (low-iron glass)	
Thickness	0.004 m
Emissivity	0.88
Extinction coefficient	26/m
Refraction index	1.526
Depth of air gap underneath	0.025 m
PV encapsulation (TPT + EVA + solar cell + EVA + TPT + silicon gel)	
Solar cell type	single-crystalline silicon
Cell area	1.11 m^2
Cell electrical efficiency at STC	13%
Solar cell temperature coefficient	0.005/K
Emissivity	0.8
Absorptivity	0.8
Packing factor (wrt glazing)	63%
Thermal absorber (Aluminum)	
No. of flat-box absorber module	15
Absorber module size	0.105 × 1.38 × 0.012 m
No. of header	2
Header size	1.575 × 0.025 (dia.) × 0.002 (thick) m
Thermal insulation layer (glass wool)	
Thickness	0.03 m
Back cover (galvanized iron)	
Thickness	0.001 m
Water tank and connecting pipes	
Water storage capacity	155 kg
Tank length	1.2 m
Tank diameter	0.21 m
Pipe diameter	0.015 m
Thickness of insulation layer at tank	0.025 m
Thickness of insulation layer on pipe	0.02 m

4.2. PV Systems. In the last decades, plenty of works have been reported on life cycle performance of PV systems in both free-stand and building-integrated manners. The

Table 2: Collector and technical design data of BiPV/T system.

Design parameters	Data
Front glazing (low-iron glass)	
Thickness	0.004 m
Surface area	1.61 m^2
Depth of air gap underneath	0.025 m
PV encapsulation (TPT + EVA + solar cell + EVA + TPT + silicon gel)	
Solar cell type	single-crystalline silicon
Cell area	0.81 m^2
Cell electrical efficiency at STC	13%
Solar cell temperature coefficient	0.005/K
Emissivity	0.8
Absorptivity	0.8
Packing factor (wrt glazing)	50%
Thermal absorber (aluminum alloy)	
Thermal capacity	903 kJ/(kg·K)
Density	2702 kg/m^3
Thermal conductivity	237 W/(m·K)
Emissivity	0.8
Absorptivity	0.9
Insulation material (glass wool)	
Thickness	0.03 m
Air gap between insulation layer and building wall	0.02 m
Building wall (brick)	
Thickness	0.15 m
Density	1600 kg/m^3
Thermal capacity	880 J/(kg·K)
Thermal conductivity	1.0 W/(m·K)
Water tank (steel) and connecting pipes (copper)	
Water storage capacity	0.46 m^3
Tank length	1.5 m
Tank diameter	0.54 m
Pipe diameter	0.055 m
Thickness of insulation layer at tank	0.025 m
Thickness of insulation layer on pipe	0.02 m

estimations of EPBT and GPBT have been kept on revising owing to the advancements in PV technology.

The production of a PV module includes the following processes:

(i) silicon purification and processing,

(ii) silicon ingot slicing, and

(iii) PV module fabrication.

Silica is first melted and manufactured into metallurgical-grade silicon (MG-Si), then into electronic silicon (EG-Si) through the Siemen's process or into solar-grade silicon (SoG-Si) through the modified Siemens process [17]. Finally, after the Czochralski process (for sc-Si) or other production process, silicon is made available for the solar cell production. The silicon ingot is needed to be sliced into wafer. The technologies of cell production include etching, doping, screen printing, and coating. The solar cells are then tested, packed, and interconnected with other components to form PV modules.

Alsema [18] studied the EPBT and the GHG emissions of grid-connected PV systems. The cumulative energy demands of sc-Si and mc-Si frameless modules were evaluated as 5700 and 4200 MJ/m^2. Further, it was pointed out that with the implementation of new manufacturing technologies, the above data could be as low as 3200 and 2600 MJ/m^2. Later on, Alsema et al. [19, 20] reviewed the important options that were available for further reduce energy consumption and environment impacts of the PV module production processes. As for BOS, Alsema and Nieuwlaar [21] presented that because of the less use of aluminum in supporting structure, the energy requirement for array support of ground-mounted PV system was about 1800 MJ/m^2, but this could be only 700 MJ/m^2 for rooftop installation; hence rooftop systems should have better potentials for EPBT reduction than ground-mounted systems.

Mason et al. [22] studied the energy contents of the BOS components used in a 3.5 MWp mc-Si PV plant. By integrating the weight of the PV modules with the supports, the embodied energy of the BOS components was found as low as 542 MJ/m^2—a sharp reduction from the previous estimations. Fthenakis and Kim [23] showed that in Japan the primary energy demand for sc-Si PV module was in the range of 4160–15520 MJ/m^2, and the life-cycle GHG emissions rate for PV systems in the United States were from 22 to 49 g CO_2-eq/kWh$_e$.

In Singapore, Kannan et al. studied a 2.7 kWp distributed PV system with sc-Si modules [24]. Specific energy consumptions for the PV modules and the inverters were estimated 16 and 0.17 MWh$_e$/kWp respectively. The manufacturing of solar PV modules accounted for 81% of the life cycle energy use. The aluminium supporting structure accounted for about 10%, and the recycling of aluminium accounted for another 7%. The EPBT was estimated to be 6.74 years. It was claimed that this can be reduced to 3.5 years if the primary energy use on PV module production is reduced by 50%.

In India, Nawaz and Tiwari [25] calculated EPBT by evaluating the energy requirement for manufacturing a sc-Si PV system for open field and rooftop conditions with BOS. Mitigation of CO_2 emissions at macrolevel (where lifetime of battery and PV system are the same) and microlevel of the PV system has also been studied. For a 1 m^2 sc-Si PV system, their estimations give an embodied energy of 666 kWh for silicon purification and processing, 120 kWh for cell fabrication, and 190 kWh for subsequent PV module production. Hence without BOS, the embodied energy was estimated 976 kWh/m^2 and the GHG emission was 27.23 kg/m^2.

In Hong Kong Lu and Yang [26] investigated the EPBT and GPBT of a roof-mounted 22 kW BiPV system. It was found that 71% of the embodied energy on the whole is from the embodied energy of the PV modules, whereas the remaining 29% is from the embodied energy of BOS. The

EPBT of the PV system was then calculated as 7.3 years. Considering the fuel mixture composition of local power stations, the corresponding GPBT is 5.2 years. Further, it was predicted that the possible range of EPBT of BiPV installations in Hong Kong is from 7.1 years (for optimal orientation) to 20 years (for west-facing vertical façade).

Bankier and Gale [27] gave a review of EPBT of roof mounted PV systems reported in the 10-year period (1996–2005). A large range of discrepancy was found. They pointed out that the limitations to the accuracy of the assessments came from the difficulties in determining realistic energy conversion factors, and in determining realistic energy values for human labor. According to their estimation, the appropriate range of EPBT for mc-Si PV module installations should be between 2–8 years. A more recent review was done by Sherwani et al. [28]. The EPBT for sc-Si, mc-Si, and a-Si PV systems have been estimated in the ranges of 3.2–15.5, 1.5–5.7, and 2.5–3.2, years, respectively. Similarly, GHG emissions are 44–280, 9.4–104, and 15.6–50 g CO_2-eq/kWh.

4.3. PV/T Systems. While there have been plenty studies of EPBT and GPBT on solar thermal and PV systems, our literature review shows that those on PV/T systems have been very few. In particular, there is so far no reported work on the assessment of PV/T collectors with channel-type absorber design.

Battisti and Corrado [29] made evaluation based on a conventional mc-Si building-integrated system located in Rome, Italy. An experimental PV/T system with heat recovery for DSHW application was examined. Evaluations were made for alternative heat recovery to replace either natural gas or electricity. Their results give the EPBT and GPBT of PV system as 3.3 and 4.1 years. On the other hand, those of the PV/T systems designed for natural gas replacement are 2.3 and 2.4 years.

Also in Italy, Tripanagnostopoulos et al. [30] evaluated the energy and environmental performance of their modified 3 kWp mc-Si PV and experimental water-cooled PV/T sheet-and-tube collector systems designed for horizontal-roof (free-stand) and tilted-roof (building integrated) installations. The application advantage of the glazed/unglazed PV/T over the PV options was demonstrated through the better LCA performances. The EPBT of the PV and BiPV system were found to be 2.9 and 3.2 years, whereas the GPBT were 2.7 and 3.1 years, respectively. For PV/T system with 35°C operating temperature, the EPBT of the PV/T and BiPV/T options were both 1.6 years, and the GPBT were 1.9 and 2.0 years respectively. The study showed that nearly the whole of the environmental impacts are due to PV module production, aluminium parts (reflectors and heat-recovery-unit) as well as copper parts (for heat-recovery-unit and hydraulic circuit), with barely significant contributions from the other system components, such as support structures or electrical/electronic devices. The disposal phase contribution is again almost negligible.

Dubey and Tiwari [31] carried out an environmental impact analysis of a hybrid PV/T solar water heater for use in the Delhi climate of India. With a glazed sheet-and-tube flat plate collector system designed for pump operation, the EPBT was found 1.3 years.

5. Environmental Analysis of Aluminum Rectangular-Channel PV/T Systems

5.1. EPBT of Free-Stand System. Skillful lamination of solar cell onto thermal absorber with layers of EVA and TPT is needed for PV/T collector production. Aluminum thermal absorber parts are made available by raw material mining and extraction, ingot melting, mechanical extrusion, machining, and assembling into whole piece. The major-component production and assembly processes include front glass (low iron), PV-laminated absorber, insulation material and aluminum frame. The supply was from the mainland. As for the BOS, the electrical BOS components include inverters, electrical wirings, and electronic devices. The mechanical BOS include water storage tank, pipe work, supporting structure, and accessories. The embodied energy to be considered in the LCA include the above during production, plus those related to the required transportation from factory to installation site, construction and testing, decommissioning and disposal, and any other end-of-life energy requirements.

Table 3 summarizes the materials used and cumulative energy of the free-stand PV/T collector system. The cumulative energy intensity of sc-Si PV module was estimated as 976 kWh/m^2, making references to [25, 26]. That of the inverter and electrical parts was taken as 5% of the PV module. The other values of cumulative energy intensity in MJ/unit was obtained from the Hong Kong government EMSD (Electrical and Mechanical Services Department) database that covers the specific (per unit quantity) impact profile due to consumption of materials in the "Cradle-to-As-built" stage [32]. The total cumulative energy comes up to 3041.8 kWh or 1728 kWh/m^2 for this free-stand system. Table 4 shows the distribution of the embodied energy in this case. It can be seen that the hybrid PV/T collector itself accounts for around 80% of the embodied energy. For the BOS, the water tank accounts for 11.4%, the other mechanical components accounts for 7%, whereas the electrical accessories accounts for only 1.8%. $\sum_{\text{pvt}}$ and $\sum_{\text{bos}}$ are then 2429 and 613 kWh, respectively.

With the installation of this PV/T system, two kinds of energy saving are involved: thermal energy for water heating and electrical energy. This will be no air-conditioning saving. A thermal energy saving of 2650 MJ/year and electricity saving of 473 MJ/year give an E_t of 736 kWh/year and an E_{pv} of 398 kWh/year. In the computation, a heat-to-electricity conversion factor of 0.33 has been used. Mainly labor costs were considered in E_{om}. This is estimated as 41 kWh/year and is therefore not significant. With (2), the EPBT is found 2.8 years. This is much shorter than the expected CPBT of 12.1 years reported in our previous work [6]. Assuming that the working life of PV/T system is similar to PV system, that is, 15–30 years in general [29], then it can be concluded that

Table 3: Cumulative energy in free-stand PV/T system.

Materials	Quantity consumed (kg)	Cumulative energy intensity (MJ/unit)	Cumulative energy (kWh)
PV/T collector			
Front glazing			
Low-iron glass ($1.76\,m^2$)	19.7	19.7	107.9
Thermal insulation			
Glass wool	1.69	31.7	14.9
Thermal absorber			
Aluminum absorber	18.3	219	1114.7
Frame and back cover			
Aluminum	1.78	219	108.0
PV Encapsulation			
PV Module	$1.11\,m^2$	976	1083.4
BOS			
Water tank			
Stainless steel tank	4.20	82.2	273.0
Tank insulation (Glass wool)	1.58	31.7	13.9
Aluminum Cladding	0.966	219	58.8
Connecting pipe			
Copper piping (15 mm dia.)	2.4 m	6.33	4.2
Pipe insulation (Glass wool)	0.0627	31.7	0.6
Structural support and accessories			
Steel stand	14.2	29.2	115.2
Pipe fittings and structural joints	7.19	140.0	93.3
Inverter + electric wiring	5% of PV module		54.2
		Total:	3041.8

Table 4: Distribution of embodied energy in PV/T collector systems.

System component description		Free-stand	BiPV/T
PV/T Collector	Mechanical components	44.2	51.8
	Electrical components	35.6	37.7
BOS	Water tank	11.4	4.9
	Pipe and structural supports	7.0	3.8
	Electrical components	1.8	1.9

the EPBT in this case study is an order of magnitude lower than its expected working life.

5.2. BiPV/T System. Table 5 summarizes the materials used and the cumulative energy in the $9.66\,m^2$ BiPV/T case. Accordingly, the values of Z_{pvt} and Z_{bos} are, respectively, 11258 and 1328 kWh. Z_{mtl} is estimated as 594 kWh, making reference to the work of Streicher et al. [10] and adjusted by the cost of living. Taking the advantage of building material replacement, the cumulative energy intensity reduces to $1241\,kWh/m^2$. The embodied energy distribution of this BiPV/T system is also given in Table 4. It can be seen that for this building integrated case the portion of the collector increases to 89%. For the BOS, the water tank accounts for 4.9%, the pipe and supporting components account for 3.8%, and the electrical components remain at less than 2%.

Table 5: Cumulative energy in BiPV/T system.

Materials	Quantity consumed (kg)	Cumulative energy intensity (MJ/unit)	Cumulative energy (kWh)
PV/T collector			
Front glazing			
Low-iron glass ($1.61\,m^2 \times 6$)	99.6	19.7	545.0
Thermal insulation			
Glass wool	9.50	31.7	83.7
Thermal absorber			
Aluminum absorber	86.7	219	5273.8
Frame and back cover			
Aluminum	10.1	219	611.8
PV Encapsulation			
PV Module	$4.86\,m^2$	976	4743.4
BOS			
Water tank			
Stainless steel tank	19.9	82.2	454.0
Insulation (Glass wool)	2.14	31.7	18.8
Aluminum Cladding	1.53	219	93.0
Connecting pipe			
Copper piping (55 mm dia.)	7 m	40.1	77.9
Pipe insulation (Glass wool)	1.07	31.7	9.4
Structural support and accessories			
Pipe fittings and structural parts	5.25	140.0	68.1
Inverter + electric wiring	5% of PV module		237.2
		Total	12585.2

With the installation of this BiPVW system, the annual energy savings include the following:

(i) thermal energy: 2258 kWh (E_t);

(ii) electrical energy: 323 kWh;

(iii) space cooling load: 206 kWh.

By taking the COP of air-conditioning plant as 3.0, E_{pv} and E_{ac} are then 979 and 208 kWh/year, respectively. In this case E_{om} is 246 kWh/year, by estimation. By (1), the EPBT is 3.8 years, which is much shorter than its CPBT of

13.8 years. A longer period of EPBT in this BiPV/T than in the free-stand case is mainly because of its vertical collector position as compared to the best angle of tilt, and also the differences in collector size and solar cell packing factor. A shorter EPBT is expected if ms-Si cell modules were used in the analyses because of the lower energy consumption during the manufacturing process. As a matter of fact, this 3.8 years for vertical-mounted BiPV/T is advantageous as compared to the 7.1 years [26] for an optimal-oriented roof-top BiPV system in Hong Kong.

5.3. GHG Emission Analysis. In our analysis, the thermal energy saving was taken as a save of town gas consumed in the building. The electrical energy saving was taken as a save in purchased electricity from the utilities. Based on the data provided by the Hong Kong government, the territory-wide emission factor of GHG coming from utility power generation is 0.7 kg CO_2-eq/kWh$_e$ including the transmission losses [33]. As for town gas, the emission factors for CO_2, CH_4, and N_2O are, respectively, 2.815 kg/unit, 0.0446 g/unit and 0.0099 g/unit, where 1 unit of town gas is equivalent to 48 MJ consumed. For the free-stand case, the above information gives an annual reduction in GHG emission of 285 kg CO_2-eq. The PV/T system itself does not produce polluting emissions during their daily operation. And in these days, most of the manufacturing activities of products consumed in Hong Kong are taking place in the Mainland, so the emission factor of China can be used in our embodied GHG assessment. In China, the primary energy consumption for power generation is 12.01 MJ/kWh$_e$ and the CO_2 emission rate for coal-fired power plant is 24.7 g CO_2-eq/MJ [34], the embodied GHG intensity of the PV/T collector in this case is therefore 0.297 kg CO_2-eq/kWh cumulative energy. The local emission factor was used for the BOS part since local acquisition was assumed. Accordingly, with (4) this approximation gives a GPBT of 3.2 years for the free-stand system.

Similarly, for the BiPVT system the saving in air-conditioning energy is converted as electricity saving based on a system COP (coefficient of performance) of 3.0. With (3) this gives a GPBT of 4.0 years. The result is again lower than the previously estimated GPBT of 5.2 years for the general performance of BiPV systems in Hong Kong [26].

For completeness, Table 6 shows the technical data in the evaluation of their CPBT. Comparing with the free-stand PV/T case, the BiPV/T system had a lower investment cost on unit collector area basis. This is because on one hand there were building materials saving and there was no requirement on the steel stands which is essential for tilt-mounting of the free-stand PV/T collector. On the other hand, it was benefitted by the economy of scale for mass handling of the system components. During operation, however, the vertical collector position of the BiPV/T system made it disadvantageous in the quantity of year-round solar radiation received by the collector surface. At the same time, there would be greater transmission loss for a centralized energy system. The simulation results showed that the annual useful heat gains of the free-stand and the building integrated cases are 418 kWh/m^2 and 233 kWh/m^2, respectively, on unit glazing area basis. And the electrical energy gains are 118 kWh/m^2 and 66.4 kWh/m^2 on unit PV cell area basis. These came out with the CPBT of 12.1 years for the free-stand case and 13.8 years for the building integrated case.

Table 6: Evaluation of cost payback time.

Investment: HK$	Free-stand PV/T [6]	BiPV/T [7]
Water storage tank	400	750
Collector frame and support	400	1800
Modular thermal absorber	600	2700
Solar cells and encapsulation	4000	17500
Inverter	700	1000
Piping, wiring and accessories	300	900
Installation costs	1500	3000
Total system costs (HK$)	7900	27650
Useful energy savings	MJ (kWh)	MJ (kWh)
Thermal energy	2650.4 (736.2)	8127.5 (2257.6)
Electrical energy	473.2 (131.4)	1162.4 (322.9)
Space cooling load	—	742.6 (206.3)
Cost savings: HK$		
Gaseous fuel at HK$0.2/MJ	530.1	1625.5
Electricity at HK$0.95/kWh	124.9	372.0
Annual saving	655.0	1997.5
Cost payback time (CPBT)	12.1 years	13.8 years

Note: USD1 is equivalent to HK$7.8.

Our above findings are generally in line with the estimations by other researchers based on their own collector designs and local applications. Nevertheless, it should be noted that the above picture is not static. It is expected that the continuing improvements in material and energy utilization and recycling will change the current environmental profiles. On the other hand, the progression in solar cell performance will also lead to better EPBT and GPBT.

6. Conclusion

An environmental life-cycle assessment has been done to evaluate the energy and environmental profiles of two cases of PV/T system application in Hong Kong. In both cases, aluminum rectangular-channel absorber in association with sc-Si PV encapsulation was adopted in the single-glazed flat-plate PV/T collector design. In our analysis, the cumulative energy inputs and the embodied GHG emissions were determined by established methodology and technical data making reference to reported research works as well as local government publications. The annual thermal and electrical energy outputs were from results of dynamic simulation based on the TMY dataset of Hong Kong and validated PV/T system models. Our estimation shows that the EPBT of the free-stand PV/T system at the best angle of tilt is around 2.8 years, which is an order of magnitude lower than the expected system working life. In the vertical-mounted BiPV/T case, this is 3.8 years which is again considerably better than the general performance of roof-top BiPV system in Hong Kong. The corresponding GPBT of 3.2 and 4.0 years as a result demonstrate the environmental superiority of this PV/T option over many other competing renewable energy systems.

Acknowledgment

The work described in this paper was supported by a Grant from the Research Grants Council of the Hong Kong Special Administrative Region, China (Project no. CityU112009).

References

[1] B. Sandnes and J. Rekstad, "A photovoltaic/thermal (PV/T) collector with a polymer absorber plate. Experimental study and analytical model," *Solar Energy*, vol. 72, no. 1, pp. 63–73, 2002.

[2] C. Cristofari, G. Notton, P. Poggi, and A. Louche, "Modelling and performance of a copolymer solar water heating collector," *Solar Energy*, vol. 72, no. 2, pp. 99–112, 2002.

[3] Y. Tripanagnostopoulos, T. Nousia, M. Souliotis, and P. Yianoulis, "Hybrid photovoltaic/thermal solar systems," *Solar Energy*, vol. 72, no. 3, pp. 217–234, 2002.

[4] T. T. Chow, J. Ji, and W. He, "Photovoltaic-thermal collector system for domestic application," *Journal of Solar Energy Engineering, Transactions of the ASME*, vol. 129, no. 2, pp. 205–209, 2007.

[5] T. T. Chow, W. He, A. L. S. Chan, K. F. Fong, Z. Lin, and J. Ji, "Computer modeling and experimental validation of a building-integrated photovoltaic and water heating system," *Applied Thermal Engineering*, vol. 28, no. 11-12, pp. 1356–1364, 2008.

[6] T. T. Chow, W. He, J. Ji, and A. L. S. Chan, "Performance evaluation of photovoltaic-thermosyphon system for subtropical climate application," *Solar Energy*, vol. 81, no. 1, pp. 123–130, 2007.

[7] T. T. Chow, A. L. S. Chan, K. F. Fong, Z. Lin, W. He, and J. Ji, "Annual performance of building-integrated photovoltaic/water-heating system for warm climate application," *Applied Energy*, vol. 86, no. 5, pp. 689–696, 2009.

[8] ISO, (International Organization for Standardization) 14040 Standard, Environmental Management-Life cycle Assessment-Principles and Framework, 1997.

[9] G. N. Tiwari and R. K. Mishra, *Advanced Renewable Energy Sources*, RSC Publishing, Cambridge, UK, 2012.

[10] E. Streicher, W. Heidemann, and H. Müller-Steinhagen, "Energy payback time—a key number for the assessment of thermal solar systems," in *Proceedings of EuroSun*, pp. 20–23, Freiburg, Germany, June 2004.

[11] G. Tsilingiridis, G. Martinopoulos, and N. Kyriakis, "Life cycle environmental impact of a thermosyphonic domestic solar hot water system in comparison with electrical and gas water heating," *Renewable Energy*, vol. 29, no. 8, pp. 1277–1288, 2004.

[12] F. Ardente, G. Beccali, M. Cellura, and V. Lo Brano, "Life cycle assessment of a solar thermal collector," *Renewable Energy*, vol. 30, no. 7, pp. 1031–1054, 2005.

[13] S. Kalogirou, "Thermal performance, economic and environmental life cycle analysis of thermosiphon solar water heaters," *Solar Energy*, vol. 83, no. 1, pp. 39–48, 2009.

[14] R. H. Crawford, G. J. Treloar, B. D. Ilozor, and P. E. D. Love, "Comparative greenhouse emissions analysis of domestic solar hot water systems," *Building Research and Information*, vol. 31, no. 1, pp. 34–47, 2003.

[15] M. Arif, "Life cycle analysis and carbon credit earned by solar water heating system," *International Journal of Research in Engineering and Applied Sciences*, vol. 2, no. 2, pp. 1884–1905, 2012.

[16] Y. Hang, M. Qu, and F. Zhao, "Economic and environmental life cycle analysis of solar hot water systems in the United States," *Energy and Buildings*, vol. 45, pp. 181–188, 2012.

[17] N. Jungbluth, "Life cycle assessment of crystalline photovoltaics in the Swiss ecoinvent database," *Progress in Photovoltaics: Research and Applications*, vol. 13, no. 5, pp. 429–446, 2005.

[18] E. A. Alsema, "Energy pay-back time and CO_2 emissions of PV systems," *Progress in Photovoltaics Research and Applications*, vol. 8, pp. 17–25, 2000.

[19] E. A. Alsema, M. J. de Wild-Scholten, and V. M. Fthenakis, "Environmental impacts of PV electricity generation, a critical comparison of energy supply options," in *Proceedings of 21st European Photovoltaic Solar Energy Conference*, Dresden, Germany, 2006.

[20] E. A. Alsema and M. J. de Wild-Scholten, "Reduction of the environmental impacts in crystalline silicon module manufacturing," in *Proceedings of the 22nd European Photovoltaic Solar Energy Conference*, Milan, Italy, 2007.

[21] E. A. Alsema and E. Nieuwlaar, "Energy viability of photovoltaic systems," *Energy Policy*, vol. 28, no. 14, pp. 999–1010, 2000.

[22] J. E. Mason, V. M. Fthenakis, T. Hansen, and H. C. Kim, "Energy payback and life-cycle CO_2 emissions of the BOS in an optimized 3.5MW PV installation," *Progress in Photovoltaics: Research and Applications*, vol. 14, no. 2, pp. 179–190, 2006.

[23] V. M. Fthenakis and H. C. Kim, "Greenhouse-gas emissions from solar electric- and nuclear power: a life-cycle study," *Energy Policy*, vol. 35, no. 4, pp. 2549–2557, 2007.

[24] R. Kannan, K. C. Leong, R. Osman, H. K. Ho, and C. P. Tso, "Life cycle assessment study of solar PV systems: an example of a 2.7 kWp distributed solar PV system in Singapore," *Solar Energy*, vol. 80, no. 5, pp. 555–563, 2006.

[25] I. Nawaz and G. N. Tiwari, "Embodied energy analysis of photovoltaic (PV) system based on macro- and micro-level," *Energy Policy*, vol. 34, no. 17, pp. 3144–3152, 2006.

[26] L. Lu and H. X. Yang, "Environmental payback time analysis of a roof-mounted building-integrated photovoltaic (BIPV) system in Hong Kong," *Applied Energy*, vol. 87, no. 12, pp. 3625–3631, 2010.

[27] C. Bankier and S. Gale, "Energy payback of roof mounted photovoltaic cells," *The Environmental Engineer*, vol. 7, no. 4, pp. 11–14, 2006.

[28] A. F. Sherwani, J. A. Usmani, and Varun, "Life cycle assessment of solar PV based electricity generation systems: a review," *Renewable and Sustainable Energy Reviews*, vol. 14, no. 1, pp. 540–544, 2010.

[29] R. Battisti and A. Corrado, "Evaluation of technical improvements of photovoltaic systems through life cycle assessment methodology," *Energy*, vol. 30, no. 7, pp. 952–967, 2005.

[30] Y. Tripanagnostopoulos, M. Souliotis, R. Battisti, and A. Corrado, "Energy, cost and LCA results of PV and hybrid PV/T solar systems," *Progress in Photovoltaics: Research and Applications*, vol. 13, no. 3, pp. 235–250, 2005.

[31] S. Dubey and G. N. Tiwari, "Life cycle cost analysis and carbon credit earned by PV/T solar water heater for Delhi climatic conditions," *Open Environmental Sciences*, vol. 2, pp. 15–25, 2008.

[32] Hong Kong Government, Life Cycle Assessment (LCA) and Life Cycle Costing (LCC) Tool for Commercial Building Developments in Hong Kong: User Manual, EMSD, Hong Kong SAR Government publication, 2005.

[33] "Hong Kong Government," Guidelines to Account for and Report on Greenhouse Gas Emission and Removals for Buildings (Commercial, Residential or Institutional Purposes) in Hong Kong, EMSD and EPD, Hong Kong SAR Government publication, 2008.

[34] M. Ito, K. Kato, K. Komoto, T. Kichimi, and K. Kurokawa, "A comparative study on cost and life-cycle analysis for 100 MW very large-scale PV (VLS-PV) systems in deserts using m-Si, a-Si, CdTe, and CIS modules," *Progress in Photovoltaics: Research and Applications*, vol. 16, no. 1, pp. 17–30, 2008.

2

Enhancement of Photocatalytic Activity of ZnO/SiO_2 by Nanosized Pt for Photocatalytic Degradation of Phenol in Wastewater

R. M. Mohamed[1,2] and M. A. Barakat[3,4]

[1] *Chemistry Department, Faculty of Science, King Abdulaziz University, P.O. Box. 80203, Jeddah 21589, Saudi Arabia*
[2] *Nanostructured Materials and Nanotechnology Division, Advanced Materials Department, Central Metallurgical Research & Development Institute (CMRDI), P.O. Box 87 Helwan, Cairo 11421, Egypt*
[3] *Department of Environmental Sciences, Faculty of Meteorology, Environment, and Arid Land Agriculture, King Abdulaziz University, Jeddah 21589, Saudi Arabia*
[4] *Minerals Technology Department, Central Metallurgical Research & Development Institute (CMRDI), Cairo 11421, Egypt*

Correspondence should be addressed to M. A. Barakat, mabarakat@gmail.com

Academic Editor: Manickavachagam Muruganandham

ZnO-SiO_2 nanoparticles were synthesized by a sol-gel technique from $Zn(NO_3)_2 \cdot 6H_2O$ and tetraethyl orthosilicate (TEOS). The synthesized samples were further modified by nanosized Pt from H_2PtCl_6 solution through photoassisted deposition (PAD) and impregnation (Img) routes. The obtained samples were characterized by a series of techniques including X-ray diffraction (XRD), UV-Vis diffuse reflectance spectroscopy, N_2 adsorption, extended X-ray absorption fine structure (EXAFS), and transmission electron microscopy (TEM). The photocatalytic activity of the Pt-ZnO/SiO_2 was evaluated by photocatalytic degradation of phenol in synthetic wastewater under UV-irradiation. Results obtained revealed that the surface area and the photocatalytic activity of the prepared samples were increased in the order ZnO/SiO_2 < PAD: Pt-ZnO/SiO_2 < img: Pt-ZnO/SiO_2. The surface area decreased from 480 to 460 and 450 m^2/g, while the efficiency of the phenol degradation increased from 80 to 85 and 100%, with the ZnO/SiO_2, Img: Pt-ZnO-SiO_2, and PAD: Pt-ZnO-SiO_2 samples, respectively.

1. Introduction

Zinc oxide (ZnO) is an n-type semiconductor with a wide direct bandgap of 3.37 eV. Recently, much effort has been devoted to study ZnO as a promising photocatalyst for photocatalytic degradation of water pollutants, owing to its high activity, low cost, and environmentally friendly feature [1–4]. ZnO microcrystal showed also good photocatalytic activity for dye wastewater treatment [5].

However, a major drawback of ZnO is the large bandgap of 3.37 eV; so, wavelengths below 400 nm are necessary for excitation. Another disadvantage of ZnO is that charge carrier recombination of photogenerated electron/hole pairs occurs within nanoseconds and the photocatalytic activity is low [6–9]. Therefore, it is necessary to improve its visible-light activities by extending its absorption threshold from the UV light region to the visible light region and also reduce the recombination of photogenerated electron/hole. Different works were performed recently to improve the activity of ZnO catalyst. Development of core/shell-structured materials on a nanometer scale has been receiving extensive attention [10, 11]. The shell can alter the charge, functionality, and reactivity of surface or improve the stability and dispersive ability of the core material. Furthermore, catalytic, optical, or magnetic functions can be imparted to the core particles by the shell material. In general, the synthesis of core/shell-structured material has the goal of obtaining a new composite material having synergetic or complementary behaviors between the core and shell materials. Many studies on the synthesis of composites such as NiO [12], V_2O_5 [13], TiO_2 [14], Fe_2O_3 [15], Pt [16], and Ag [17, 18] coated with SiO_2 shells have been reported. SiO_2 is one of the most studied shell candidates due to its relative ease in preparation, good environmental stability, and compatibility with other materials, which motivated us to prepare the core/shell-structured composite of ZnO and SiO$_2$ and expected to achieve novel properties resulting from the synergic interaction of these two chemical components. One of the most promising methods to increase the photocatalytic efficiency is surface modification of ZnO, this can be achieved by metal doping into the ZnO catalyst. Dopant can act as a sink to collect photogenerated electrons from the conduction band of the semiconductor. Thus, it hinders the recombination of photogenerated electrons and holes through increasing the charge separation [19–22]. The surface modification of ZnO nanoparticles by preparing charge-transfer catalysts with mixing multicomponent oxides can enhance the surface chemical and physical properties and be considered as the key for the successful photocatalytic applications of such nanoparticles. Several metal ions such as Fe [23] and Ag [24, 25] have been used as dopants for ZnO to improve its photocatalytic activity. Photocatalytic degradation of phenolic compounds with semiconducting oxides holds promise for the purification and treatment of both drinking and industrial wastewater [26–28]. The presence of such organic pollutants in aquatic environments has caused several environmental pollution problems.

In the present work, Pt/ZnO-SiO$_2$ nanoparticles with large specific surface areas had been synthesized by the application of a photoassisted deposition (PAD) and impregnation (Img) methods, and the properties of the nanoparticles were characterized by XRD, TEM, EXAFS, UV-Vis/DRS, and BET analysis. The photocatalytic activity of the synthesized nanoparticles was evaluated by the photodegradation of phenol (as a model for pollutants in wastewater) under UV irradiation.

2. Experimental Methods

2.1. Chemicals. $Zn(NO_3)_2 \cdot 6H_2O$, H_2PtCl_6, and tetraethyl orthosilicate (TEOS) were purchased from Aldrich and selected as the precursors of zinc, platinum, and silica, respectively. Standard grade phenol was purchased from Merck with 99.5% purity. 1 M NaOH and 1 M HCl solutions were used for pH adjustment. All chemicals used in this work were of reagent-grade quality. The water used was deionized water.

2.2. Preparation of ZnO/SiO$_2$. ZnO/SiO$_2$ nanoparticles were synthesized via a sol-gel technique and calcination. In a typical procedure, 20 mL TEOS was mixed with ethyl alcohol (C_2H_5OH), ultrapure water (H_2O), and nitric acid (HNO_3) under vigorous stirring for 1 hr. The overall molar ratio of $TEOS : C_2H_5OH : H_2O : HNO_3$was 1 : 4 : 8 : 0.02. Subsequently, after 60 min, an aqueous solution of $Zn(NO_3)_2 \cdot 6H_2O$ was added in to the above solution under vigorous stirring for 60 min. The reaction was performed at room temperature. The prepared samples were aged for 24 h. Finally, the samples were evaporated and dried at 80°C, followed by calcination at 550°C for 3 h in air.

2.3. Preparation of Pt Loaded on ZnO/SiO$_2$ Using PAD Method. Pt metal (3 wt%) was deposited on ZnO-SiO$_2$ from aqueous solution of H_2PtCl_6 under UV-light irradiation. The samples were dried at 100°C and reduced by H_2 (20 mL/min) at 350°C for 4 h.

2.4. Preparation of Pt Loaded on ZnO/SiO$_2$ Using Img Method. In a typical impregnation (Img) method, the 3 wt% of Pt metal was deposited by a simple impregnation of ZnO-SiO$_2$ in the absence of light with aqueous solution of H_2PtCl_6. The samples were dried at 100°C and reduced by H_2 (20 mL/min) at 350°C for 4 h.

2.5. Characterization Techniques. To determine the crystallite sizes and identities of the Pt loaded on ZnO/SiO$_2$ nanocomposite photocatalyst, X-ray diffraction (XRD) analysis was carried out at room temperature using Rigaku X-ray diffractometer with Cu Kα radiation over a 2θ collection range of 10–80°. The shape of the samples was tested using Hitachi H-9500 transmission electron microscope (TEM), the prepared samples were prepared by suspending the prepared samples in ethanol, followed by ultrasonication for 30 min, then a small amount of this solution onto a carbon-coated copper grid and drying for TEM. Specific surface area was calculated from measurements of N_2-adsorption using Nova 2000 series chromatech apparatus at 77 K. Prior to the measurements, all samples were treated under vacuum at 200°C for 2 h. The band gap of the samples was identified by UV-visible diffuse reflectance spectra (UV-Vis-DRS) in air at room temperature in the wavelength range of 200–800 nm using Shimadzu UV-2450 spectrophotometer. The extended X-ray absorption fine structure (EXAFS) is performed at BL-7C facility [29] of the Photon Factory at the National Laboratory for High Energy Physics, Tsukuba, Tokyo, Japan. A Si (1 1 1) double-crystal was used to monochromatize the X-rays from the 2.5 GeV electron storage ring. The K-edge EXAFS spectra of Fe were measured in the fluorescence mode at 25°C. The Fourier transformation was performed on K^3-weighted EXAFS oscillations in effective range from 0–5 Å.

2.6. Photocatalytic Activity Measurements. Photodegradation experiments were performed with a photocatalytic reactor system at 25°C. This bench-scale system consisted of a cylindrical Pyrex-glass cell with 1 L. A 150-Watts mercury lamp was placed in a 5 cm diameter quartz tube with one end tightly sealed by a Teflon stopper. The photoreactor was filled with 1 L of 100 mg/L aqueous phenol solution with 1 g/L of each of the three prepared nanoparticles samples. The whole reactor was cooled with a water-cooled jacket on its outside to the temperature. Compressed air was purged into the solution by bubbling compressed air from the bottom to maintain an aerobic condition [30]. Magnetic stirrer was also used to keep the solution chemically uniform. The pH of the solutions was adjusted and kept constant at 7 during the experiments by adding NaOH (1 M) and HCl (1 M) using an Orion Model 801 A pH meter. The experiments were carried out for 60 minutes. The liquid samples were filtered for analysis through 0.2 μm syringe

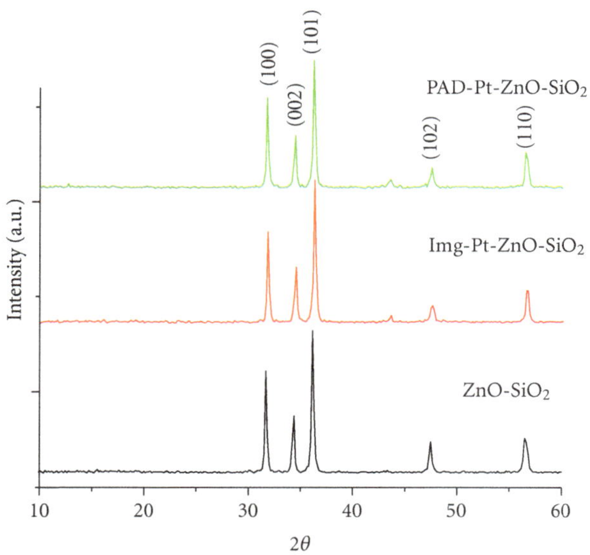

FIGURE 1: XRD patterns of the ($ZnO\text{-}SiO_2$, Img: $Pt\text{-}ZnO\text{-}SiO_2$ and PAD: $Pt\text{-}ZnO\text{-}SiO_2$).

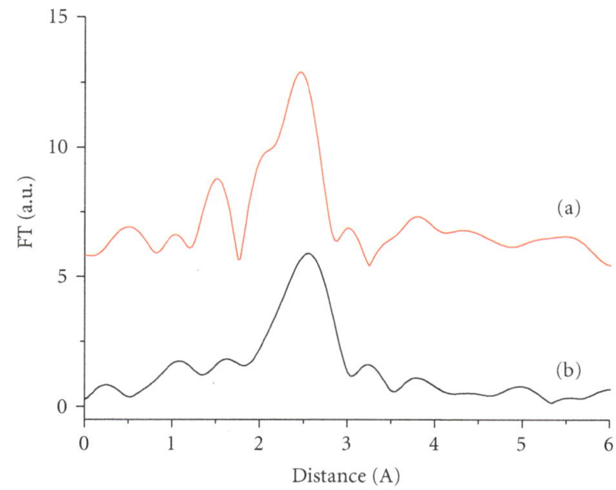

FIGURE 2: Fourier transforms of the Pt L_{III}-edge EXAFS spectra for Img: $Pt\text{-}ZnO\text{-}SiO_2$ (a) and PAD: $Pt\text{-}ZnO\text{-}SiO_2$ (b).

filters. The residual phenol was analyzed by the reaction with 4-aminoantipyrine [31]. Diluted sample of phenol was treated with 2 mL of 4 N NH_4OH, 1 mL of 1.5% 4-aminoantipyrine, and 1 mL of 4% K_3 $[Fe(CN)_6]$ and quantitatively diluted to 100 mL with H_2O. After 5 min, the reaction was determined colorimetrically at 510 nm. The photodegradation efficiency of phenol has been calculated applying the following equation:

$$\% \text{ Photodegradation efficiency} = \frac{C_0 - C}{C_0} \times 100, \quad (1)$$

where C_0 is the initial phenol concentration; C is the retained phenol in solution.

3. Results and Discussions

3.1. Phase Analysis. The XRD patterns of the $ZnO\text{-}SiO_2$ and Pt-doped $ZnO\text{-}SiO_2$ nanoparticles prepared by (Img) and (PAD) routes are shown in Figure 1. It can be seen that the diffraction patterns of $ZnO\text{-}SiO_2$ sample and all Pt-doped $ZnO\text{-}SiO_2$ are mainly composed of ZnO phase which still exists after applying both mentioned preparation methods. While in the Pt-doped samples, no diffraction peaks of Pt were observed, this is probably attributed to the low Pt-doping content (ca. 3 wt%). Moreover, it is obvious that Pt is well dispersed over the $ZnO\text{-}SiO_2$ surface.

3.2. Nanostructure Characterization. Figure 2 displays the Fourier transforms of Pt L_{III}-edge EXAFS spectra of the Pt-loaded catalysts. It can be noticed that the presence of the peak assigned to the Pt–Pt bond of Pt metal at around 2.5 Å is an indication of the formation of nanosized Pt metal [32]. In addition, the intensity of the Pt–Pt peak of the PAD: $Pt\text{-}ZnO\text{-}SiO_2$ catalyst is smaller than that prepared by impregnation route. Therefore, Pt metal particles formed on (PAD: $Pt\text{-}ZnO\text{-}SiO_2$) showed smaller particle size than (Img: $Pt\text{-}ZnO\text{-}SiO_2$). The grain sizes of PAD: $Pt\text{-}ZnO\text{-}SiO_2$ and Img: $Pt\text{-}ZnO\text{-}SiO_2$ nanocomposite photocatalysts were displayed in TEM images as shown in Figure 3. The particle size distribution obtained from the analysis of TEM images is shown in Figure 4. The result reveal that the nano-sized Pt metal with a mean diameter (d) of ca. 3 nm having a narrow size distribution was found on the PAD: $Pt\text{-}ZnO\text{-}SiO_2$ catalyst, whereas the aggregated Pt metal within various sizes is observed on Img: $Pt\text{-}ZnO\text{-}SiO_2$ catalyst (d = 15 nm) which is in agreement with the results of EXAFS measurement. These findings suggest that the size of Pt metal particles depends on the preparation method. Also, TEM micrographs showed the homogenous distribution of Pt over $ZnO\text{-}SiO_2$ matrix which was prepared by PAD method.

3.3. Surface Area Analysis. Specific surface area (S_{BET}) of ZnO/SiO_2, PAD: $Pt\text{-}ZnO\text{-}SiO_2$, and Img: $Pt\text{-}ZnO\text{-}SiO_2$ powder samples were determined. The S_{BET} values were 480, 460, and 450 m^2/g for the $ZnO\text{-}SiO_2$, PAD: $Pt\text{-}ZnO\text{-}SiO_2$, and Img: $Pt\text{-}ZnO\text{-}SiO_2$, respectively. The parameters of surface area and the data calculated from the t-plot are collected in Table 1. Furthermore, the total pore volume of $Pt\text{-}ZnO\text{-}SiO_2$ is higher than that of $ZnO\text{-}SiO_2$. The values of S_{BET} and S_t are generally close in most samples indicating the presence of mesopores. The values of S_{micro} are high compared to that of S_{meso} implying that the main surface is mesoporous solid. The surface texture data are correlated with the catalytic activity as will be mentioned later on. The N_2 adsorption-desorption isotherm of the prepared samples is shown in Figure 5. It is clear that the isotherm shows a typical type IV sorption behavior, which confirms the presence of mesoporous form.

3.4. Band-Gap Analysis (UV-Vis-DRS). Study of the UV-visible radiation absorption is an important tool for evaluating the changes in the absorbance spectra of the prepared semiconductor materials. This is expressed by the band-gap

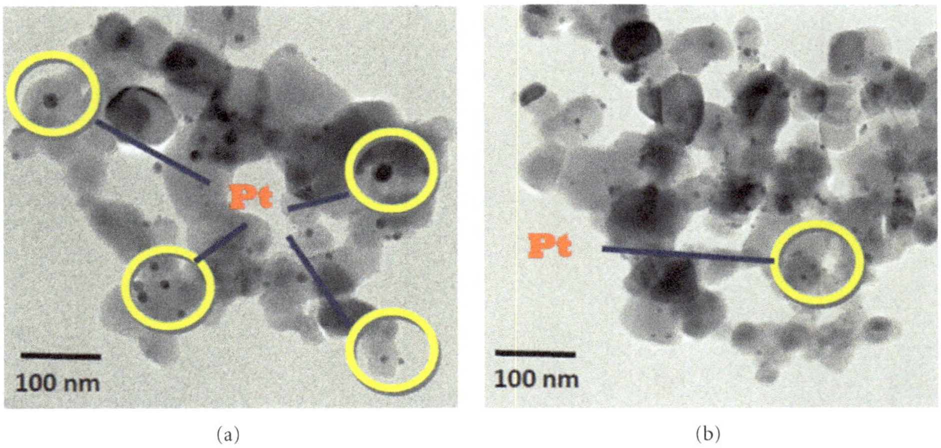

FIGURE 3: The TEM images of the Img: Pt-ZnO-SiO_2 (a) and PAD: Pt-ZnO-SiO_2 (b) catalysts.

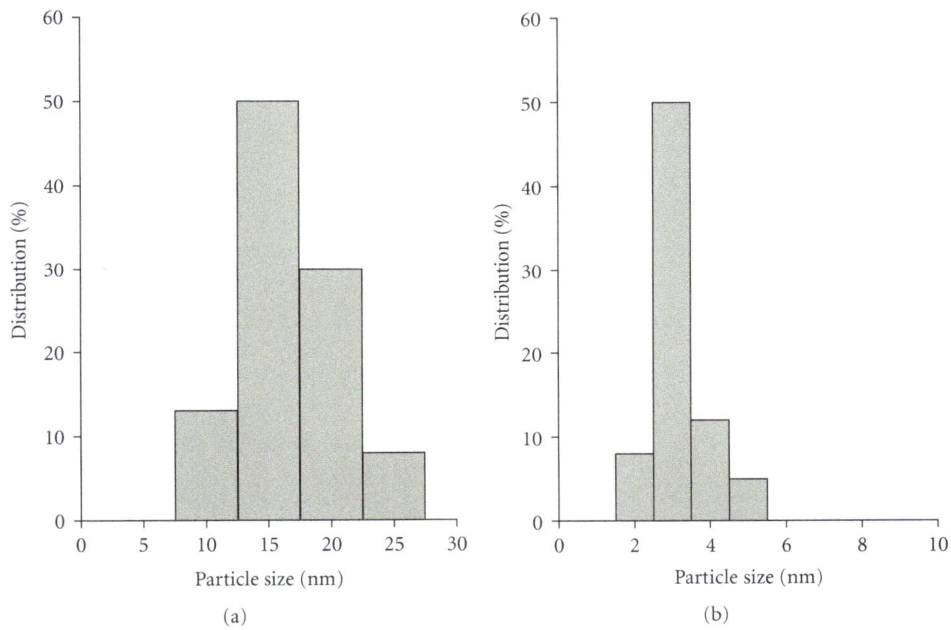

FIGURE 4: Size distribution diagrams of Pt metal obtained from the TEM images of the Img: Pt-ZnO-SiO_2 (a) and PAD: Pt-ZnO-SiO_2 (b) catalysts.

(E_g) measurement which can be altered by different parameters. For instance, E_g value for pure ZnO phase is usually reported 3.37 [7]; however, these values are influenced by the synthesis method, and also affected by the existence of impurities doping the crystalline network and also the average crystal size of the semiconductor. In a previous study, different methods for calculating the E_g from the UV-Vis reflectance spectra were used. For example, some authors calculated the E_g values by a direct extrapolation of the $F(R)$ spectrum, whereas others reported the wavelength corresponding to the maximum absorption [33]. As a consequence, quite different E_g values for ZnO samples are found in the literature. For instance, threshold wavelength values of 240 nm [34], 290 nm [35], and 360 nm [36] correspond to bandgaps 5.15, 4.28, and 3.45, respectively. Figure 6 gives UV-Vis-DRS of (ZnO-SiO_2, Img: Pt-ZnO-SiO_2, and PAD: Pt-ZnO-SiO_2). The results showed an increase in the absorbance in the visible light region with the Pt doping. The values of E_g for the synthesized semiconductors can be derived from the spectra by plotting $(F(R)\cdot h\nu)^{1/2}$ against $h\nu$ [37, 38] as shown in Figure 7 and tabulated as shown in Table 2. The results revealed that the calculated values of E_g for (ZnO-SiO_2, Img: Pt-ZnO-SiO_2, and PAD: Pt-ZnO-SiO_2) were 3.4, 3.25, and 3.05 eV, respectively. This indicates shifting the spectra of the PAD: Pt-ZnO-SiO_2 sample to the visible light area.

3.5. Evaluation of Photocatalytic Activity. The photocatalytic activity of the synthesized nanoparticles samples was evaluated by degradation of phenol in solution under UV light. Figure 8 displays the photocatalytic degradation of

Table 1: Texture parameters of $ZnO\text{-}SiO_2$, Img: $Pt\text{-}ZnO\text{-}SiO_2$, and PAD: $Pt\text{-}ZnO\text{-}SiO_2$.

Sample	S_{BET} (m^2/g)	S_t (m^2/g)	S_{micro} (cm^2/g)	S_{ext} (cm^2/g)	V_p (cm^3/g)	V_{micro} (cm^3/g)	V_{meso} (cm^3/g)	r (Å)
$ZnO\text{-}SiO_2$	480	485	380	100	0.866	0.069	0.803	38.00
Img: $Pt\text{-}ZnO\text{-}SiO_2$	450	455	500	365	0.726	0.022	0.704	36.00
PAD: $Pt\text{-}ZnO\text{-}SiO_2$	460	465	650	385	0.668	0.066	0.602	34.00

Note: (S_{BET}) BET-Surface area, (S_t) surface area derived from V_{l-t} plots, (S_{mic}) surface area of micropores, (S_{ext}) external surface area, (V_p) total pore volume, (V_{mic}) pore volume of micropores, (V_{mes}) pore volume of mesopores, and (r^-) mean pore radius.

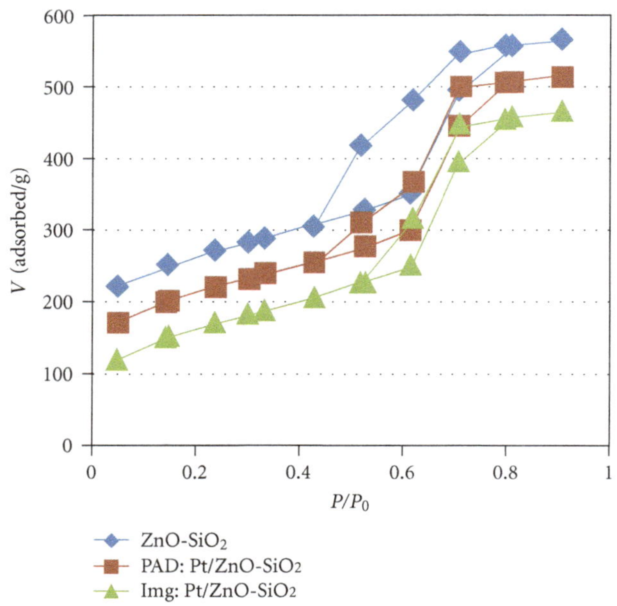

Figure 5: N_2 sorption isotherms of $ZnO\text{-}SiO_2$, PAD-$Pt/ZnO\text{-}SiO_2$, and Img-$Pt/ZnO\text{-}SiO_2$.

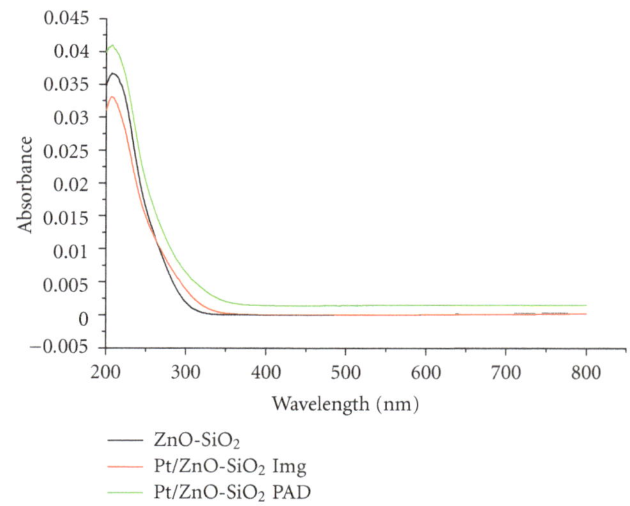

Figure 6: Diffuse reflectance UV-Vis absorption spectra of $ZnO\text{-}SiO_2$, Img: $Pt\text{-}ZnO\text{-}SiO_2$, and PAD: $Pt\text{-}ZnO\text{-}SiO_2$.

Table 2: The calculated band gap energy of $ZnO\text{-}SiO_2$, Img: $Pt\text{-}ZnO\text{-}SiO_2$, and PAD: $Pt\text{-}ZnO\text{-}SiO_2$.

Sample	Band gap energy (eV)
$ZnO\text{-}SiO_2$	3.40
Img: $Pt\text{-}ZnO\text{-}SiO_2$	3.25
PAD: $Pt\text{-}ZnO\text{-}SiO_2$	3.05

100 mg/L phenol pertaining to ($ZnO\text{-}SiO_2$, Img: $Pt\text{-}ZnO\text{-}SiO_2$, and PAD: $Pt\text{-}ZnO\text{-}SiO_2$) at solution pH 7 for different time intervals. It can be seen that the rate of phenol photodegradation increased gradually with time reaching maximum efficiency values of 80, 85, and 99.9% after 1 h with parent $ZnO\text{-}SiO_2$, Img: $Pt\text{-}ZnO\text{-}SiO_2$, and PAD: $Pt\text{-}ZnO\text{-}SiO_2$ samples, respectively. Considering that the pure Pt oxides have no photocatalytic oxidation properties, such change in the photodegradation activity may be explained in terms of the differences in interaction between Pt and $ZnO\text{-}SiO_2$ that led to several modifications in physical properties such as bandgap, particle size, and surface texture. Also, one could observe that the catalytic activity of $ZnO\text{-}SiO_2$ generally increased with the addition of Pt promoters. Figure 9 shows the good correlation between the physical properties of the synthesized samples, such as bandgap, surface area, and pore volume, with their catalytic activity. It was obvious that the photodegradation activity was gradually increased with the decrease of bandgap and the increase of the surface area and pore volume. The maximum photocatalytic degradation of phenol was achieved in the case of PAD: $Pt\text{-}ZnO\text{-}SiO_2$ in which the surface area and pore volume were maximum with lower bandgap value. It is believed that the lack of electron scavengers (surface Zn^{2+}) and hole traps (surface hydroxyl groups) is responsible for the rapid recombination rate of e^-/h^+, which leads to lower photocatalytic activity with the parent $ZnO\text{-}SiO_2$ sample [6–9]. The photocatalytic activities of the Pt-doped $ZnO\text{-}SiO_2$ nanoparticles increased due to that Pt plays two important roles; noble metal incorporation into TiO_2 dielectric provides an absorption feature due to the surface plasmon resonance (SPR) occurring over the visible range of the spectrum. In particular, Ag, Pt, and Au metals are the most popular materials due to the strong SPR character [39, 40]. On the other hand, the superior driven photocatalytic efficiency of the Pt/TiO_2 nanocomposite photocatalyst can be ascribed to the high efficiency of charge-pair separation due to the presence of deposited Pt serving as electron sinks to retard the rapid e^--h^+ couple recombination, the good photoabsorption capacity in the visible light region, and the higher concentration of surface hydroxyl groups, which are able to effectively scavenge photogenerated valence band holes [41].

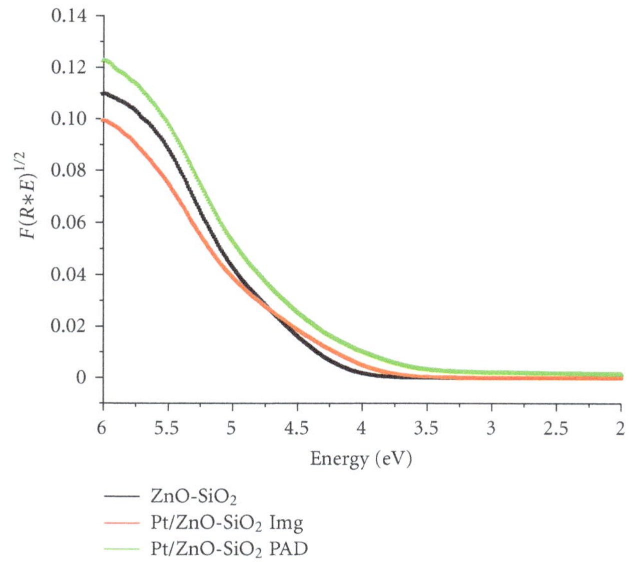

FIGURE 7: Bandgap calculated from the DR-UV-Vis spectra of $ZnO-SiO_2$, Img: $Pt-ZnO-SiO_2$, and PAD: $Pt-ZnO-SiO_2$.

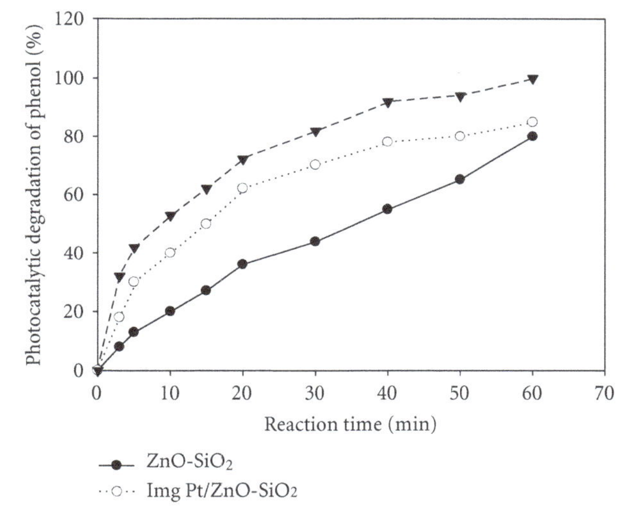

FIGURE 8: Photocatalytic degradation of phenol (%) for $ZnO-SiO_2$, Img: $Pt-ZnO-SiO_2$, and PAD: $Pt-ZnO-SiO_2$. (a) PAD: $Pt-ZnO-SiO_2$, (b) Img: $Pt-ZnO-SiO_2$, and (c) $ZnO-SiO_2$.

Accordingly, the highest activity was more clear in the PAD: $Pt-ZnO-SiO_2$ sample in which the Pt diffuses in the lattice of the semiconductor. Thus, it lowered both the bandgap (as confirmed from the UV-Vis-DRS spectra analysis) and particle size (as confirmed from the TEM analysis). This resulted in increase in the surface area and pore volume, and consequently showed the best photoactivity with phenol degradation.

4. Conclusions

Pt doping, through photoassisted deposition (PAD) and impregnation (Img) routes, can greatly enhance the performance of $ZnO-SiO_2$ as a photocatalyst. The nano-sized

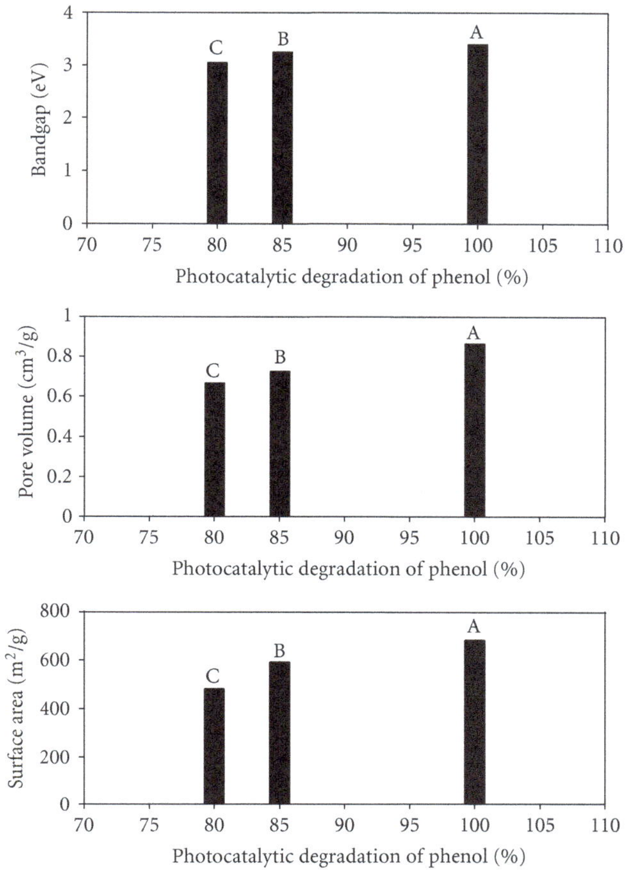

FIGURE 9: Effect of physical parameters of the materials on their photocatalytic activity.

Pt metal with a mean diameter (d) of ca. 3 nm having a narrow size distribution was found on the PAD: $Pt-ZnO-SiO_2$ catalyst, whereas the aggregated Pt metal with various sizes is observed on Img: $Pt-ZnO-SiO_2$ catalyst ($d = 15$ nm). The surface area of the synthesized samples was decreased from 480 to 460 and 450 m^2/g with $ZnO-SiO_2$, PAD: $Pt-ZnO-SiO_2$, and Img: $Pt-ZnO-SiO_2$, respectively. The UV-Vis-DRS spectra analysis confirmed the lowest bandgab of PAD: $Pt-ZnO-SiO_2$ with a value of 3.05, comparing to 3.25 and 3.4 with Img: $Pt-ZnO-SiO_2$ and $ZnO-SiO_2$, respectively. The photocatalytic degradation of phenol in synthetic wastewater solution was found to be much more effective in the PAD: $Pt-ZnO-SiO_2$. The degradation efficiency increased from 80 to 85 and 99.9%, with the ZnO/SiO_2, Img: $Pt-ZnO-SiO_2$, and PAD: $Pt-ZnO-SiO_2$ samples, respectively. The smallest particle size and lowest bandgab of the PAD: $Pt-ZnO-SiO_2$ sample resulted in a high increase in the surface area and pore volume and consequently showed the best photoactivity with phenol degradation.

Acknowledgments

This paper was funded by the Deanship of Scientific Research (DSR), King Abdulaziz University (KAU), Jeddah, under Grant no. (265/155/1431). The authors, therefore, acknowledge with thanks DSR and KAU for technical and financial support.

References

[1] C. Hariharan, "Photocatalytic degradation of organic contaminants in water by ZnO nanoparticles: revisited," *Applied Catalysis A*, vol. 304, no. 1-2, pp. 55–61, 2006.
[2] M. Mrowetz and E. Selli, "Photocatalytic degradation of formic and benzoic acids and hydrogen peroxide evolution in TiO_2 and ZnO water suspensions," *Journal of Photochemistry and Photobiology A*, vol. 180, no. 1-2, pp. 15–22, 2006.
[3] T. Pauporté and J. Rathouský, "Electrodeposited mesoporous ZnO thin films as efficient photocatalysts for the degradation of dye pollutants," *Journal of Physical Chemistry C*, vol. 111, no. 21, pp. 7639–7644, 2007.
[4] J. Yu and X. Yu, "Hydrothermal synthesis and photocatalytic activity of zinc oxide hollow spheres," *Environmental Science and Technology*, vol. 42, no. 13, pp. 4902–4907, 2008.
[5] J. H. Sun, S. Y. Dong, Y. K. Wang, and S. P. Sun, "Preparation and photocatalytic property of a novel dumbbell-shaped ZnO microcrystal photocatalyst," *Journal of Hazardous Materials*, vol. 172, no. 2-3, pp. 1520–1526, 2009.
[6] M. Zhou, J. Yu, and B. Cheng, *Journal of Hazardous Materials*, vol. 137, 2006.
[7] C. F. Klingshirn, "ZnO: material, physics and applications," *ChemPhysChem*, vol. 8, no. 6, pp. 782–803, 2007.
[8] A. Singhal, S. N. Achary, A. K. Tyagi, P. K. Manna, and S. M. Yusuf, "Colloidal Fe-doped ZnO nanocrystals: facile low temperature synthesis, characterization and properties," *Materials Science and Engineering B*, vol. 153, no. 1–3, pp. 47–52, 2008.
[9] C. R. Estrellan, C. Salim, and H. Hinode, "Photocatalytic activity of sol-gel derived TiO_2 co-doped with iron and niobium," *Reaction Kinetics and Catalysis Letters*, vol. 98, no. 1, pp. 187–192, 2009.
[10] J. J. Schneider, "Magnetic core/shell and quantum-confined semiconductor nanoparticles via chimie douce organometallic synthesis," *Advanced Materials*, vol. 13, no. 7, pp. 529–533, 2001.
[11] H. B. Lee, Y. M. Yoo, and Y. H. Han, "Characteristic optical properties and synthesis of gold-silica core-shell colloids," *Scripta Materialia*, vol. 55, no. 12, pp. 1127–1129, 2006.
[12] R. M. Mohamed and E. S. Aazam, "Photocatalytic oxidation of carbon monoxide over NiO/SnO_2 nanocomposites under UV irradiation," *Journal of Nanotechnology*, vol. 2012, Article ID 794874, 9 pages, 2012.
[13] A. A. Ismail, I. A. Ibrahim, and R. M. Mohamed, "Sol-gel synthesis of vanadia-silica for photocatalytic degradation of cyanide," *Applied Catalysis B*, vol. 45, no. 2, pp. 161–166, 2003.
[14] J. Zhang, Z. Liu, B. Han et al., "Preparation of silica and TiO_2-SiO_2 core-shell nanoparticles in water-in-oil microemulsion using compressed CO_2 as reactant and antisolvent," *Journal of Supercritical Fluids*, vol. 36, no. 3, pp. 194–201, 2006.
[15] V. Maurice, T. Georgelin, J. M. Siaugue, and V. Cabuil, "Synthesis and characterization of functionalized core-shell γFe_2O_3-SiO_2 nanoparticles," *Journal of Magnetism and Magnetic Materials*, vol. 321, no. 10, pp. 1408–1413, 2009.
[16] R. M. Mohamed, "Characterization and catalytic properties of nano-sized Pt metal catalyst on TiO_2-SiO_2 synthesized by photo-assisted deposition and impregnation methods," *Journal of Materials Processing Technology*, vol. 209, no. 1, pp. 577–583, 2009.
[17] K. S. Chou and C. C. Chen, "Fabrication and characterization of silver core and porous silica shell nanocomposite particles," *Microporous and Mesoporous Materials*, vol. 98, no. 1–3, pp. 208–213, 2007.
[18] R. M. Mohamed and I. A. Mkhalid, "Characterization and catalytic properties of nano-sized Ag metal catalyst on TiO_2-SiO_2 synthesized by photo-assisted deposition and impregnation methods," *Journal of Alloys and Compounds*, vol. 501, no. 2, pp. 301–306, 2010.
[19] M. J. Height, S. E. Pratsinis, O. Mekasuwandumrong, and P. Praserthdam, "Ag-ZnO catalysts for UV-photodegradation of methylene blue," *Applied Catalysis B*, vol. 63, no. 3-4, pp. 305–312, 2006.
[20] T.-Y. Han, C.-F. Wu, and C.-T. Hsieh, "Hydrothermal synthesis and visible light photocatalysis of metal-doped titania nanoparticles," *Journal of Vacuum Science and Technology B*, vol. 25, no. 2, article 430, 6 pages, 2007.
[21] R. Ullah and J. Dutta, "Photocatalytic degradation of organic dyes with manganese-doped ZnO nanoparticles," *Journal of Hazardous Materials*, vol. 156, no. 1–3, pp. 194–200, 2008.
[22] R. Slama, F. Ghribi, A. Houas, C. Barthou, and L. El Mir, *International Journal of Nanoelectronics and Materials*, vol. 3, p. 133, 2010.
[23] R. M. Mohamed, M. A. Al-Rayyani, E. S. Baeissa, and I. A. Mkhalid, "Nano-sized Fe-metal catalyst on ZnO-SiO_2: (photo-assisted deposition and impregnation) Synthesis routes and nanostructure characterization," *Journal of Alloys and Compounds*, vol. 509, no. 24, pp. 6824–6828, 2011.
[24] C. Ren, B. Yang, M. Wu et al., "Synthesis of Ag/ZnO nanorods array with enhanced photocatalytic performance," *Journal of Hazardous Materials*, vol. 182, no. 1–3, pp. 123–129, 2010.
[25] Y. Lai, M. Meng, and Y. Yu, "One-step synthesis, characterizations and mechanistic study of nanosheets-constructed fluffy ZnO and Ag/ZnO spheres used for Rhodamine B photodegradation," *Applied Catalysis B*, vol. 100, no. 3-4, pp. 491–501, 2010.
[26] M. A. Barakat, H. Schaeffer, G. Hayes, and S. Ismat-Shah, "Photocatalytic degradation of 2-chlorophenol by Co-doped TiO_2 nanoparticles," *Applied Catalysis B*, vol. 57, no. 1, pp. 23–30, 2005.
[27] M. A. Barakat, J. M. Tseng, and C. P. Huang, "Hydrogen peroxide-assisted photocatalytic oxidation of phenolic compounds," *Applied Catalysis B*, vol. 59, no. 1-2, pp. 99–104, 2005.
[28] A. Abdel Aal, M. A. Barakat, and R. M. Mohamed, "Electrophoreted Zn-TiO2-ZnO nanocomposite coating films for photocatalytic degradation of 2-chlorophenol," *Applied Surface Science*, vol. 254, no. 15, pp. 4577–4583, 2008.
[29] M. Nomura and A. Koyama, "Design of an XAFS beamline at the photon factory: possibilities of bent conical mirrors," *Journal of Synchrotron Radiation*, vol. 6, no. 3, pp. 182–184, 1999.
[30] M. A. Barakat, Y. T. Chen, and C. P. Huang, "Removal of toxic cyanide and Cu(II) Ions from water by illuminated TiO_2 catalyst," *Applied Catalysis B*, vol. 53, no. 1, pp. 13–20, 2004.
[31] J. R. Pemberton, "Retention of preservative levels of formaldehyde in desiccated biological products," *Journal of Clinical Microbiology*, vol. 2, no. 2, pp. 144–146, 1975.
[32] A. Caballero, F. Villain, H. Dexpert, F. Lepeltier, B. Didillon, and J. Lynch, "In situ EXAFS studies of modifications to supported metallic catalysts under reactive atmospheres," *Catalysis Letters*, vol. 20, no. 1-2, pp. 1–13, 1993.
[33] R. J. Davis and Z. Liu, "Titania-silica: a model binary oxide catalyst system," *Chemistry of Materials*, vol. 9, no. 11, pp. 2311–2324, 1997.

[34] H. Yoshida, T. Shimizu, C. Murata, and T. Hattori, "Highly dispersed zinc oxide species on silica as active sites for photoepoxidation of propene by molecular oxygen," *Journal of Catalysis*, vol. 220, no. 1, pp. 226–232, 2003.

[35] M. Huang, P. E. J. Rivera-Díaz-Del-Castillo, O. Bouaziz, and S. Van Der Zwaag, "Irreversible thermodynamics modelling of plastic deformation of metals," *Materials Science and Technology*, vol. 24, no. 4, pp. 495–500, 2008.

[36] G. D. Mihai, V. Meynen, M. Mertens, N. Bilba, P. Cool, and E. F. Vansant, "ZnO nanoparticles supported on mesoporous MCM-41 and SBA-15: a comparative physicochemical and photocatalytic study," *Journal of Materials Science*, vol. 45, no. 21, pp. 5786–5794, 2010.

[37] M. Anpo and M. Che, "Applications of photoluminescence techniques to the characterization of solid surfaces in relation to adsorption, catalysis, and photocatalysis," *Advances in Catalysis*, vol. 44, pp. 119–257, 1999.

[38] G. Lassaletta, A. Fernández, J. P. Espinós, and A. R. González-Elipe, "Spectroscopic characterization of quantum-sized TiO_2 supported on silica: influence of size and TiO_2-SiO_2 interface composition," *Journal of Physical Chemistry*, vol. 99, no. 5, pp. 1484–1490, 1995.

[39] J. F. Rivadulla, M. C. Vergara, M. C. Blanco, M. A. López-Quintela, and J. Rivas, "Optical properties of platinum particles synthesized in microemulsions," *Journal of Physical Chemistry B*, vol. 101, no. 44, pp. 8997–9004, 1997.

[40] N. A. M. Barakat, K. D. Woo, M. A. Kanjwal, K. E. Choi, M. S. Khil, and H. Y. Kim, "Surface plasmon resonances, optical properties, and electrical conductivity thermal hysteresis of silver nanofibers produced by the electrospinning technique," *Langmuir*, vol. 24, no. 20, pp. 11982–11987, 2008.

[41] D. Zhang, "Visible light-induced photocatalysis through surface plasmon excitation of platinum-metallized titania for photocatalytic bleaching of rhodamine B," *Monatshefte für Chemie*, vol. 143, no. 5, pp. 729–738, 2012.

3

Influence of Electrolyte Refreshing on the Photoelectrochemical Performance of Fiber-Shaped Dye-Sensitized Solar Cells

Zhibin Lv, Hongwei Wu, Xin Cai, Yongping Fu, Dan Wang, Zengze Chu, and Dechun Zou

Beijing National Laboratory for Molecular Sciences, Key Laboratory of Polymer Chemistry and Physics of Ministry of Education, College of Chemistry and Molecular Engineering, Peking University, Beijing 100871, China

Correspondence should be addressed to Dechun Zou, dczou@pku.edu.cn

Academic Editor: Gerrit Boschloo

Given the convenient sealing of fiber-shaped dye-sensitized solar cells (FDSSCs), the electrolyte refreshing effect on the photoelectrochemical performance of FDSSCs was studied. The electron transport and interfacial recombination kinetics were also systematically investigated by electrochemical impedance spectroscopy. With increased electrolyte refreshing times from 0 to 10, the open-circuit voltage (V_{oc}) and fill factor (FF) increased, whereas the photocurrent density (J_{sc}) and power conversion efficiency (PCE) significantly decreased. The increased V_{oc} was mainly ascribed to the electron recombination resistance ($R_{ct,\,WE}$) at the TiO_2/electrolyte interface and electron lifetime. The decreased J_{sc} and PCE were due to dye desorption and the increase of series resistance. Further investigation proved that Li^+ played a vital role in increasing V_{oc} as electrolyte refreshing and Li^+ had more significant impact than TBP (tert-butyl pyridine) on maintaining high V_{oc}.

1. Introduction

Given their high-energy conversion efficiency and simple fabrication process, dye-sensitized solar cells (DSSCs) have been attracting worldwide scientific and technological interest for decades [1–3]. Research on the optimization of materials and fabrication procedures is progressive.

Consequently, current planar DSSCs using I_3^-/I^- electrolytes, have achieved up to 11%-12% conversion efficiency [4–6], and the efficiency of large-scale modules has also reached 8% with aperture areas of more than 20 cm^2 [7]. DSSCs are considered as promising alternatives to conventional silicon-based photovoltaic devices for sustainable, low-cost, and environment-friendly energy supply. However, there are still obstacles to overcome, such as cost, limited efficiency in the large-scale module, electrolyte leakage, cell sealing, and stability.

Fiber solar cells have received great attention in recent years [8–22]. Their advantages over DSSCs having traditional sandwiched planar structure include the following. (1) Nonflat structured solar cells get rid of the dependence on transparent conductive oxides (TCOs) [8]. Traditional solar cells and the reported tubular solar cells still use TCOs, which usually account for 30%–50% of total cell costs [23]. (2) The fiber cell is a macro-one-dimensional structure, and has a smaller package area ratio than the two-dimensional structure. Making a larger cell only requires increasing the length of the cell. When the cell is increased to a certain size, the package area of the cell remains basically unchanged [20]. (3) The flexible fiber cell can directly adopt the traditional preparation technology. In contrast, special technologies must be adopted to create all kinds of traditional flexible flat cells (like OPVs and planar DSSCs) to ensure that the flexible substrate will not be damaged during the preparation process, such as low temperatures. (4) Existing textile techniques can be directly used in weaving for the mass production of fiber cells. This method is another further improvement of the traditional roll-to-roll technology for producing flexible electronic devices [18, 24].

The advancements in fiber solar cells, including fiber-shaped organic solar cells and fiber-shaped DSSCs (FDSSCs), have thus far been impressive [8, 9, 11]. The highest power

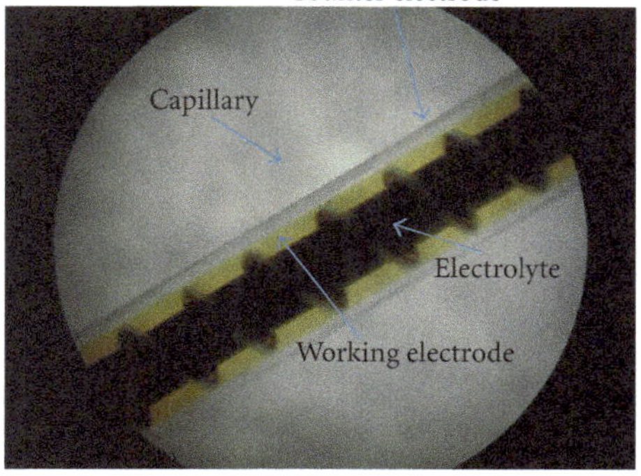

FIGURE 1: An optical photo of FDSSC structure details.

efficiency of FDSSCs ever reported is above 7% [18]. FDSSCs not only address the problems of traditional flat cells, but also provide some new features to elucidate the factors that affect their performances. For example, Fu et al. [18] have studied the three-dimensional light collecting ability of FDSSCs. FDSSCs are found to be able to harvest surrounding light efficiently to double their power output. Under high light intensity, FDSSCs combined with a condenser also still has a relatively high output power and fill factor, with little performance decline. O'Connor et al. [14] have proposed a new type of solar cell that combines a series of fiber cell units with narrow-band spectrum absorption. A proper arrangement in space is adopted to realize full-spectrum absorption of the sun spectra, resulting in 17% theoretical power convention efficiency. All these above results are difficult to achieve when using planar DSSCs.

FDSSCs act as units, and they are easily encapsulated because only one end or both ends of the capillary need to be sealed during the sealing process [20]. This fact enables the possibility of more conveniently renewing the electrolyte, in contrast to planar DSSCs whose electrolytes cannot be refreshed once they are sealed tightly. In the present paper, the effect of electrolyte refreshing on the photovoltaic performance of FDSSCs was studied. The electron transport and interfacial recombination kinetics were also systematically investigated using electron impedance spectroscopy (EIS).

2. Experimental

2.1. Materials. All chemicals and solvents used were of pure quality. The Ti filament (99.7% pure, diameter = 0.25 mm) was purchased from Alfa Aesar. LiI, I_2, and 4-tetrabutylpyridine (TBP) were purchased from Aldrich. Acetonitrile (CH_3CN, GR: guaranteed reagent) was purchased from Siyou Inc., Tianjin, China. N719 dye ($[(C_4H_9)_4N]_2[Ru(II)L_2(NCS)_2]$, where L = 2,2-bipyridyl-4,4 dicarboxylic acid; ruthenium TBA535) and 1-butyl-3-methylimidazolium iodide (BMII) were from the Dalian Rainbow Light Solar Technology Development Co., Ltd. The TiO_2 colloid was prepared as follows. About 12.0 mL of tetra-butyl titanate (Aldrich) and 2.0 mL of acetic acid were mixed with vigorous stirring. The mixture was slowly dropped into 60.0 mL of water, and 0.5 mL of 65% nitric acid was added to the system. After stirring for 2 h in an oil bath at 80°C, 35 mL of the mixture was transferred into the autoclave and left undisturbed at 220°C for 12 h. After removing part of the solvent, 0.6 g of polyethylene oxide (M_w = 4000; Aldrich) was added and the TiO_2 colloid was ready.

2.2. Fabrication of FDSSCs. The Ti wires were cleaned using acetone, isopropanol, and then ethanol. The Ti wires were immersed into the prepared TiO_2 colloid and dipped in and out several times. Sintering at 400°C for 5 min followed. A thin layer of TiO_2 film was formed on the Ti wire. The above step was repeated until the required thickness of the TiO_2 film was obtained. The TiO_2-coated electrode was sintered at 450°C for 30 min. The electrode was sensitized in a 10^{-4} M N719 ethanol solution for 24 h to obtain the sensitized working electrode. The counterelectrode was a Pt wire 40 μm in diameter. The Pt wire was uniformly and spirally wound onto the photoanode and then inserted into the sealing capillary (0.5 mm inner diameter and 0.85 mm outer diameter) filled with the electrolyte. This electrolyte contained 0.04 M iodine, 0.6 M BMII, 0.05 M LiI, and 0.30 M TBP in acetonitrile. The capillary was sealed with wax to prevent the electrolyte leakage. Each time refreshing the electrolyte, the wax was melted to decapsulate the FDSSC. The used electrolyte was removed, and the unsealed FDSSCs were allowed to suck the fresh electrolyte by capillarity. Finally, the cells were sealed again using wax.

2.3. Measurement and Methods. All device tests and short circuit current responses were conducted on a Keithley Model 2000, with sunlight simulator (YS-55, Japan). The illumination intensity is 100 mW/cm^2. The aperture area of the cell was obtained from the project area A_{pron}, which was equal to twice the sum of the Ti wire diameter and TiO_2 film thickness, multiplied by the length of the device. The short circuit current density J_{sc} was I/A_{proj} (I is the short circuit current). The power conversion efficiency PCE was $J_{sc} \times V_{oc} \times FF/P_{in}$, where V_{oc} is the open-circuit voltage, FF is the fill factor, and P_{in} is the intensity of incident light. The electrochemical impedance spectrum was obtained using PGSTAT302N (Autolab Corp., Switzerland) within the frequency range of 0.05 Hz to 300 kHz at room temperature and was fitted using Nova 1.7 software.

3. Results and Discussion

Generally, the electrolytes of FTO-glass-based planar DSSCs could not be or difficultly refreshed once there is already an electrolyte leakage. FDSSCs overcome this problem because of their simple and detachable sealing methods [20]. The new issue is whether the electrolyte refreshing times affect the performance of FDSSCs. Figure 1 is an optical photo of FDSSC, in which the details of FDSSC are demonstrated. The FDSSC consists a counterelectrode wrapped around the working electrode, which then encapsulated in a capillary filled with electrolytes. Figure 2 shows the photocurrent density-voltage (J-V) characteristics of devices with electrolyte refreshing

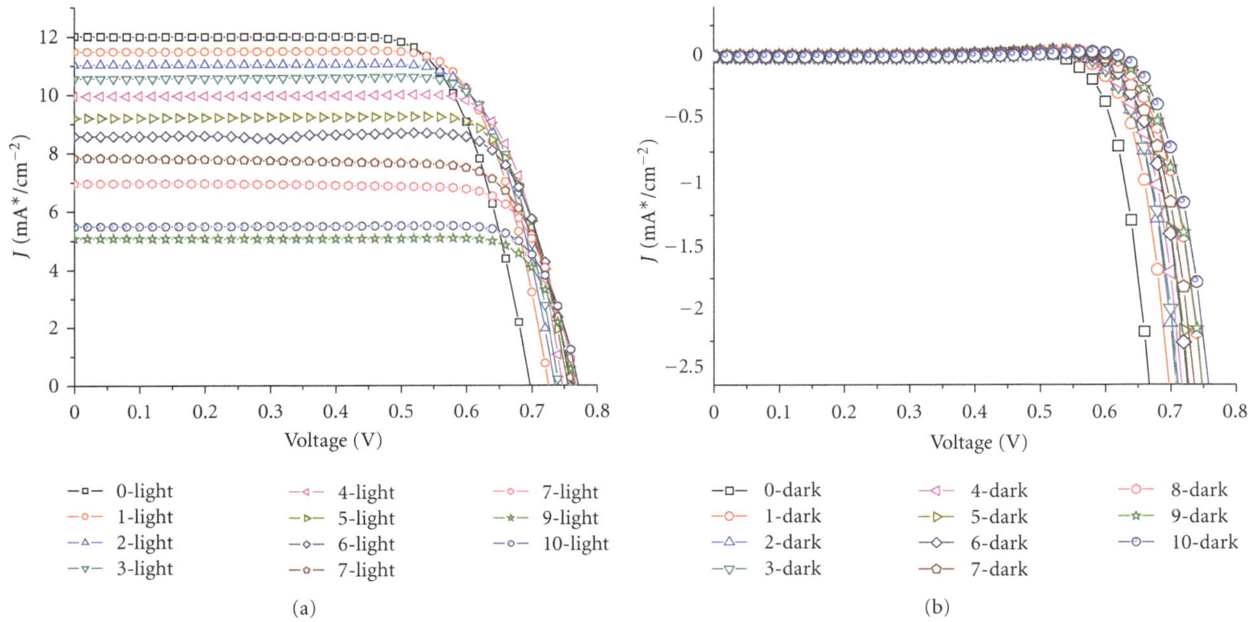

FIGURE 2: (a) Photocurrent density-voltage (J-V) characteristics of devices with electrolyte refreshing times under AM 1.5 illumination (100 mW/cm). (b) J-V curves of the device with electrolyte refreshing times under dark condition.

times under air mass (AM) 1.5 illumination (100 mW/cm^2) (Figure 2(a)) and under dark conditions (Figure 2(b)). The dependency of the photovoltaic performance parameters (J_{sc}, V_{oc}, FF, and PCE) on the electrolyte refreshing times is shown in Figure 3. With increased electrolyte refreshing times from 0 to 10, V_{oc} gradually increased from 0.697 to 0.760 V. FF also increased from 0.730 to 0.820, whereas J_{sc} decreased from 12.01 to 5.07 mA/cm^2. The PCE was at the apex when the electrolyte was refreshed for the first time. At this point, V_{oc}, J_{sc}, FF, and PCE of the FDSSCs were 0.725 V, 11.47 mA/cm^2, 0.749, and 6.23%, respectively. During the second electrolyte refreshing, the PCE began to drop. All these results indicated that certain phenomena occurred inside the FDSSCs when the electrolytes were frequently refreshed.

In a renewable electrochemical system [25, 26]

$$V_{oc} = \left(\frac{kT}{e}\right) \ln\left(\frac{I_{inj}}{n_{cb} k_{et} [I_3^-]}\right), \quad (1)$$

where I_{inj} is the photocurrent injected to the TiO_2 from the excited dye, n_{cb} is the photoelectron density of the conductive band, and k_{et} is the constant of the reaction, and

$$I_3^- + 2e_{cb}(TiO_2) \longrightarrow 3I^-, \quad (2)$$

where I_3^- can enter into tiny cavities inaccessible to dye molecules. The second equation is related to the dark reaction. As shown in Figure 1(b), the dark current density decreased with increased electrolyte refreshing times, especially during the initial stage. This finding meant that, with electrolyte refreshing, the dark reaction was restrained and k_{et} decreased. Consequently, there was less electron recombination between the electrons at the TiO_2 conduction band and I_3^- in the electrolyte. According to (1), V_{oc} increases for the same solar cell if weakened by the dark reaction.

As presented in Figure 3, J_{sc} and PCE also both decreased with increased electrolyte refreshing times. EIS was used to investigate these tendencies. EIS measured in the dark is a well-known technique for studying charge-transfer processes [27–30]. Wide ranges of electrochemical processes and internal resistances that govern the performance of DSSCs could be analyzed by this method. Important general conclusions have been made based on EIS studies. The impedance spectra of DSSCs consist of three components [29, 30], namely, charge transfers at the Pt/electrolyte and TiO_2/electrolyte interfaces, as well as the diffusion of I_3^- in the electrolyte. Each time the electrolyte is refreshed, the impedance data of the FDSSs were acquired at a forward bias of 0.75 V in the dark. Figure 4 shows the Nyquist plots of the corresponding FDSSCs with various electrolyte refreshing times in the dark at an applied forward bias of 0.75 V. The inset in Figure 3 is the equivalent circuit used to fit the experimental data [30, 31]. From left to right in Figure 3, the first arc is the interface between the counterelectrode and electrolyte; the second arc is the interface between TiO_2 and the electrolyte. In the equivalent circuit, R_s is the ohmic serial resistance. R_s represented the transmission resistance of the conducting substrate as well as the interface resistance between the conducting substrate and TiO_2. $R_{ct,CE}$: is the charge transfer resistance, by which the regeneration of I^- at the counterelectrode is characterized. Parallel to $R_{ct,CE}$, the constant phased element (CPE) describes the double-layer capacitance C_{Pt} at the counterelectrode/electrolyte interface.

$$\begin{aligned} CPE &= Y_0 (j\omega)^{-n}, \quad 0 < n \le 1,\ n = 1, \\ CPE &= \frac{1}{(j\omega C)}, \end{aligned} \quad (3)$$

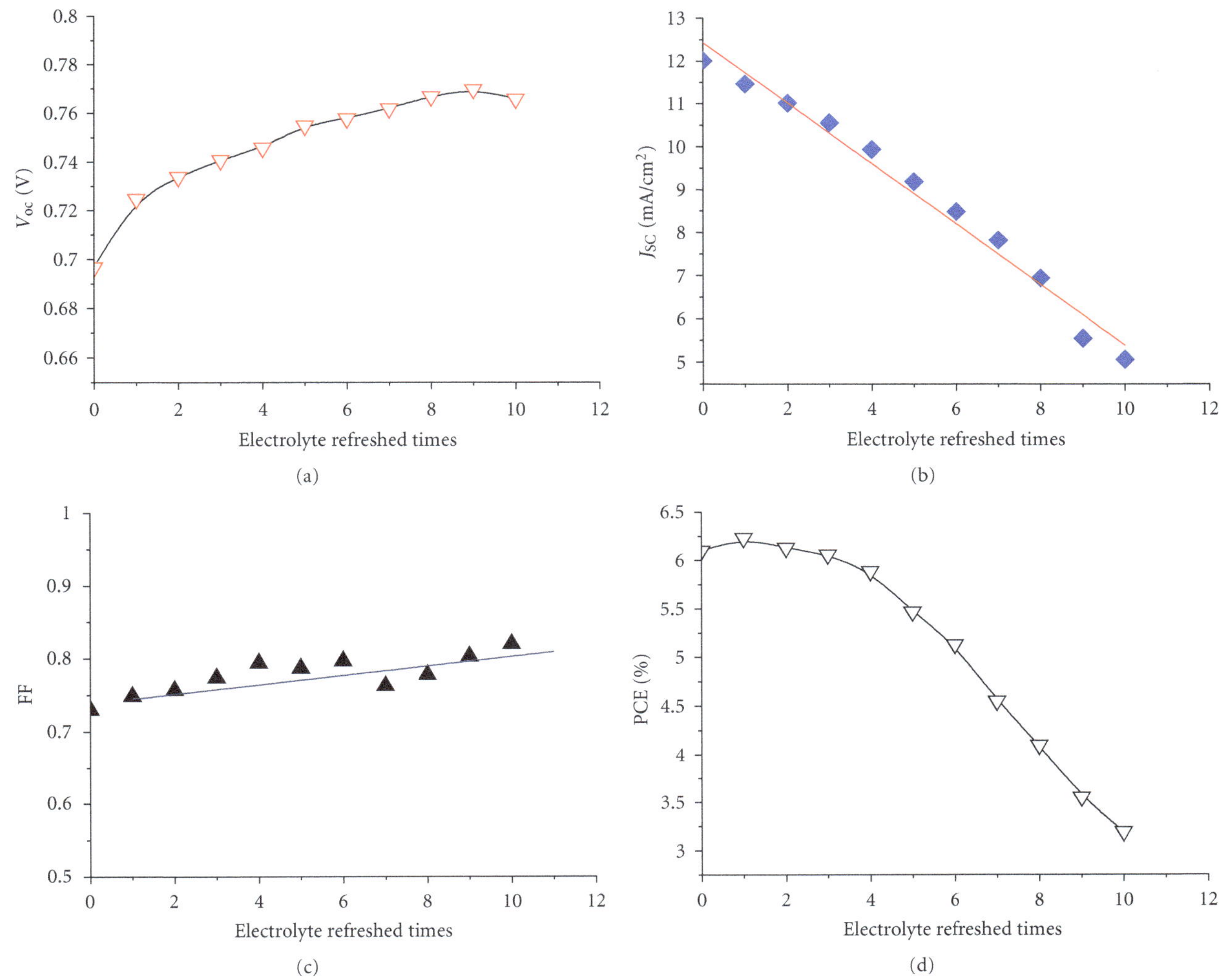

FIGURE 3: Dependency of V_{oc} (a), J_{sc} (b), FF (c), and PCE (d) on electrolyte refreshing times.

where C is the interface capacitance. $R_{ct,WE}$ and its parallel CPE describe the interfacial charge recombination reaction of electrons with the I_3^- ions and chemical capacitance at TiO_2/electrolyte interface, respectively. W is the impedance for the ion diffusion of electrolytes. For $R_{ct,WE}$ [32, 33],

$$R_{CT,WE} = \frac{1}{e}\left(\frac{\partial j_{rec}}{\partial F_n}\right)^{-1} = R_0 \exp\left[-\frac{\beta}{k_B T}\left(E_{F_n} - E_{redox}\right)\right] = \frac{k_B T \tau}{n_{cb} e^2}, \tag{4}$$

$$\tau = \sqrt{R_{CT,WE} C_\mu}, \tag{5}$$

$$C_\mu = e^2 \frac{\partial n_{cb}}{\partial E_{F_n}} = e^2 \frac{\overline{n}_{cb}}{k_B T}, \tag{6}$$

where j_{rec} is the recombination (or loss) current, E_{Fn} is the position of the Fermi level of electrons, E_{redox} is the potential of the redox couple, $E_{Fn} - E_{redox} = eV_{bias}$, V_{bias} is the potential at the electrode, β is the transfer coefficient, R_0 is a constant, k_B is the Boltzmann constant, T is the absolute temperature, n_{cb} is the electron density of the TiO_2 conduction band, C_μ is the TiO_2/electrolyte interface capacitance, τ is the electron life time, e is the unit charge, F_n is the Fermi level of TiO_2, and $\overline{n}_{cb}$ is the average electron density of the TiO_2 conduction band.

By fitting the experimental data, the useful parameters for understanding the underlying mechanisms of electron transport and recombination were attained, as indicated in Figure 5. The electron combination $R_{ct,WE}$ (Figure 5(a)) increased from 47.8 Ω at the initial to 340 Ω at the tenth electrolyte refreshing. According to (4), at the very same $V_{bias}(E_{Fn} - E_{redox} = eV_{bias})$, the recombination (or loss) current j_{rec} decreases with increased electrolyte refreshing time and electron time τ, as shown in Figure 5(b). Consequently, (7) could be derived as follows [32, 33]:

$$\frac{1}{e}\left(\frac{\partial j_{rec}}{\partial F_n}\right)^{-1} = R_0 \exp\left[-\frac{e\beta}{k_B T} V_{oc}\right]. \tag{7}$$

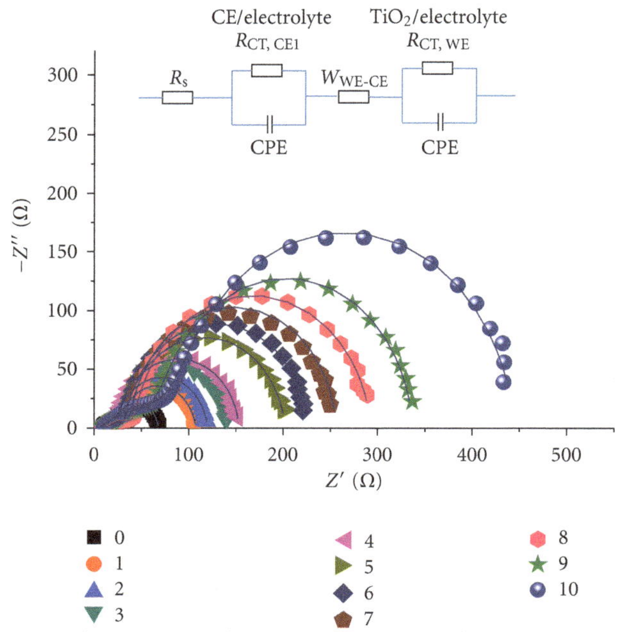

FIGURE 4: The Nyquist plots of the corresponding DSSCs under various electrolyte refreshing times in the dark at an applied forward bias of −0.75 V. The dot is the experiment data and the line is the fitting data via the inset equivalent circuit.

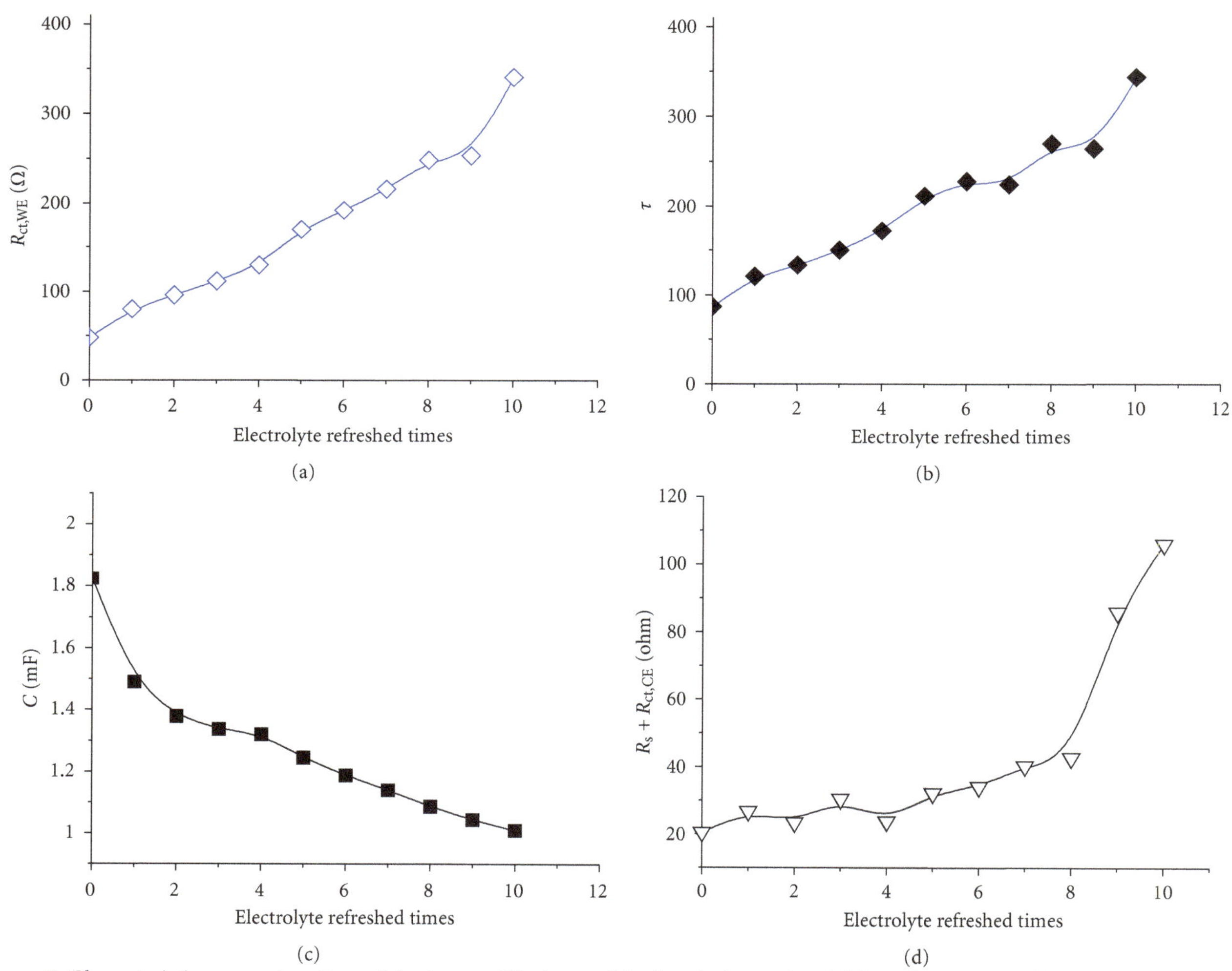

FIGURE 5: Characteristic parameters $R_{ct,WE}$ (a), electron life time τ (b), chemical capacity C (c), and $R_s + R_{ct,CE}$ (d) of the arc impedance spectra of DSSCs with various electrolyte refreshing times in the dark at an applied forward bias of −0.75 V.

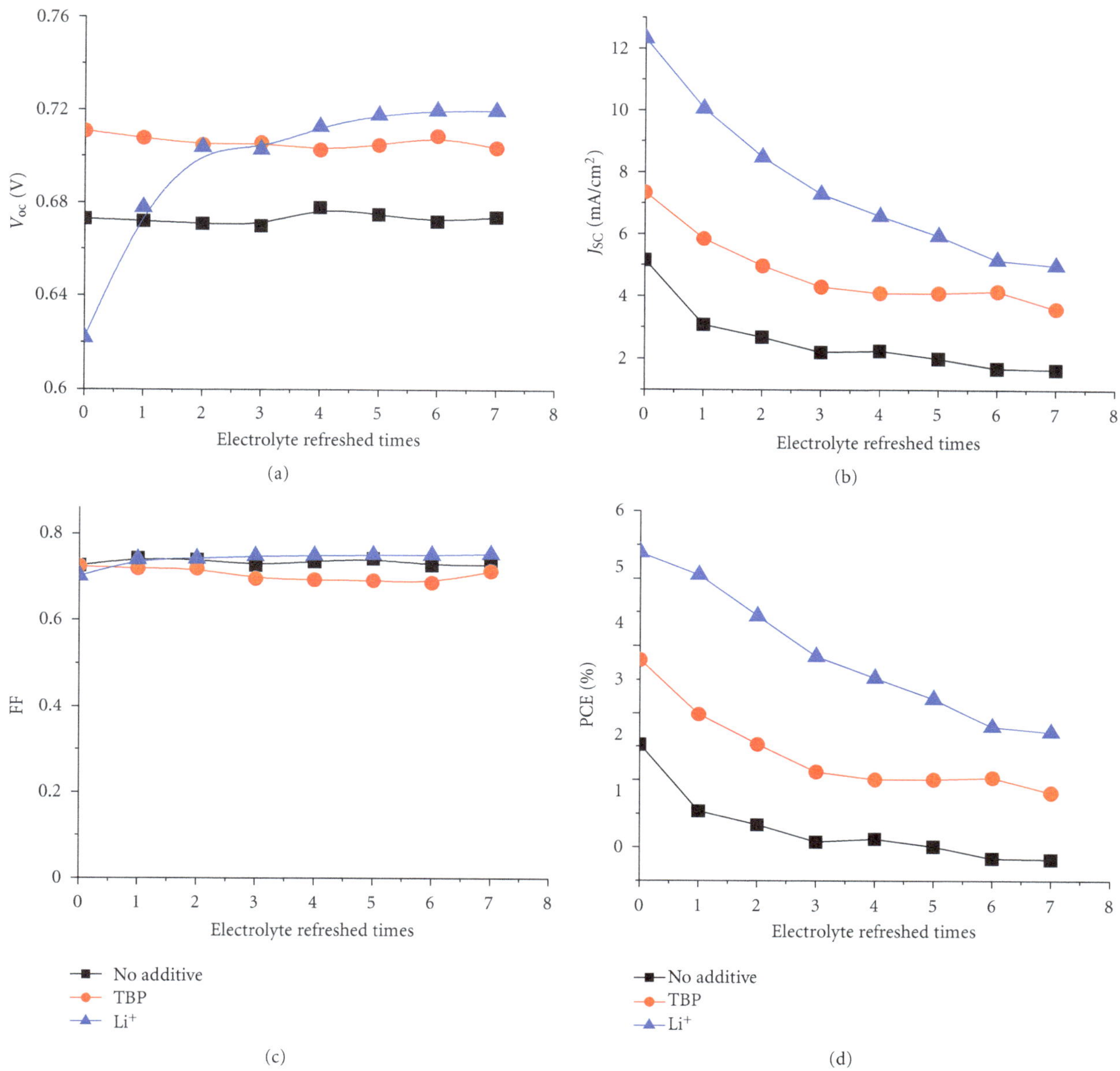

FIGURE 6: *I-V* parameters (V_{oc} (a), J_{sc} (b), FF (c), and PCE (d)) of devices based on different electrolytes varied with electrolyte refreshing times. No additive: 0.04 M I_2 and 0.6 M BMII; TBP: 0.04 M I_2, 0.6 M BMII, and 0.03 M TBP; Li^+: 0.04 M I_2, 0.6 M BMII, and 0.05 M LiI.

Therefore, under the same illumination, there would be less electron recombination and larger V_{oc}. This conclusion was consistent with the dark current shown in Figure 1(b) and with the conclusions derived from (1) and (2).

As aforementioned, J_{sc} declined with electrolyte refreshing. This finding may be ascribed to several aspects. On one hand, it was due to dye desorption. Figure 5(c) shows that the TiO_2/electrolyte interface capacitance C_μ decreased with electrolyte refreshing. According to (6), $\overline{n}_{cb}$ decreases with electrolyte refreshing. This tendency indicates less photocurrent injected into the TiO_2 from the excited dye, which may be due to dye desorption. On the other hand, the increased $R_s + R_{ct,CE}$ with electrolyte refreshing (Figure 5(d)) decreased the J_{sc} [31], which was likely to the electrolyte corrosion of the counter electrolyte [34].

The additive TBP and Li^+ play important roles in maintaining a high V_{oc} [35, 36]. The tendency shown in Figure 3 may be attributed to the remaining additive in the solar cell with electrolyte refreshing. To prove this assumption, the electrolyte refreshing effects on devices based on different electrolytes A, B, and C were investigated. The contents of electrolytes A, B, and C were detailed as follows: A: 0.04 M I_2 and 0.6 M BMII; B: 0.04 M I_2, 0.6 M BMII, and 0.03 M TBP; C: 0.04 M I_2, 0.6 M BMII, and 0.05 M LiI. All the *I-V* parameters of the devices based on A, B, and C are shown in Figure 6. In Figure 5(a), V_{oc} of the devices based on A and B changed little, whereas that of the devices based on C obviously increased with electrolyte refreshing. FF of all the devices only slightly changed with electrolyte refreshing. J_{sc} and PCE of all the devices decline in a similar tendency, which can be ascribed to the interface charge transport and recombination analyzed in the previous parts. Therefore, we could conclude from Figure 6 that the remaining Li^+ significantly influenced V_{oc} with electrolyte refreshing, the remaining TBP little influenced V_{oc} with electrolyte refreshing and the factors that decreased J_{sc} were mainly related to the decreased PCE. Comparing Figures 3 and 6, we could find that V_{oc} of devices based on electrolyte with both TBP and Li^+ was larger than that of A, B, C, which were based on electrolyte with no additive, with TBP only and with Li^+ only, respectively. It is confirmed that TBP and Li^+ were in synergism to increase V_{oc}.

4. Conclusions

The effect of electrolyte refreshing on the photovoltaic performance of FDSSCs was studied. The electron transport and interfacial recombination kinetics were also systematically investigated by EIS. With increased electrolyte refreshing times from 0 to 10 times, FF and especially V_{oc} increased, whereas J_{sc} and PCE significantly decreased. The increase in V_{oc} was mainly ascribed to less electron recombination with the I_3^- ions as the electrolytes were refreshed according to EIS. Li^+ also played a critical role in V_{oc}, increase and it worked with TBP synergistically to improve V_{oc}. The decline in J_{sc} and PCE was primarily brought about by dye desorption and the increase of series resistance.

Acknowledgments

This work was jointly supported by the NSFC (50833001), MOST (2011CB933300), and MOE (309001) of China.

References

[1] B. O'Regan and M. Grätzel, "A low-cost, high-efficiency solar cell based on dye-sensitized colloidal TiO_2 films," *Nature*, vol. 353, no. 6346, pp. 737–740, 1991.

[2] M. Grätzel, "Photoelectrochemical cells," *Nature*, vol. 414, no. 6861, pp. 338–344, 2001.

[3] M. K. Nazeeruddin, A. Kay, I. Rodicio et al., "Conversion of light to electricity by cis-X2bis(2, 2′-bipyridyl-4, 4′-dicarboxylate)ruthenium(II) charge-transfer sensitizers (X = Cl-, Br-, I-, CN-, and SCN-) on nanocrystalline TiO_2 electrodes," *Journal of the American Chemical Society*, vol. 115, no. 14, pp. 6382–6390, 1993.

[4] J. E. Moser, P. Bonnôte, and M. Grätzel, "Molecular photovoltaics," *Coordination Chemistry Reviews*, vol. 171, no. 1, pp. 245–250, 1998.

[5] L. M. Goncalves, V. De Zea Bermudez, H. A. Ribeiro, and A. M. Mendes, "Dye-sensitized solar cells: a safe bet for the future," *Energy and Environmental Science*, vol. 1, no. 6, pp. 655–667, 2008.

[6] Q. Yu, Y. Wang, Z. Yi et al., "High-efficiency dye-sensitized solar cells: the influence of lithium ions on exciton dissociation, charge recombination, and surface states," *ACS Nano*, vol. 4, no. 10, pp. 6032–6038, 2010.

[7] L. Han, A. Fukui, Y. Chiba et al., "Integrated dye-sensitized solar cell module with conversion efficiency of 8.2%," *Applied Physics Letters*, vol. 94, no. 1, Article ID 013305, 2009.

[8] X. Fan, Z. Chu, F. Wang, L. Chen, Y. Tang, and D. Zou, "Wire-shaped flexible dye-sensitized solar cells," *Advanced Materials*, vol. 20, no. 3, pp. 592–595, 2008.

[9] M. R. Lee, R. D. Eckert, K. Forberich, G. Dennler, C. J. Brabec, and R. A. Gaudiana, "Solar power wires based on organic photovoltaic materials," *Science*, vol. 324, no. 5924, pp. 232–235, 2009.

[10] H. Wang, Y. Liu, M. Li, H. Huang, M. Zhong, and H. Shen, "Hydrothermal growth of large-scale macroporous TiO_2 nanowires and its application in 3D dye-sensitized solar cells," *Applied Physics A*, vol. 97, no. 1, pp. 25–29, 2009.

[11] B. Weintraub, Y. G. Wei, and Z. L. Wang, "Optical fiber/nanowire hybrid structures for efficient three-dimensional dye-sensitized solar cells," *Angewandte Chemie*, vol. 48, no. 47, pp. 8981–8985, 2009.

[12] S. Huang, Q. Zhang, X. Huang et al., "Fibrous CdS/CdSe quantum dot co-sensitized solar cells based on ordered TiO_2 nanotube arrays ," *Nanotechnology*, vol. 21, no. 37, Article ID 375201, 2010.

[13] Y. Liu, M. Li, H. Wang et al., "Synthesis of TiO_2 nanotube arrays and its application in mini-3D dye-sensitized solar cells," *Journal of Physics D*, vol. 43, no. 20, 2010.

[14] B. O'Connor, D. Nothern, K. P. Pipe, and M. Shtein, "High efficiency, broadband solar cell architectures based on arrays of volumetrically distributed narrowband photovoltaic fibers," *Optics Express*, vol. 18, no. 19, pp. A432–A443, 2010.

[15] D. Zou, D. Wang, Z. Chu, Z. Lv, and X. Fan, "Fiber-shaped flexible solar cells," *Coordination Chemistry Reviews*, vol. 254, no. 9-10, pp. 1169–1178, 2010.

[16] J. Bae, Y. J. Park, M. Lee et al., "Single-fiber-based hybridization of energy converters and storage units using graphene as electrodes," *Advanced Materials*, vol. 23, no. 30, pp. 3446–3449, 2011.

[17] T. Chen, S. Wang, Z. Yang et al., "Flexible, light-weight, ultrastrong, and semiconductive carbon nanotube fibers for a highly efficient solar cell," *Angewandte Chemie*, vol. 50, no. 8, pp. 1815–1819, 2011.

[18] Y. Fu, Z. Lv, S. Hou et al., "Conjunction of fiber solar cells with groovy micro-reflectors as highly efficient energy harvesters," *Energy and Environmental Science*, vol. 4, no. 9, pp. 3379–3383, 2011.

[19] S. Huang, X. Guo, X. Huang et al., "Highly efficient fibrous dye-sensitized solar cells based on TiO_2 nanotube arrays," *Nanotechnology*, vol. 22, no. 31, Article ID 315402, 2011.

[20] Z. Lv, Y. Fu, S. Hou et al., "Large size, high efficiency fiber-shaped dye-sensitized solar cells," *Physical Chemistry Chemical Physics*, vol. 13, no. 21, pp. 10076–10083, 2011.

[21] D. Wang, S. Hou, H. Wu, C. Zhang, Z. Chu, and D. Zou, "Fiber-shaped all-solid state dye sensitized solar cell with remarkably enhanced performance via substrate surface engineering and TiO_2 film modification," *Journal of Materials Chemistry*, vol. 21, no. 17, pp. 6383–6388, 2011.

[22] J. Yu, D. Wang, Y. Huang et al., "A cylindrical core-shell-like TiO_2 nanotube array anode for flexible fiber-type dye-sensitized solar cells," *Nanoscale Research Letters*, vol. 6, article 94, 2011.

[23] K. Okada, H. Matsui, T. Kawashima, T. Ezure, and N. Tanabe, "$100\,mm \times 100\,mm$ large-sized dye sensitized solar cells," *Journal of Photochemistry and Photobiology A*, vol. 164, no. 1–3, pp. 193–198, 2004.

[24] M. A. Green, "Consolidation of thin-film photovoltaic technology: the coming decade of opportunity," *Progress in Photovoltaics*, vol. 14, no. 5, pp. 383–392, 2006.

[25] M. L. Rosenbluth and N. S. Lewis, "'Ideal' behavior of the open circuit voltage of semiconductor/liquid junctions," *The Journal of Physical Chemistry*, vol. 93, no. 9, pp. 3735–3740, 1989.
[26] A. Kumar, P. G. Santangelo, and N. S. Lewis, "Electrolysis of water at SrTiO3 photoelectrodes: distinguishing between the statistical and stochastic formalisms for electron-transfer processes in fuel-forming photoelectrochemical systems," *The Journal of Physical Chemistry*, vol. 96, no. 2, pp. 834–842, 1992.
[27] J. Bisquert, "Theory of the impedance of electron diffusion and recombination in a thin layer," *The Journal of Physical Chemistry B*, vol. 106, no. 2, pp. 325–333, 2002.
[28] J. Bisquert, A. Zaban, and P. Salvador, "Analysis of the mechanisms of electron recombination in nanoporous TiO_2 dye-sensitized solar cells. Nonequilibrium steady-state statistics and interfacial electron transfer via surface states," *The Journal of Physical Chemistry B*, vol. 106, no. 34, pp. 8774–8782, 2002.
[29] I. Morá-Seró and J. Bisquert, "Effect of reduced selectivity of contacts on the current-potential characteristics and conversion performance of solar cells," *Solar Energy Materials and Solar Cells*, vol. 85, no. 1, pp. 51–62, 2005.
[30] Q. Wang, J. E. Moser, and M. Grätzel, "Electrochemical impedance spectroscopic analysis of dye-sensitized solar cells," *The Journal of Physical Chemistry B*, vol. 109, no. 31, pp. 14945–14953, 2005.
[31] Q. Wang, S. Ito, M. Grätzel et al., "Characteristics of high efficiency dye-sensitized solar cells," *The Journal of Physical Chemistry B*, vol. 110, no. 50, pp. 25210–25221, 2006.
[32] F. Fabregat-Santiago, J. Bisquert, E. Palomares et al., "Correlation between photovoltaic performance and impedance spectroscopy of dye-sensitized solar cells based on ionic liquids," *The Journal of Physical Chemistry C*, vol. 111, no. 17, pp. 6550–6560, 2007.
[33] I. Mora-Seró, J. Bisquert, F. Fabregat-Santiago et al., "Implications of the negative capacitance observed at forward bias in nanocomposite and polycrystalline solar cells," *Nano Letters*, vol. 6, no. 4, pp. 640–650, 2006.
[34] T. N. Murakami, S. Ito, Q. Wang et al., "Highly efficient dye-sensitized solar cells based on carbon black counter electrodes," *Journal of the Electrochemical Society*, vol. 153, no. 12, Article ID 047612JES, pp. A2255–A2261, 2006.
[35] S. Rühle, M. Greenshtein, S. G. Chen et al., "Molecular adjustment of the electronic properties of nanoporous electrodes in dye-sensitized solar cells," *The Journal of Physical Chemistry B*, vol. 109, no. 40, pp. 18907–18913, 2005.
[36] K. Ryuzi, K. Motohiro, S. Kodate, A. Furube, N. Fuke, and N. Koide, "Effects of 4-tert-butylpyridine and li ions on photoinduced electron injection efficiency in black-dye-sensitized Nanocrystalline TiO_2 films," *The Journal of Physical Chemistry C*, vol. 113, no. 48, pp. 20738–20744, 2009.

4

Reliable and Damage-Free Estimation of Resistivity of ZnO Thin Films for Photovoltaic Applications Using Photoluminescence Technique

N. Poornima, T. V. Vimalkumar, V. G. Rajeshmon, C. Sudha Kartha, and K. P. Vijayakumar

Department of Physics, Cochin University of Science and Technology, Cochin, Kerala 682022, India

Correspondence should be addressed to K. P. Vijayakumar; kpv@cusat.ac.in

Academic Editor: Mahmoud M. El-Nahass

This work projects photoluminescence (PL) as an alternative technique to estimate the order of resistivity of zinc oxide (ZnO) thin films. ZnO thin films, deposited using chemical spray pyrolysis (CSP) by varying the deposition parameters like solvent, spray rate, pH of precursor, and so forth, have been used for this study. Variation in the deposition conditions has tremendous impact on the luminescence properties as well as resistivity. Two emissions could be recorded for all samples—the near band edge emission (NBE) at 380 nm and the deep level emission (DLE) at ~500 nm which are competing in nature. It is observed that the ratio of intensities of DLE to NBE (I_{DLE}/I_{NBE}) can be reduced by controlling oxygen incorporation in the sample. *I-V* measurements indicate that restricting oxygen incorporation reduces resistivity considerably. Variation of I_{DLE}/I_{NBE} and resistivity for samples prepared under different deposition conditions is similar in nature. I_{DLE}/I_{NBE} was always less than resistivity by an order for all samples. Thus from PL measurements alone, the order of resistivity of the samples can be estimated.

1. Introduction

Extensive research is being promoted on ZnO owing to the unique optoelectronic properties such as wide band gap of 3.3 eV, large room-temperature exciton binding energy of 60 meV, and piezoelectricity [1–5]. The wide band gap makes it an excellent candidate for use in UV sensors [6], light emitting diodes [7, 8], solar cells [9–11], and so forth. It serves its purpose as antireflection coating in solar cells [12] and nanorod layers of ZnO can be coupled with compound semiconductor absorber layers as in extremely thin absorber (ETA) cells [13–15] or can be used in organic photovoltaics [16, 17].

A wide range of deposition techniques have been reported for ZnO [18–35]. For device applications, films with good optical quality are required on a large scale. This is where cost-effective deposition techniques as wet chemical methods and spray pyrolysis play a vital role [36, 37]. We fabricated ZnO thin films using the chemical spray pyrolysis (CSP) technique with an intention to develop them as an affordable substitute to ITO. The focus of the industry is mainly on developing cost-effective deposition as well as characterization techniques so that the end product is consumer friendly. In the present work, with the method for depositing TCO being CSP, the requirement for cost-effective technique is fulfilled. ZnO is a candidate which exhibits strong excitonic as well as defect-assisted luminescence. Any change in composition by varying the stoichiometry, incorporating dopants, and varying the deposition parameters can cause variation in the PL spectrum. Electrical properties, mainly the resistivity, can be varied desirably by bringing about such variations. Many groups have come up with the observation that PL can be used to relate to the electrical properties [38, 39]. Our group also has reported such an observation on SnS thin films [40].

Thin films of ZnO were deposited by spraying alcoholic solution of zinc salt and various parameters such as spray rate, solvent used, pH of solution, and dopants used were varied; its effect on PL spectrum and resistivity was analyzed. In order to substantiate our findings that PL can be used to analyze the resistivity of ZnO thin films, we have also included our own similar observations with indium sulphide (In_2S_3) thin films.

2. Experimental Technique

ZnO thin films were deposited on soda lime glass substrates maintained at 450°C by spraying 0.3 M solution of zinc acetate dissolved in a mixture of alcohol and water. We had already reported the effectiveness of employing postdeposition treatment (inversion) on improving the resistivity of ZnO thin films by two orders [41]. Hence all the samples prepared were given this inversion treatment in order to have the added benefit of the treatment in addition to the variation in deposition parameters themselves. The alphabet letter "I" has been included in all the sample names to indicate that the samples have been subjected to "inversion process." Initially films were prepared using both "propanol-water mixture" and "ethanol-water mixture" as a base solvent. Films prepared with propanol-water mixture as a solvent were named ZP-I and those prepared with ethanol-water mixture as a solvent were named ZE-I. Alcohol-to-water ratio in the solvent is an important parameter. After identifying propanol as the better solvent, films were prepared for different propanol-to-water ratios, namely, 0.25 : 1, 0.42 : 1, 1 : 1, 1.5 : 1, and 2.3 : 1, and these were named ZP0.25 : 1-I, ZP0.42 : 1-I, and so forth. With the optimum propanol-to-water ratio in the solvent, pH of the spray solution was varied from 3 to 6. Films prepared were named as 3-I, 4-I, 5-I, and 6-I for simplicity. By fixing the propanol-to-water ratio and pH at the optimum value, films were prepared for three different spray rates—3, 7, and 10 mL/min, which were named as Z3-I, Z7-I, and Z10-I, respectively. With the main motivation being bringing down the resistivity of ZnO for TCO applications, doping with indium was tried and samples were prepared with various doping percentages of indium after optimizing all other deposition parameters. These were named as Z0.5In-I, Z1In-I, Z1.5In-I, Z2In-I, and Z2.5In-I, respectively, based on the doping percentages of indium. PL measurements were done on all samples prepared. The output of He-Cd laser (Kimmon Koha Co., Ltd.) at 325 nm was used as the excitation source and the emission spectrum was recorded using USB2000 Ocean Optics spectrophotometer coupled to the computer via the custom-made software OOIbase32. Thickness of all samples was measured using Dektak 6M stylus profiler. Resistivity of the samples was obtained from "two probe" *I*-*V* measurements using Keithley 236 Source Measure Unit. For resistivity measurements, ohmic contacts of dimension 5 mm × 1 mm using silver paint were given on the sample. The correlation observed between PL measurements and resistivity of the samples has been explained in subsequent sections.

3. Results and Discussion

The room-temperature PL spectrum of ZnO has two bands—the near band edge emission (NBE) arising out of the recombination of bound excitons and the deep level emission (DLE). The origin of DLE is still being debated over; many have attributed the emission as arising due to oxygen vacancies, zinc interstitials, and so forth [42–44], while others have related this emission to the O_{Zn} antisite defect [45, 46]. The effect of doping various transition metals on the broad defect-related luminescence of ZnO has also been reported [47]. Studies on the behavioural patterns of defects with excitation power are also reported. For systems with excellent optical quality, NBE is prominent whereas when the defect concentration is high, DLE gains prominence [48]. It has been generally observed that when the metal-to-oxygen ratio is high, the NBE gains strength. The effect of various treatments on PL spectrum has been described in subsequent sections. For all our samples, we could observe two emissions—the NBE and the DLE centered in the blue green. Assigning the emission as a transition from conduction band to antisite oxygen has been substantiated using irradiation studies on ZnO [49].

We begin our discussion by throwing light on the postdeposition treatment that we have given to all films. Still now there is not much consensus and understanding on the formation and annihilation of point defects on annealing. Any postdeposition treatment given definitely affects the extent of oxygen incorporation and it can be observed that there is an interesting correlation between the extent of oxygen incorporation, PL spectrum, and resistivity. Figure 1(a) depicts the variation of the PL spectrum for various treatments given to the sample. For convenience, samples left to cool on their own after deposition have been named ZR and samples which were subjected to the postdeposition treatment have been named ZI. We have presented the results on ZI, ZR, ZI-v (sample subjected to vacuum annealing after giving postdeposition treatment), and ZI-va (ZI-v subjected to air annealing) in Figure 1(a).

The reduction in oxygen incorporation results in reduction of the defect O_{Zn} antisite, resulting in reduction of the intensity of DLE. Upon postdeposition treatment, the Zn/O ratio increases (which has been verified using XPS and has already been reported [41]) and the intensity of NBE increases. The resistivity is also reduced by two orders. Results from XPS studies complement those from PL and rightly indicate that PL can also be used for an indirect analysis of composition. ZI was subjected to vacuum annealing and the vacuum annealed sample was subjected to air annealing solely to confirm that oxygen was the factor that decides the intensity of DLE as well as the resistivity. It can be seen that for samples of ZI-v, the fine structure of NBE appears which indicates the enhanced crystallinity of the sample. Resistivity is lowered further by an order. For samples of ZI-va, NBE has broadened and the intensity of DLE has increased. Resistivity has gone up by an order on air annealing. We observe that there is a relation between the ratio of intensities of DLE to NBE (I_{DLE}/I_{NBE}) and the resistivity. The smaller the ratio, the lower the resistivity (or the better the conductivity) and vice versa. The ratio I_{DLE}/I_{NBE} is a measure of the concentration of the defects present in the sample. Thus the choice of this ratio as an indicator of sample quality can be justified. Figure 1(b) represents the effect of postdeposition treatment on I_{DLE}/I_{NBE} and resistivity. The graph has been plotted on a log scale so that the variation can be clearly understood. It can be seen from the graph that the minimum value for both I_{DLE}/I_{NBE} and resistivity is for sample of ZI-v and the variation of the two parameters is similar.

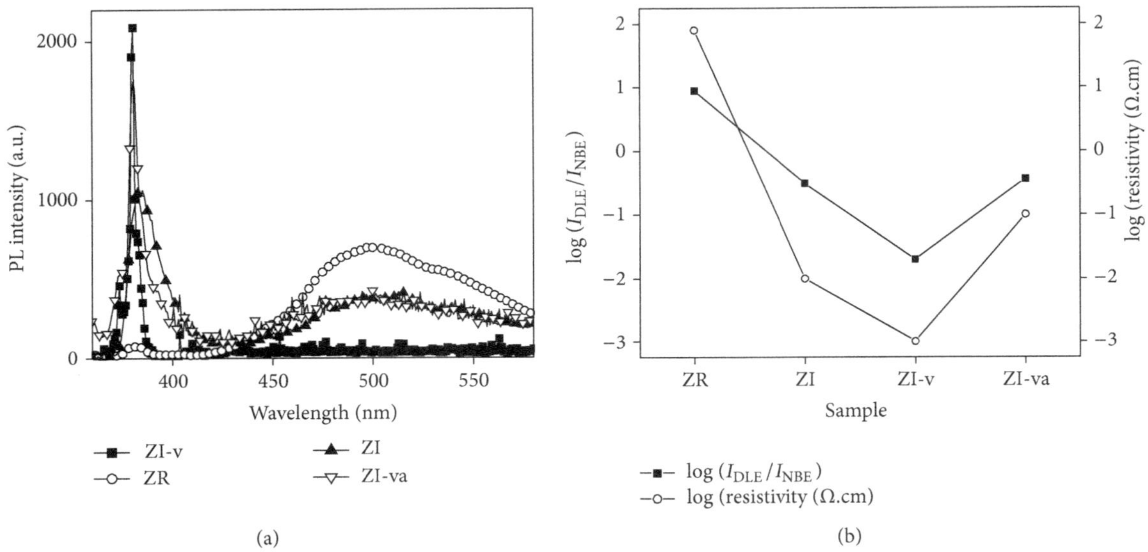

FIGURE 1: (a) PL spectrum of samples ZR, ZI, ZI-v, and ZI-va. (b) Variation of I_{DLE}/I_{NBE} and resistivity for various treatments to a sample.

3.1. Effect of Variation of Deposition Parameters

3.1.1. Effect of Solvent Variation on PL Spectrum. Transparency of the films was not good and resistivity was also high for films prepared using distilled water as a solvent. The effect of different alcoholic medium on obtaining crystalline transparent ZnO has also been an area of interest to many [50]. The benefit of using alcoholic solution in obtaining homogeneous films as well as clearly defined nanostructures has already been reported by many groups [51]. In order to improve the optical quality of the films, alcohol-water mixture was opted as a solvent.

The effect of using two different types of alcoholic solutions on the PL spectrum of ZnO thin films is shown in Figure 2.

It can be seen that for ZP-I, NBE is much enhanced in intensity and the ratio of I_{DLE}/I_{NBE} is smaller when compared to ZE-I. The defect concentration is higher in ZE-I. The high vapour pressure of alcohol helps in the formation of a uniform spray cone of finer droplets [52]. With alcohol being highly volatile, it escapes easily once it falls on the heated substrate. The lower I_{DLE}/I_{NBE} ratio as well as the lower resistivity indicates that propanol is better than ethanol. The variation in I_{DLE}/I_{NBE} and resistivity of ZnO thin films for different propanol-to-water ratios in the solvent was studied and has been plotted in Figure 3(a).

Resistivity is lowest [2.5×10^{-2} Ω·cm] for a propanol-to-water ratio of 1 : 1. Here also we observe that the variation of I_{DLE}/I_{NBE} for all samples matches closely with the variation of resistivity. Supporting evidence from structural, optical, and electrical studies, and so forth has been reported elsewhere [53].

3.1.2. Effect of Variation of Spray Solution pH on PL Spectrum. The effect of variation of pH of spray solution on the PL spectrum is shown in Figure 4.

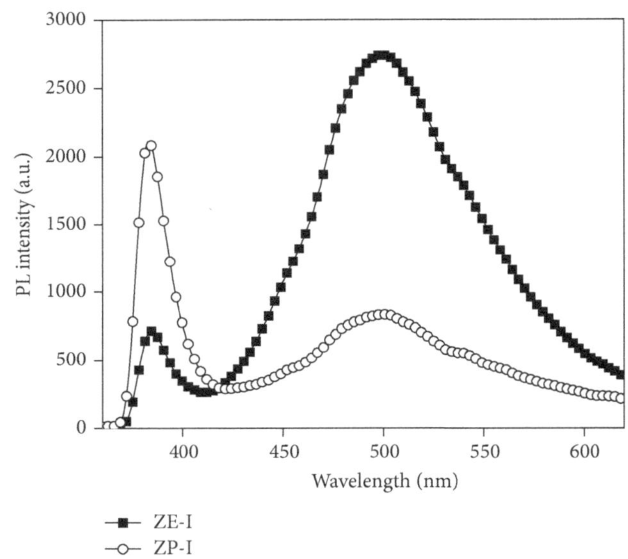

FIGURE 2: PL spectrum of ZP-I (propanol-based films) and ZE-I (ethanol-based films).

pH control was attained by adding a few drops of acetic acid to the precursor solution. This also inhibits hydroxide formation. Acetic acid is not added at all in sample 6-I (pH of spray solution = 6), and in sample 4-I a few drops of acetic acid has been added so as to make the pH of the spray solution equal to 4. pH determines the complexes formed at the beginning of the reaction, the enhancement in crystallinity, morphology, and so forth of the films formed [54].

We have observed that films with excellent optical (maximum transmittance) and electrical (lowest resistivity) properties were obtained when the spray solution was acidic (pH = 4). The variation of I_{DLE}/I_{NBE} and resistivity is plotted

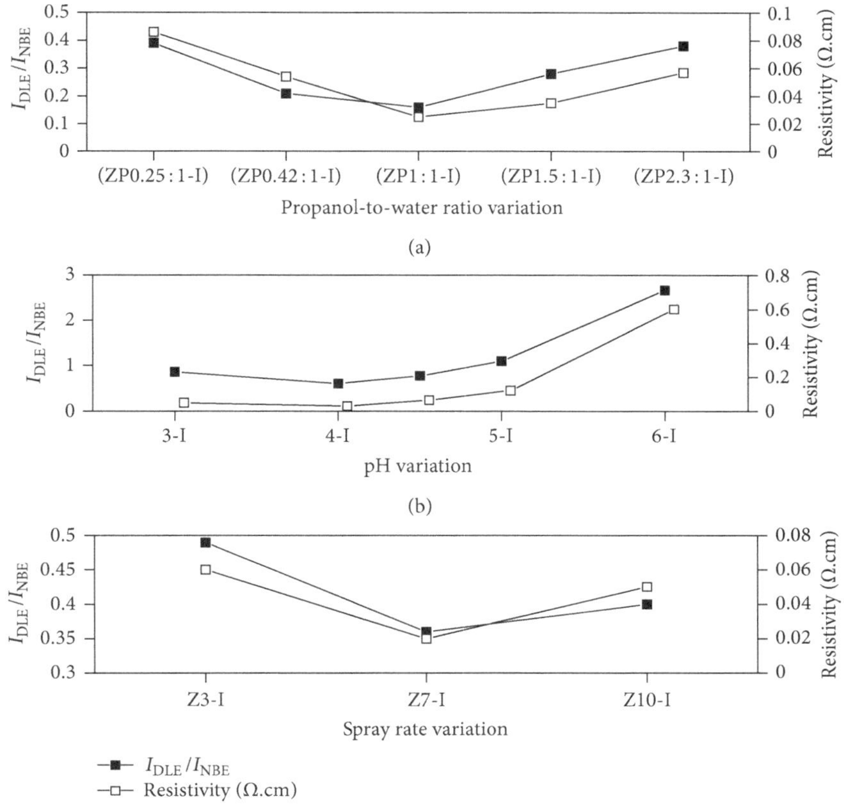

FIGURE 3: Variation of I_{DLE}/I_{NBE} and resistivity for (a) different propanol-to-water ratios in the solvent, (b) various pH of spray solution, and (c) different spray rates.

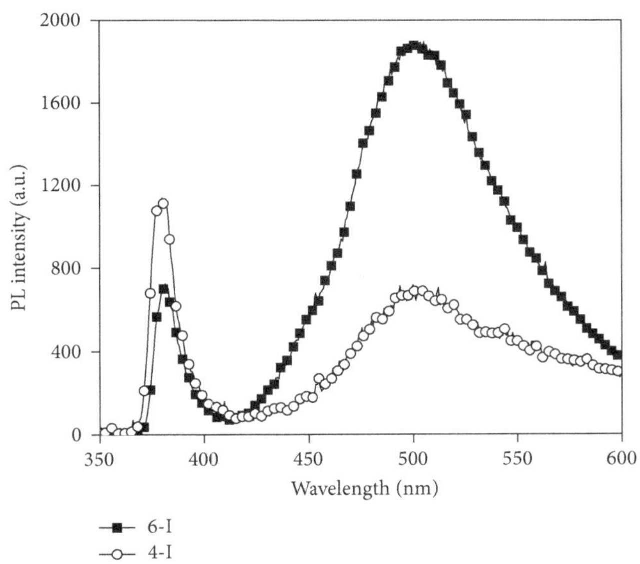

FIGURE 4: PL spectrum of ZnO thin films prepared using a spray solution of different pH.

in Figure 3(b). They show a similar variation and least values for both have been obtained when the pH of spray solution is maintained at 4. It has been well studied that pH determines the route of the reaction by controlling the intermediates formed as well as the rate of the reaction. Acidic pH favours Zn^{2+} ion formation which in turn facilitates ZnO formation. At pH other than 4, there might be unreacting intermediates which tend to deteriorate the properties of the film.

3.1.3. Effect of Spray Rate on PL Spectrum. In spray deposition, the solution in the form of fine droplets falls on the preheated substrate and the desired compound forms on the substrate. The droplets that strike the substrate initiate the formation of disk-shaped structures which later coalesce and form a continuous film. The continuity and roughness of the film depend largely on the spray rate. Figure 5(a) depicts the effect of spray rate on the PL spectrum and Figure 3(c) depicts its effect on the ratio of I_{DLE}/I_{NBE} and the resistivity. It can be observed that the ratio as well as the resistivity is minimum for Z7-I.

The full width at the half maximum (FWHM) of NBE is minimum and the intensity is the maximum when spray rate is 7 mL/min as opposed to 3 and 10 mL/min. Such low values of FWHM and intense NBE peak indicate the fine optical quality of the prepared films. In addition to finding out the ratio of I_{DLE}/I_{NBE}, the FWHM of the NBE peak at 3.26 eV was also found out which is another strong indicator of crystalline

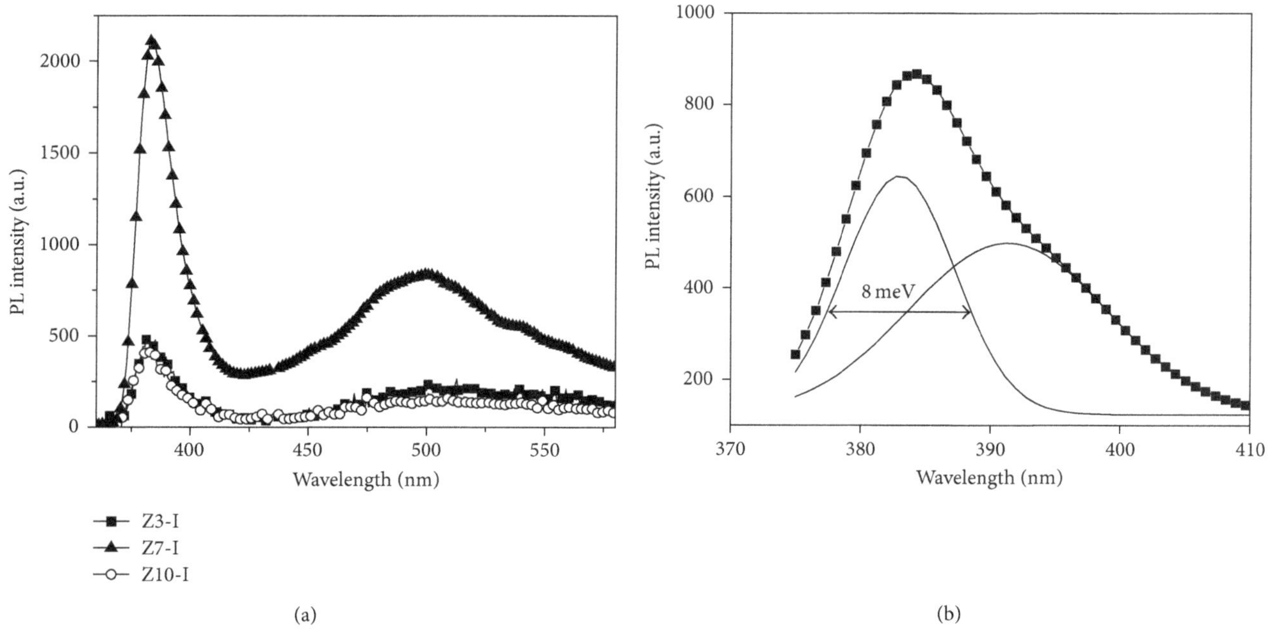

FIGURE 5: (a) PL spectrum of ZnO thin films prepared at different spray rates. (b) Gaussian fitting and FWHM of the NBE peak at 3.26 eV.

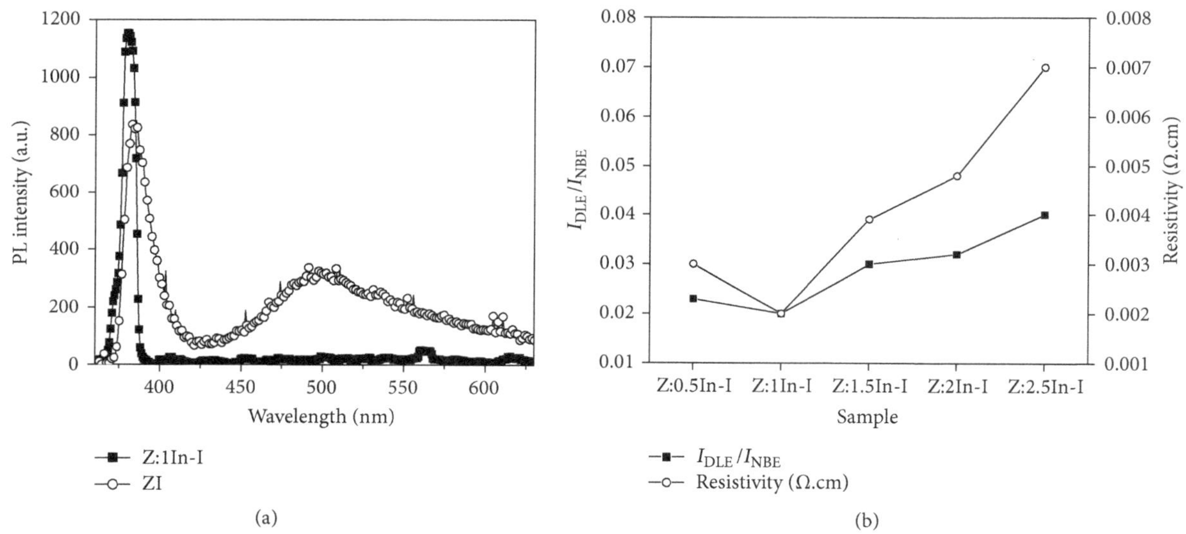

FIGURE 6: (a) Effect of indium doping on ZnO thin films. (b) Variation of I_{DLE}/I_{NBE} and resistivity with indium doping in ZnO thin films.

quality of the material. It could be observed that for films prepared at 7 mL/min, using a propanol-based precursor solution of pH 4 and given a postdeposition treatment, the FWHM of NBE was the least (≈8 meV) (Figure 5(b)).

3.2. Effect of Indium Doping on the PL Spectrum. With all the deposition conditions being optimized, the next step for reducing the resistivity was by doping with a trivalent metal. The effect of indium doping on the PL spectrum of ZnO is shown in Figure 6(a).

It can be seen that DLE is quite subdued for indium-incorporated films and the ratio of I_{DLE}/I_{NBE} has been observed to be the least for ZnO : 1In. Electrical measurements indicate that resistivity is also minimum for ZnO : 1In. The reason for reduction in resistivity may be because when indium is introduced into ZnO thin films, it accommodates itself in vacant Zn lattice sites thereby reducing the probability of oxygen occupying zinc site. This argument can be justified as the ionic radii of Zn^{2+} and In^{3+} are close-lying (0.74 and 0.75 Å, resp.), which in turn reduces the concentration of O_{Zn} antisite defect centres which explains the reduction in intensity of DLE and thereby the ratio I_{DLE}/I_{NBE}. But this process cannot take place indefinitely. So beyond a particular quantity of indium incorporated, the resistivity starts

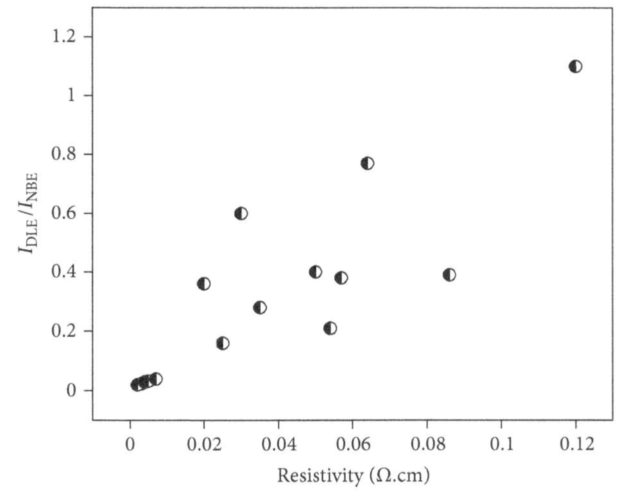

FIGURE 7: Variation of I_{DLE}/I_{NBE} as a function of resistivity for all ZnO samples chosen for study.

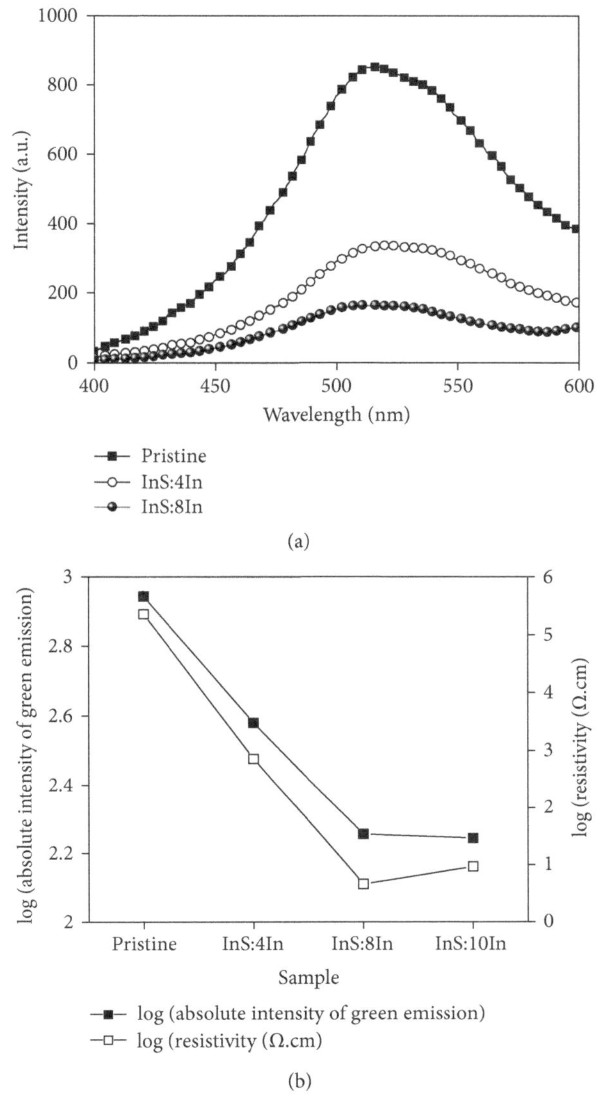

FIGURE 8: (a) Variation in intensity of green emission in the PL spectrum of In_2S_3 thin films on indium incorporation. (b) Variation of absolute intensity of green emission and resistivity of In_2S_3 thin films on indium incorporation drawn in a log scale.

increasing. This might be due to scattering of carriers from excess indium occupying sites other than lattice positions [55]. Figure 6(b) depicts the effect of indium doping on I_{DLE}/I_{NBE} and resistivity.

Based on our discussion in the previous sections, variation of I_{DLE}/I_{NBE} with resistivity has been plotted (Figure 7) in order to analyze the correlation between the two parameters.

The extent of correlation is usually expressed in terms of "correlation coefficient." Generally, if the two parameters are perfectly correlated, the correlation coefficient assumes the value 1. For perfectly uncorrelated data, it is zero. Another point to note is whether the parameters are positively or negatively correlated. We say that two parameters are positively correlated when both parameters exhibit a similar trend in variation, that is, when one parameter is high, if the other is also high and vice versa. For perfect positive correlation, the plot would be linear with positive slope. From our observations, resistivity and I_{DLE}/I_{NBE} exhibit positive correlation and the correlation coefficient was found to be 0.844. This is a reasonably good value for the correlation coefficient. From our observations, we arrive at an empirical relation between the order of resistivity (Y) and the order of I_{DLE}/I_{NBE} (X) as $Y = 10^{-1}X$, which holds true for all the ZnO samples we have tried so far.

In order to substantiate the results on ZnO, we performed a trial to check whether PL can be used to qualitatively analyze the resistivity of samples by conducting PL studies on spray-deposited In_2S_3 thin films. In_2S_3 thin films were prepared by spraying a solution containing indium chloride and thiourea in the ratio 1.2 : 12 at a spray rate of 6 mL/min onto a substrate maintained at 350°C. By birth, In_2S_3 is formed with a large number of vacancies ordering at tetrahedral sites [56] and is also highly resistive (as-deposited films had resistivity of 2.3×10^5 Ω·cm). To bring down the resistivity for certain specific applications, we diffused the metal indium into as-deposited In_2S_3 thin films. Different quantities of indium were diffused by vacuum evaporation followed by subsequent annealing and samples were named as InS : 4In, InS : 6In, InS : 8In, and InS : 10In, respectively. With indium diffusion, the resistivity of the samples could be reduced by five orders (4.7 Ω·cm for InS : 8In). The PL spectrum of In_2S_3 is comprised of a green band (arising from V_S-V_{In} transition) and a red band (arising from In_i-O_{V_S} transition) [57]. The green emission in the PL spectrum is focussed in Figure 8(a). The diffused "In" occupies the vacant indium sites, which reduces the density of states of the defect V_{In} which explains the drastic reduction in the intensity of green emission. "In" incorporation reduces the resistivity by lowering the scattering of carriers from defects.

The variation of absolute intensity of GB in the PL spectrum and resistivity of the films on indium incorporation has been depicted in Figure 8(b). It can be seen that the variation of both parameters follows a similar pattern.

Though we could not arrive at an empirical relation connecting PL and resistivity for In_2S_3, the results on In_2S_3 complement our observations on ZnO thin films. From our results on ZnO as well as In_2S_3 thin films, it is clear that any manipulation of deposition parameters, stoichiometry, or doping to enhance the conductivity is reflected clearly in the PL spectrum and it is possible to assess the electrical properties using PL.

4. Conclusion

The major success of this work lies in asserting PL as the best available tool for damage-free analysis of the sample. We have been able to establish the direct connection of resistivity with the luminescence properties of ZnO thin films prepared by varying different deposition conditions. Using the empirical relation that we have proposed connecting the order of resistivity and the order of I_{DLE}/I_{NBE}, the idea of the order of resistivity can be estimated solely from PL measurements. This technique is of benefit to the industry as well as for analysis of samples as in situ nondestructive analysis on a large scale during processing is possible, employing PL.

Conflict of Interests

The authors of the paper do not have direct financial relation with any commercial identity mentioned in their paper that might lead to a conflict of interests for any of the authors.

Acknowledgments

N. Poornima is grateful to UGC for the fellowship under the UGC-RFSMS scheme. The authors are grateful to MNRE (Government of India), for many of the experimental setup.

References

[1] U. Özgür, Ya. I. Alimov, C. Liu et al., "A comprehensive review of ZnO materials and devices," *Journal of Applied Physics*, vol. 98, no. 4, Article ID 041301, 103 pages, 2005.

[2] M. Wraback, H. Shen, S. Liang, C. R. Gorla, and Y. Lu, "High contrast, ultrafast optically addressed ultraviolet light modulator based upon optical anisotropy in ZnO films grown on R-plane sapphire," *Applied Physics Letters*, vol. 74, no. 4, pp. 507–509, 1999.

[3] D. C. Reynolds, D. C. Look, and B. Jogai, "Optically pumped ultraviolet lasing from ZnO," *Solid State Communications*, vol. 99, no. 12, pp. 873–875, 1996.

[4] D. C. Look, C. Coşkun, B. Claflin, and G. C. Farlow, "Electrical and optical properties of defects and impurities in ZnO," *Physica B*, vol. 340–342, pp. 32–38, 2003.

[5] T. T. Chen, C. L. Cheng, S. P. Fu, and Y. F. Chen, "Photoelastic effect in ZnO nanorods," *Nanotechnology*, vol. 18, no. 22, Article ID 225705, 4 pages, 2007.

[6] C. Jagadish and S. J. Pearton, Eds., *ZnO Bulk, Thin Films and Nanostructures*, Elsevier, Oxford, UK, 2006.

[7] R. Könenkamp, R. C. Word, and M. Godinez, "Ultraviolet electroluminescence from ZnO/polymer heterojunction light-emitting diodes," *Nano Letters*, vol. 5, no. 10, pp. 2005–2008, 2005.

[8] D. C. Kim, W. S. Han, H. K. Cho, B. H. Kong, and H. S. Kim, "Multidimensional ZnO light-emitting diode structures grown by metal organic chemical vapor deposition on p-Si," *Applied Physics Letters*, vol. 91, no. 23, Article ID 231901, 3 pages, 2007.

[9] P. Jackson, D. Hariskos, E. Lotter et al., "New world record efficiency for $Cu(In,Ga)Se_2$ thin-film solar cells beyond 20%," *Progress in Photovoltaics: Research and Applications*, vol. 19, no. 7, pp. 894–897, 2011.

[10] A. Gupta and A. D. Compaan, "14% CdS/CdTe thin film cells with ZnO:Al TCO," in *Proceedings of the MRS Spring Meeting*, pp. 1–6, San Francisco, Calif, USA, 2003, paper no. B3.9.

[11] U. Rau and M. Schmidt, "Electronic properties of ZnOCd-SCu(In, Ga)Se_2 solar cells—aspects of heterojunction formation," *Thin Solid Films*, vol. 387, no. 1-2, pp. 141–146, 2001.

[12] J. Y. Chen and K. W. Sun, "Growth of vertically aligned ZnO nanorod arrays as antireflection layer on silicon solar cells," *Solar Energy Materials and Solar Cells*, vol. 94, no. 5, pp. 930–934, 2010.

[13] R. Tena-Zaera, M. A. Ryan, A. Katty, G. Hodes, S. Bastide, and C. Lévy-Clément, "Fabrication and characterization of ZnO nanowires/CdSe/CuSCN eta-solar cell," *Comptes Rendus Chimie*, vol. 9, no. 5-6, pp. 717–729, 2006.

[14] M. Krunks, E. Kärber, A. Katerski et al., "Extremely thin absorber layer solar cells on zinc oxide nanorods by chemical spray," *Solar Energy Materials and Solar Cells*, vol. 94, no. 7, pp. 1191–1195, 2010.

[15] D. Kieven, T. Dittrich, A. Belaidi et al., "Effect of internal surface area on the performance of ZnO In_2 S_3 CuSCN solar cells with extremely thin absorber," *Applied Physics Letters*, vol. 92, no. 15, Article ID 153107, 3 pages, 2008.

[16] W. J. E. Beek, L. H. Slooff, M. M. Wienk, J. M. Kroon, and R. A. J. Janssen, "Hybrid solar cells using a zinc oxide precursor and a conjugated polymer," *Advanced Functional Materials*, vol. 15, no. 10, pp. 1703–1707, 2005.

[17] W. J. E. Beek, M. M. Wienk, M. Kemerink, X. Yang, and R. A. J. Janssen, "Hybrid zinc oxide conjugated polymer bulk heterojunction solar cells," *Journal of Physical Chemistry B*, vol. 109, no. 19, pp. 9505–9516, 2005.

[18] J. S. Huang and C. F. Lin, "Influences of ZnO sol-gel thin film characteristics on ZnO nanowire arrays prepared at low temperature using all solution-based processing," *Journal of Applied Physics*, vol. 103, no. 1, Article ID 014304, 5 pages, 2008.

[19] D. Bera, L. Qian, S. Sabui, S. Santra, and P. H. Holloway, "Photoluminescence of ZnO quantum dots produced by a sol-gel process," *Optical Materials*, vol. 30, no. 8, pp. 1233–1239, 2008.

[20] T. Hiramatsu, M. Furuta, H. Furuta, T. Matsuda, and T. Hirao, "Influence of thermal annealing on microstructures of zinc oxide films deposited by RF magnetron sputtering," *Japanese Journal of Applied Physics*, vol. 46, no. 6A, pp. 3319–3323, 2007.

[21] J. Lu, Y. Zhang, Z. Ye, L. Wang, B. Zhao, and J. Huang, "p-type ZnO films deposited by DC reactive magnetron sputtering at different ammonia concentrations," *Materials Letters*, vol. 57, no. 22-23, pp. 3311–3314, 2003.

[22] E. S. Jung, J. Y. Lee, H. S. Kim, and N. W. Jang, "Structural and optical characteristics of ZnO films with oxygen content," *Journal of the Korean Physical Society*, vol. 47, no. 3, pp. S480–S484, 2005.

[23] H. Lin, S. Zhou, J. Zhou et al., "Structural and optical properties of a-plane ZnO thin films synthesized on γ-$LiAlO_2$ (302) substrates by low pressure metal-organic chemical vapor deposition," *Thin Solid Films*, vol. 516, no. 18, pp. 6079–6082, 2008.

[24] E. Guziewicz, I. A. Kowalik, M. Godlewski et al., "Extremely low temperature growth of ZnO by atomic layer deposition," *Journal of Applied Physics*, vol. 103, no. 3, Article ID 033515, 6 pages, 2008.

[25] A. Wójcik, M. Godlewski, E. Guziewicz, R. Minikayev, and W. Paszkowicz, "Controlling of preferential growth mode of ZnO thin films grown by atomic layer deposition," *Journal of Crystal Growth*, vol. 310, no. 2, pp. 284–289, 2008.

[26] E. Przeździecka, T. Krajewski, L. Wachnicki et al., "Characterization of ZnO films grown at low temperature," *Acta Physica Polonica A*, vol. 114, no. 5, pp. 1303–1310, 2008.

[27] K. Miyamoto, M. Sano, H. Kato, and T. Yao, "High-electron-mobility ZnO epilayers grown by plasma-assisted molecular beam epitaxy," *Journal of Crystal Growth*, vol. 265, no. 1-2, pp. 34–40, 2004.

[28] S. H. Park, H. Suzuki, T. Minegishi et al., "Low-temperature growth of high-quality ZnO layers by surfactant-mediated molecular-beam epitaxy," *Journal of Crystal Growth*, vol. 309, no. 2, pp. 158–163, 2007.

[29] B. J. Norris, J. Anderson, J. F. Wager, and D. A. Keszler, "Spin-coated zinc oxide transparent transistors," *Journal of Physics D*, vol. 36, no. 20, pp. L105–L107, 2003.

[30] P. F. Carcia, R. S. McLean, M. H. Reilly, and G. Nunes, "Transparent ZnO thin-film transistor fabricated by rf magnetron sputtering," *Applied Physics Letters*, vol. 82, no. 7, pp. 1117–1119, 2003.

[31] J. Siddiqui, E. Cagin, D. Chen, and J. D. Phillips, "ZnO thin-film transistors with polycrystalline (Ba,Sr) TiO_3 gate insulators," *Applied Physics Letters*, vol. 88, no. 21, Article ID 212903, 3 pages, 2006.

[32] K. K. Kim, J. H. Song, H. J. Jung, W. K. Choi, S. J. Park, and J. H. Song, "The grain size effects on the photoluminescence of ZnO/α-Al_2O_3 grown by radio-frequency magnetron sputtering," *Journal of Applied Physics*, vol. 87, no. 7, pp. 3573–3575, 2000.

[33] P. Fons, K. Iwata, S. Niki, A. Yamada, and K. Matsubara, "Growth of high-quality epitaxial ZnO films on α-Al_2O_3," *Journal of Crystal Growth*, vol. 201-202, pp. 627–632, 1999.

[34] Y. Liu, C. R. Gorla, S. Liang et al., "Ultraviolet detectors based on epitaxial ZnO films grown by MOCVD," *Journal of Electronic Materials*, vol. 29, no. 1, pp. 69–74, 2000.

[35] R. D. Vispute, V. Talyansky, S. Choopun et al., "Heteroepitaxy of ZnO on GaN and its implications for fabrication of hybrid optoelectronic devices," *Applied Physics Letters*, vol. 73, no. 3, pp. 348–350, 1998.

[36] V. V. Ursaki, I. M. Tiginyanu, V. V. Zalamai et al., "Photoluminescence of ZnO layers grown on opals by chemical deposition from zinc nitrate solution," *Semiconductor Science and Technology*, vol. 19, no. 7, pp. 851–854, 2004.

[37] K. Krunks, O. Bijakina, V. Mikli, T. Varema, and E. Mellikov, "Zinc oxide thin films by spray pyrolysis method," *Physica Scripta*, vol. 79, pp. 209–212, 1999.

[38] V. G. Rajeshmon, C. Sudha Kartha, K. P. Vijayakumar, C. Sanjeeviraja, T. Abe, and Y. Kashiwaba, "Role of precursor solution in controlling the opto-electronic properties of spray pyrolysed Cu_2ZnSnS_4 thin films," *Solar Energy*, vol. 85, no. 2, pp. 249–255, 2011.

[39] T. T. John, C. Sudha Kartha, K. P. Vijayakumar, T. Abe, and Y. Kashiwaba, "Preparation of indium sulfide thin films by spray pyrolysis using a new precursor indium nitrate," *Applied Surface Science*, vol. 252, no. 5, pp. 1360–1367, 2005.

[40] T. H. Sajeesh, C. Sudha Kartha, C. Sanjeeviraja, T. Abe, Y. Kashiwaba, and K. P. Vijayakumar, "Ex situ Sn diffusion: a well-suited technique for enhancing the photovoltaic properties of a SnS absorber layer," *Journal of Physics D*, vol. 43, no. 44, Article ID 445102, 6 pages, 2010.

[41] T. V. Vimalkumar, N. Poornima, C. Sudha Kartha, K. P. Vijayakumar, T. Abe, and Y. Kashiwaba, "Enhancement of electrical conductivity in sprayed ZnO thin film through zero-energyprocess," *Physica B*, vol. 405, no. 24, pp. 4957–4960, 2010.

[42] K. Vanheusden, W. L. Warren, C. H. Seager, D. R. Tallant, J. A. Voigt, and B. E. Gnade, "Mechanisms behind green photoluminescence in ZnO phosphor powders," *Journal of Applied Physics*, vol. 79, no. 10, pp. 7983–7990, 1996.

[43] F. A. Kroger, *The Chemistry of Imperfect Crystals*, North-Holland, Amsterdam, The Netherlands, 1974.

[44] H. J. Egelhaaf and D. Oelkrug, "Luminescence and nonradiative deactivation of excited states involving oxygen defect centers in polycrystalline ZnO," *Journal of Crystal Growth*, vol. 161, no. 1–4, pp. 190–194, 1996.

[45] B. Lin, Z. Fu, Y. Jia, and G. Liao, "Defect photoluminescence of undoping ZnO films and its dependence on annealing conditions," *Journal of the Electrochemical Society*, vol. 148, no. 3, pp. G110–G113, 2001.

[46] P. M. Ratheesh Kumar, C. Sudha Kartha, K. P. Vijayakumar, F. Singh, and D. K. Avasthi, "Effect of fluorine doping on structural, electrical and optical properties of ZnO thin films," *Materials Science and Engineering B*, vol. 117, no. 3, pp. 307–312, 2005.

[47] C. Y. Lin, W. H. Wang, C. S. Lee, K. W. Sun, and Y. W. Suen, "Magnetophotoluminescence properties of Co-doped ZnO nanorods," *Applied Physics Letters*, vol. 94, no. 15, Article ID 151909, 3 pages, 2009.

[48] L. Kumari and W. Z. Li, "Synthesis, structure and optical properties of zinc oxide hexagonal microprisms," *Crystal Research and Technology*, vol. 45, no. 3, pp. 311–315, 2010.

[49] P. M. Ratheesh Kumar, K. P. Vijayakumar, and C. Sudha Kartha, "On the origin of blue-green luminescence in spray pyrolysed ZnO thin films," *Journal of Materials Science*, vol. 42, no. 8, pp. 2598–2602, 2007.

[50] A. Maldonado, M. D. L. Olvera, R. Asomoza, and S. Tirado-Guerra, "Highly conductive and transparent in-doped zinc oxide thin films deposited by chemical spray using $Zn(C_5H_7O_2)_2$," *Journal of Materials Science: Materials in Electronics*, vol. 12, no. 11, pp. 623–627, 2001.

[51] J. S. Cho, S. K. Hong, D. S. Jung, and Y. C. Kang, "Effects of solvent on the properties of nano-sized glass powders prepared by flame spray pyrolysis," *Journal of the Ceramic Society of Japan*, vol. 116, no. 1350, pp. 334–340, 2008.

[52] E. Karber, T. Raadik, T. Dedova et al., "Photoluminescence of spray pyrolysis deposited ZnO nanorods," *Nanoscale Research Letters*, vol. 6, no. 1, article 359, pp. 1–7, 2011.

[53] T. V. Vimalkumar, N. Poornima, C. Sudha Kartha, and K. P. Vijayakumar, "Effect of precursor medium on structural, electrical and optical properties of sprayed polycrystalline ZnO thin films," *Materials Science and Engineering B*, vol. 175, no. 1, pp. 29–35, 2010.

[54] A. Maldonado, R. Asomoza, J. Cañetas-Ortega et al., "Effect of the pH on the physical properties of ZnO : in thin films

deposited by spray pyrolysis," *Solar Energy Materials and Solar Cells*, vol. 57, no. 4, pp. 331–344, 1999.

[55] A. Maldonado, M. D. L. Olvera, R. Asomoza, and S. Tirado-Guerra, "Highly conductive and transparent in-doped zinc oxide thin films deposited by chemical spray using $Zn(C_5H_7O_2)_2$," *Journal of Materials Science: Materials in Electronics*, vol. 12, no. 11, pp. 623–627, 2001.

[56] K. Kambas, J. Spyridelis, and M. Balkanski, "Far infrared and Raman optical study of α- and β-In_2S_3 compounds," *Physica Status Solidi (B)*, vol. 105, no. 1, pp. 291–296, 1981.

[57] R. Jayakrishnan, T. T. John, C. Sudha Kartha, K. P. Vijayakumar, T. Abe, and Y. Kashiwaba, "Defect analysis of sprayed β-In2S3 thin films using photoluminescence studies," *Semiconductor Science and Technology*, vol. 20, no. 12, pp. 1162–1167, 2005.

Bromate Formation Characteristics of UV Irradiation, Hydrogen Peroxide Addition, Ozonation, and Their Combination Processes

Naoyuki Kishimoto[1] and Eri Nakamura[2]

[1] *Faculty of Science and Technology, Ryukoku University, 1-5 Yokotani, Setaoe-cho, Otsu 520-2194, Japan*
[2] *Graduate School of Science and Technology, Ryukoku University, 1-5 Yokotani, Setaoe-cho, Otsu 520-2194, Japan*

Correspondence should be addressed to Naoyuki Kishimoto, naoyuki@rins.ryukoku.ac.jp

Academic Editor: Manickavachagam Muruganandham

Bromate formation characteristics of six-physicochemical oxidation processes, UV irradiation, single addition of hydrogen peroxide, ozonation, UV irradiation with hydrogen peroxide addition (UV/H_2O_2), ozonation with hydrogen peroxide addition (O_3/H_2O_2), and ozonation with UV irradiation (O_3/UV) were investigated using 1.88 μM of potassium bromide solution with or without 6.4 μM of 4-chlorobenzoic acid. Bromate was not detected during UV irradiation, single addition of H_2O_2, and UV/H_2O_2, whereas ozone-based treatments produced BrO_3^-. Hydroxyl radicals played more important role in bromate formation than molecular ozone. Acidification and addition of radical scavengers such as 4-chlorobenzoic acid were effective in inhibiting bromate formation during the ozone-based treatments because of inhibition of hydroxyl radical generation and consumption of hydroxyl radicals, respectively. The H_2O_2 addition was unable to decompose 4-chlorobenzoic acid, though O_3/UV and O_3/H_2O_2 showed the rapid degradation, and UV irradiation and UV/H_2O_2 showed the slow degradation. Consequently, if the concentration of organic contaminants is low, the UV irradiation and/or UV/H_2O_2 are applicable to organic contaminants removal without bromate formation. However, if the concentration of organic contaminants is high, O_3/H_2O_2 and O_3/UV should be discussed as advanced oxidation processes because of their high organic removal efficiency and low bromate formation potential at the optimum condition.

1. Introduction

Nowadays, the world demand for water is growing because of the rapid population growth. Furthermore, pollution of freshwater resources proceeds in all over the world. For instance, China encounters severe water pollution caused by industrial chemicals, heavy metals, and algal toxin with an extraordinary economic growth [1]. Gadgil [2] reported that about half the population in the developing world is suffering from one or more of the six main diseases, diarrhea, ascaris dracunculisis, hookworm, schistosomiasis, and trachoma, which are associated with water supply and sanitation. In the industrialized countries, micropollutants like pharmaceuticals gather much concern as potential contaminants in drinking water [3] and surface water [4]. As a result of this situation, water supply section has made efforts to supply a plenty of safe drinking water. In this context, various advanced water treatment like UV disinfection, ozonation, and adsorption processes [5–7] have been introduced to water purification plants.

UV irradiation and ozone-based chemical oxidation are widely used as advanced water purification processes. These processes can achieve higher level of disinfection and organic pollutants removal [9, 10]. However, bromate (BrO_3^-) formation in these chemical oxidation processes may bring a potential health risk, because BrO_3^- is a possibly carcinogenic to human [11]. Therefore, it is important to understand BrO_3^- formation potential in these processes.

Various knowledge of the BrO_3^- formation during UV and ozone-based chemical oxidation processes has been accumulated for past a few decades. For instance, ozonation of bromide-containing water produces BrO_3^- via ozone and hydroxyl radical pathways [8], but pH depression [12] and ammonia addition [12, 13] successfully decrease BrO_3^- formation. The pH depression decreased 50–63% in BrO_3^- formation per a decline in one-pH-unit [12] because of a depression of hydroxyl radical generation and a decrease in hypobromite (BrO^-), which is a key intermediate of BrO_3^- formation. The inhibition effect of ammonia addition on BrO_3^- formation is caused by bromamines formation from the reaction of HOBr with ammonia [14]. The effective ammonia dose for BrO_3^- depression was limited to 200 μg/L and further increase in ammonia addition did not enhance the BrO_3^- minimization [14]. Effect of hydrogen peroxide (H_2O_2) in ozonation on BrO_3^- formation is complicated. As H_2O_2 can reduce BrO^- into bromide (Br^-) [15], it seems to be useful to depress the BrO_3^- formation. However, ozone reacts with hydroperoxide anion (HO_2^-) and produces hydroxyl radicals ($^\bullet OH$) [16], which promotes oxidation of Br^- [8]. Ozekin et al. [13] reported that an increase in H_2O_2 dose in ozonized water enhanced BrO_3^- formation at pH 6.5, though the BrO_3^- formation at pH 8.5 did not depended on the H_2O_2 dose and was smaller than that of ozone alone. Kim et al. [17] pointed out the importance of molar ratio of H_2O_2/O_3 for BrO_3^- formation during ozonation with H_2O_2 addition; the molar ratio of H_2O_2/O_3 above 0.5 and the dissolved ozone concentration below 0.1 mg/L successfully depress the BrO_3^- formation. The inconsistent results between Ozekin et al. [13] and Kim et al. [17] may be caused by the difference in their experimental designs: H_2O_2 was injected during ozonation [17] or after ozonation [13]. But, both research groups suggested that the $^\bullet OH$ generation under a certain concentration of dissolved ozone enhanced the BrO_3^- formation during ozone/H_2O_2 treatment [13, 17]. UV light is known to decompose BrO_3^- into BrO^- and/or Br^- [18]. As the result, Collivignarelli and Sorlini [19] reported that BrO_3^- formation during ozonation with UV 254 nm irradiation (O_3/UV) was about 40% lower than that during conventional ozonation. However, Ratpukdi et al. [20] showed that the BrO_3^- formation potential of O_3/UV was similar to ozonation alone, though ozonation combined with vacuum UV irradiation could decrease the BrO_3^- formation. Thus, the BrO_3^- formation mechanisms of UV and ozone-based chemical oxidation processes have been explored extensively. However, each research was performed using different reactors, different procedures, and different water matrices. Therefore, it is not easy to judge which process should be selected for BrO_3^- control.

In this study, BrO_3^- formation in UV irradiation, H_2O_2 addition, ozonation, and their combination processes, UV irradiation with H_2O_2 addition (UV/H_2O_2), ozonation with H_2O_2 addition (O_3/H_2O_2), and ozonation with UV irradiation (O_3/UV), were discussed using the same reactor and the same water matrix to provide comparable information of their features of BrO_3^- formation and its control.

2. Experimental

2.1. Material and Experimental Conditions. A low-pressure mercury vapor lamp (20W, UVL20PH-6, Sen Lights, Japan) was used as a UV light source. Ozone gas was generated from analytical grade oxygen gas with a silent discharge ozonizer (ED-OG-R3Lt, Eco Design, Japan). Hydrogen peroxide was purchased from Nacalai Tesque, Japan as about 35% aqueous solution (extra pure grade) and used without further purification. The accurate H_2O_2 concentration was checked just before an experiment and final concentration was set at 10, 100, or 1,000 μM. Figure 1 shows the experimental setup. The reactor was made of glass with a volume of 1.9 L. The UV lamp in a duplex quartz jacket was installed in the center of the reactor. Ozone was injected through two gas diffusers made of glass at the injection rate of about 20 mg/min. Inlet and outlet ozone gas concentration was monitored with two ozone monitors (EG-600, Ebara Jitsugyo, Japan). The exhaust ozone gas was dried with a gas dryer (DH106-1, Komatsu Electronics, Japan) before ozone monitoring, because water vapor biases the ozone concentration. Oxygen gas flow rate was regulated with a mass flow controller (CMQ9200, Yamatake, Japan) at 500 mL/min. Test solution was 1.9 L of 1.88 μM potassium bromide (KBr, Nacalai Tesque, Japan) solution with or without 6.4 μM of 4-chlorobenzoic acid (4-CBA, Wako Chemicals, Japan). The 4-CBA was used as a model compound of organic scavengers of hydroxyl radical ($^\bullet OH$), because it was unreactive with ozone [21]. The solution pH was adjusted by addition of sulfuric acid or sodium hydroxide at around 2.5 or 7. An experimental run continued for 10 or 30 minutes and solution in the reactor was sampled every two or five minutes for chemical analyses of BrO_3^-, bromide ion (Br^-), dissolved ozone, H_2O_2, 4-CBA, and pH.

2.2. Chemical Analysis. The Br^- concentration was analyzed using an ion chromatography system with a conductivity detector (DX-500, Dionex, USA). Analytical conditions were as follows. Column: Dionex IonPac AS12A with a suppressor (Dionex ASRS-ULTRA 4 mm); mobile phase: aqueous solution with 2.7 mM sodium carbonate and 0.3 mM sodium bicarbonate; flow rate: 1.0 mL/min; sample injection volume: 100 μL; oven temperature: 40°C. The BrO_3^- concentration was determined by the ion chromatography coupled with a postcolumn system (Dionex BRS-500) [22]. Reaction conditions were as follows, reactant A: 1.5 M potassium bromide and 1.0 M sulfuric acid; reactant B: 1.2 mM sodium nitrite; flow rate: 0.4 mL/min for reactant A and 0.2 mL/min for reactant B; reaction temperature: 40°C; detection: absorbance at 268 nm. The determination limit was estimated to be 0.050 μM. Dissolved ozone and H_2O_2 were analyzed by indigo-colorimetric method [23] and DMP method [24], respectively. The 4-CBA concentration was determined by the high-performance liquid chromatography (LV-10ADVP, Shimadzu, Japan) [21]. Analytical conditions were as follows: column: ODS-80TM (4.6 × 250 mm, Tosoh, Japan); mobile phase: acetonitrile (70%) and 0.1% phosphoric acid (30%); flow rate: 1.0 mL/min; sample injection volume: 200 μL; oven temperature: 40°C; detection: absorbance at 234 nm. The solution pH was measured with a pH meter (Twin pH B-212, Horiba, Japan).

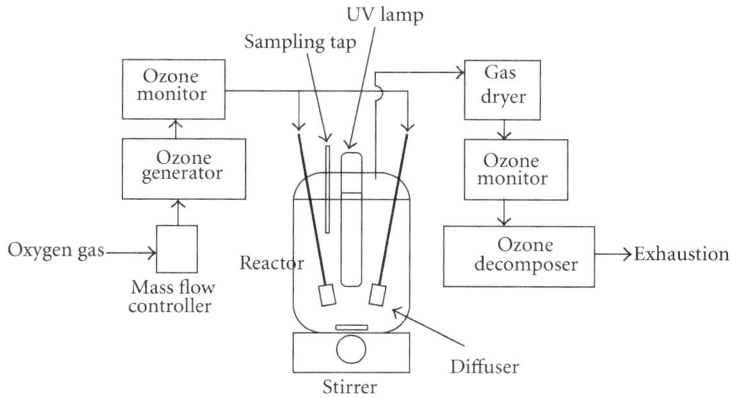

FIGURE 1: Experimental setup.

3. Results and Discussion

3.1. UV Irradiation, H_2O_2 Addition, and UV/H_2O_2 Process. The H_2O_2 addition did not change concentrations of BrO_3^-, Br^-, and 4-CBA under both acid and neutral pH conditions (data not shown). Although H_2O_2 is an oxidant, H_2O_2 is nucleophilic too. Therefore, H_2O_2 can oxidize Br^- into bromine (Br_2), but Br_2 is reduced into Br^- by H_2O_2 as follows [25]:

$$\begin{aligned} 2Br^- + H_2O_2 + 2H^+ &\longrightarrow Br_2 + 2H_2O \\ Br_2 + H_2O_2 &\longrightarrow 2H^+ + 2Br^- + O_2 \end{aligned} \quad (1)$$

Accordingly, Br^- concentration did not change because of the catalytic behavior of Br^- as shown in reactions (1). On the reactivity of 4-CBA with H_2O_2, Dionysiou et al. [26] also observed that H_2O_2 did not decompose 4-CBA under the dark condition. Since the standard electrode potential of H_2O_2 (1.736 V versus standard hydrogen electrode (SHE)) is lower than that of ozone (2.07 V versus SHE) and $^\bullet OH$ (2.38 V versus SHE) [25], the low oxidation potential of H_2O_2 may be responsible for the low reactivity with 4-CBA.

Figure 2 shows the time-course changes in Br^-, BrO_3^-, and 4-CBA concentrations during UV irradiation and UV/H_2O_2 at neutral pH. The concentration changes at acidic condition were almost the same at neutral pH, though the H_2O_2 accumulation was enhanced at acidic condition. The low-pressure mercury vapor lamp emits vacuum UV light of 185 nm, which can photolyze water molecules into hydrogen atoms and $^\bullet OH$ [27]. Therefore, H_2O_2 accumulation was caused by H_2O_2 production via the combination of two $^\bullet OH$ [28]. The concentrations of Br^- and 4-CBA declined during the UV irradiation and UV/H_2O_2, though BrO_3^- was not generated (Figure 2). No BrO_3^- formation during UV irradiation and UV/H_2O_2 was also reported by Kruithof et al. [29]. The H_2O_2 concentration in the both treatment increased with the passage of time, and the final concentration in UV irradiation reached over 10 μM, which was the initial concentration in UV/H_2O_2. The 4-CBA degradation in UV irradiation slightly delayed in comparison with that in UV/H_2O_2, but the degradation was enhanced with the accumulation of H_2O_2. Accordingly, $^\bullet OH$ generation via UV photolysis of H_2O_2 [27] was believed to contribute to the 4-CBA degradation during the UV irradiation and UV/H_2O_2. The decline in Br^- concentration without BrO_3^- accumulation indicates the formation of intermediates. Von Gunten and Oliveras [8] reported that ozone and $^\bullet OH$ oxidized Br^- to BrO_3^- (Figure 3). In this mechanism, hypobromite ion (BrO^-) and bromite ion (BrO_2^-) are the critical intermediates, which participate in all BrO_3^- formation pathways. Accordingly, UV irradiation and UV/H_2O_2 were thought to produce BrO^- and/or BrO_2^-. The H_2O_2 can reduce hypobromous acid (HOBr) and BrO^- as follows [15]:

$$\begin{aligned} &BrO^- + H_2O_2 \longrightarrow Br^- + H_2O + O_2 \\ &\qquad k = 1.3 \times 10^6\ M^{-1}\,s^{-1}, \\ &HOBr + HO_2^- \longrightarrow Br^- + H_2O + O_2 \\ &\qquad k = 7.6 \times 10^8\ M^{-1}\,s^{-1}, \\ &HOBr + H_2O_2 \longrightarrow Br^- + H^+ + H_2O + O_2 \\ &\qquad k = 1.5 \times 10^4\ M^{-1}s^{-1}. \end{aligned} \quad (2)$$

Therefore, the accumulation of H_2O_2 was inferred to contribute partly to the prevention of BrO_3^- formation in UV irradiation and UV/H_2O_2. Phillip et al. [30] reported that low-pressure mercury vapor lamps decayed free bromine into Br^- (major) and BrO_3^- (minor). Thus, the photo-degradation of HOBr/BrO^- might conduce to the prevention of BrO_3^- formation too.

3.2. Ozonation. In ozonation, BrO_3^- formation was correspondent to a decrement in Br^- at neutral pH without 4-CBA. However, the BrO_3^- formation was much lower than Br^- removal at acidic pH or coexistence of 4-CBA (Figure 4). Although both ozone and $^\bullet OH$ promote the oxidation of Br^- to BrO_3^- via BrO^- and BrO_2^- (Figure 3), our experimental results shown in Figure 4 indicated that contribution of $^\bullet OH$ to BrO_3^- evolution was relatively large. Because acidic pH restrains $^\bullet OH$ generation via self-decomposition of ozone [31], and 4-CBA is a $^\bullet OH$ radical scavenger with low reactivity with ozone [21]. Since HOBr has a pKa of 8.8–9.0 [32],

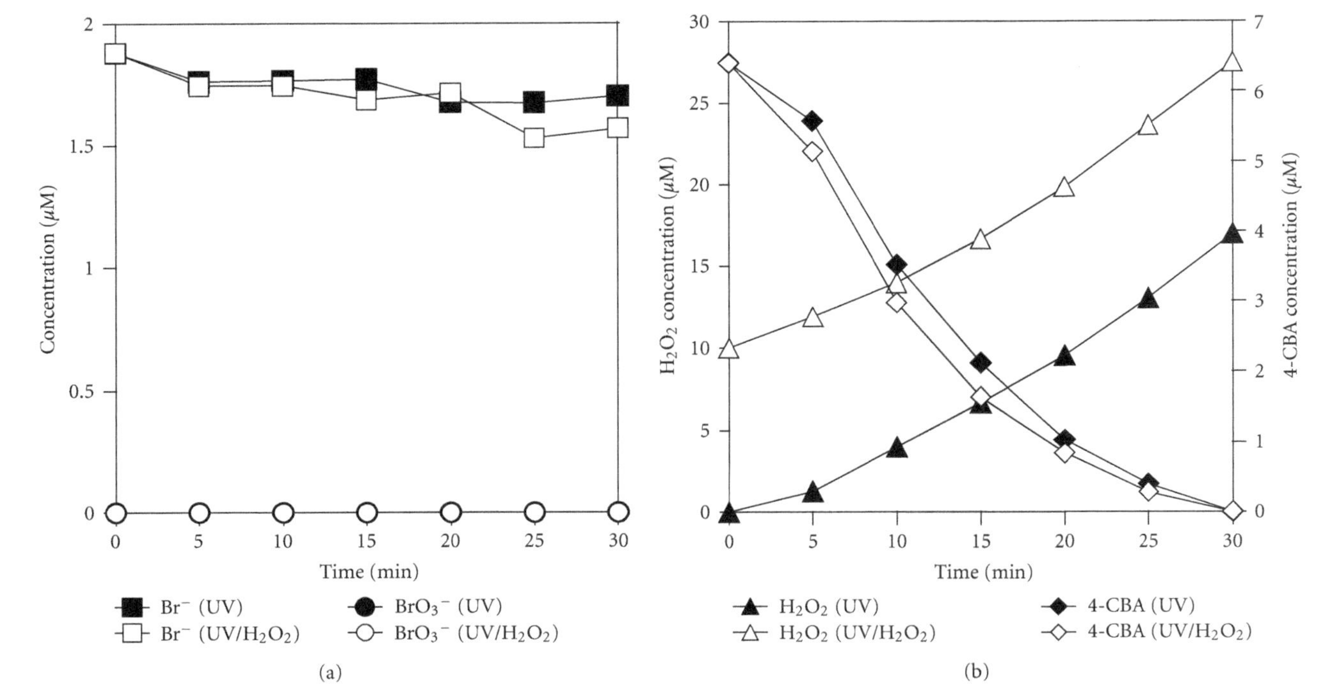

FIGURE 2: Changes in concentrations of Br^-, BrO_3^-, H_2O_2, and 4-CBA during UV irradiation (UV) and UV/H_2O_2 treatment at neutral pH. The initial H_2O_2 concentration for UV/H_2O_2 was 10 μM.

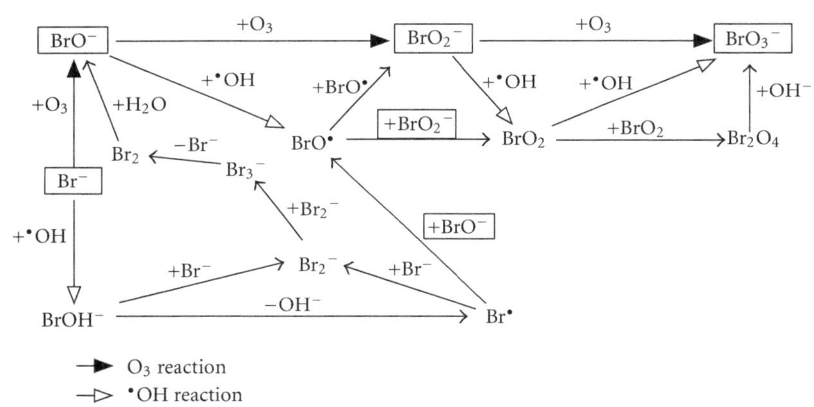

FIGURE 3: BrO_3^- formation pathways illustrated on the basis of Von Gunten and Oliveras [8].

acidification decreases the percentage of BrO^-. The decrease in BrO^- at acidic pH also contributed to the decrease in BrO_3^- formation [33]. In addition, the discordance of a decrement in Br^- and an increment in BrO_3^- at acidic pH in Figure 4 suggested the accumulation of HOBr.

3.3. O_3/H_2O_2. Figure 5 shows the changes in concentrations of Br^-, BrO_3^-, and 4-CBA during O_3/H_2O_2 at various H_2O_2 doses and pHs. The O_3/H_2O_2 processes showed additional effect on lowering BrO_3^- formation, especially at higher H_2O_2 dose. Bromate ion was not detected in O_3/H_2O_2 at the H_2O_2 dose of 1,000 μM. When the H_2O_2 dose was lowered to 100 or 10 μM, rapid BrO_3^- formation was observed (Figure 5). As H_2O_2 can reduce HOBr/BrO^- into Br^-, the excess H_2O_2 was believed to restrain BrO_3^- formation. The behavior of BrO_3^- formation in O_3/H_2O_2 at the H_2O_2 dose of 10 μM (Figure 5) was similar to that in ozonation (Figure 4), since H_2O_2 concentration in ozonation increased to around 10 μM, which was nearly equal to H_2O_2 concentration in O_3/H_2O_2 at the H_2O_2 dose of 10 μM (Figure 6). Thus, the initial H_2O_2 dose of 10 μM was too low to restrain the BrO_3^- formation in this study.

When acidification was applied, BrO_3^- was not detected in O_3/H_2O_2 at the H_2O_2 dose of 100 μM or higher. This effect was caused by both the HOBr/BrO^- reduction by H_2O_2

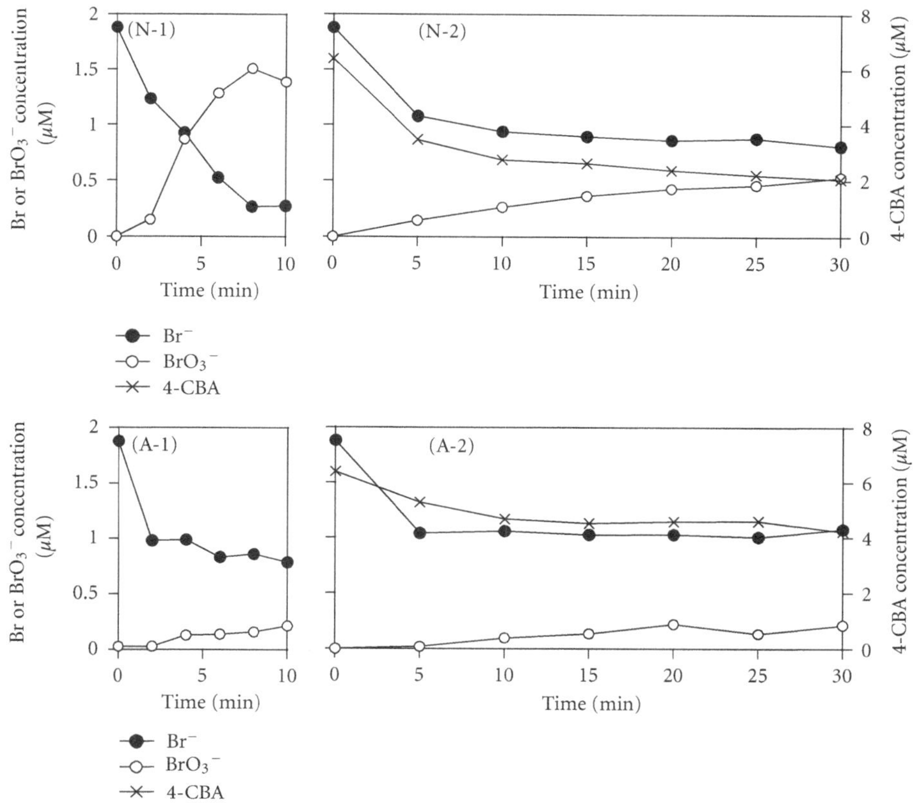

FIGURE 4: Changes in concentrations of Br^-, BrO_3^-, and 4-CBA during ozonation at neutral pH without 4-CBA (N-1), ozonation at neutral pH with 4-CBA (N-2), ozonation at acidic pH without 4-CBA (A-1), and ozonation at acidic pH with 4-CBA (A-2).

and inhibition of $^{\bullet}OH$ generation in O_3/H_2O_2. Because $^{\bullet}OH$ generation in O_3/H_2O_2 is expressed by the following reactions [16] and the acidification inhibits the dissociation of H_2O_2 and HO_2 (reactions (3) and (5)) as follows:

$$H_2O_2 \longleftrightarrow H^+ + HO_2^- \quad (3)$$

$$O_3 + HO_2^- \longrightarrow {}^{\bullet}O_3^- + HO_2^{\bullet} \quad (4)$$

$$HO_2^{\bullet} \longleftrightarrow H^+ + {}^{\bullet}O_2^- \quad (5)$$

$$O_3 + {}^{\bullet}O_2^- \longrightarrow {}^{\bullet}O_3^- + O_2 \quad (6)$$

$${}^{\bullet}O_3^- + H^+ \longleftrightarrow HO_3 \rightarrow O_2 + {}^{\bullet}OH \quad (7)$$

The inhibition of $^{\bullet}OH$ generation in O_3/H_2O_2 was also confirmed by a slow decrease in 4-CBA at acidic pH (Figure 5).

3.4. O_3/UV. Figure 7 shows changes in Br^-, BrO_3^-, and 4-CBA concentrations during O_3/UV. The O_3/UV increased BrO_3^- concentration rapidly, even at the acidic pH. However, the addition of 4-CBA successfully decreased the BrO_3^- formation regardless of the pH condition. Collivignarelli and Sorlini [19] also observed lower BrO_3^- formation in O_3/UV than that in ozonation. As mentioned in the Section 3.2, the acidification decreases BrO_3^- formation by the inhibition of $^{\bullet}OH$ generation via the self-decomposition of ozone. Accordingly, it was thought that $^{\bullet}OH$ generated by the self-decomposition of ozone did not contribute to BrO_3^- formation very much in O_3/UV. This discussion was supported by the lower concentration of dissolved ozone in O_3/UV (Figure 8). The low dissolved ozone concentration also brought the negligible contribution of molecular ozone to BrO_3^- formation. The decrease in BrO_3^- formation by the addition of 4-CBA indicated the contribution of $^{\bullet}OH$ to BrO_3^- formation. Accordingly, it is suggested that the main oxidant in O_3/UV was $^{\bullet}OH$, which mainly generated via UV photolysis of ozone [16]. The first step of $^{\bullet}OH$ generation in O_3/UV is the production of H_2O_2 [16]. Then the H_2O_2 generates $^{\bullet}OH$ through UV photolysis [16] and the same reactions as O_3/H_2O_2 (reactions (3)–(7)). As the coexistence of dissolved ozone and $^{\bullet}OH$ favors BrO_3^- formation [13, 17], low dissolved ozone concentration in O_3/UV was thought to be advantageous to the depression of BrO_3^- formation. Moreover, strong H_2O_2 accumulation was observed during O_3/UV (Figure 9). Therefore, the reduction of intermediates by H_2O_2 [15] and UV photolysis [30] was also inferred to contribute to the decline in the BrO_3^- formation potential.

3.5. Strategy for Organic Contaminants Removal with Preventing BrO_3^- Formation. The aim of advanced oxidation processes is organic contaminants removal from a water stream. Therefore, it is important to remove organic contaminants without BrO_3^- formation. In this context, H_2O_2 addition

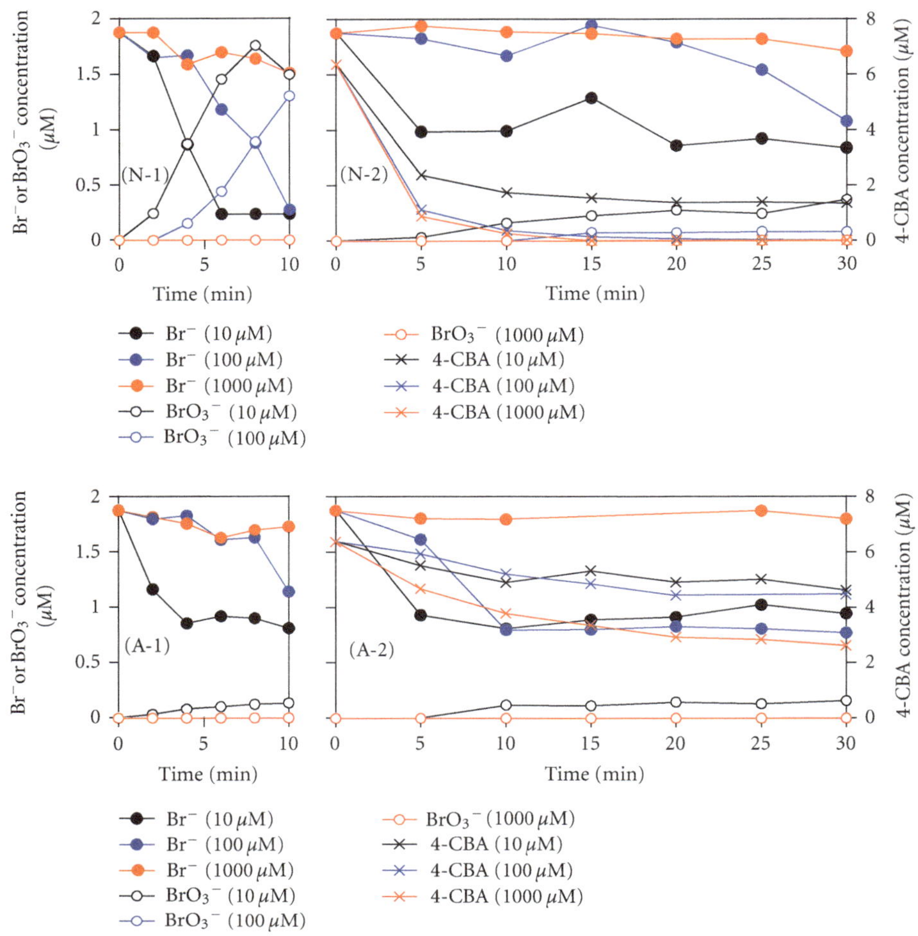

Figure 5: Changes in concentrations of Br^-, BrO_3^-, and 4-CBA during O_3/H_2O_2 at various H_2O_2 doses. The concentrations in the legend mean the H_2O_2 dose. The symbols in the bracket mean O_3/H_2O_2 at neutral pH without 4-CBA (N-1), O_3/H_2O_2 at neutral pH with 4-CBA (N-2), O_3/H_2O_2 at acidic pH without 4-CBA (A-1), and O_3/H_2O_2 at acidic pH with 4-CBA (A-2).

is inapplicable to advanced water treatment, because it is not effective to degrade refractory organic matters like 4-CBA. Ozonation is also difficult to apply to the organic contaminants removal, because it has higher BrO_3^- formation potential at the neutral pH than at the acidic pH as shown in Figure 4. Although the acidification successfully decreases the BrO_3^- formation potential of ozonation, it decreases the removal rate of organic contaminants too.

Contrary to H_2O_2 addition and ozonation, UV-based processes and O_3/H_2O_2 are potentially applicable to advanced water treatment with inhibiting BrO_3^- formation. The UV irradiation and UV/H_2O_2 successfully decompose organic contaminants without BrO_3^- formation. But their degradation rate of organic contaminants is not high, and H_2O_2 dose of 10 μM is too low to enhance the degradation rate of UV irradiation. The O_3/H_2O_2 is characterized by rapid degradation of organic contaminants and low BrO_3^- formation rate at high H_2O_2 dose. Although the acidification effectuates further decrease in BrO_3^- formation, it spoils the degradation of organic contaminants. Therefore, the acidification should not apply to O_3/H_2O_2. The feature of O_3/UV is rapid degradation of organic contaminants, low dissolved ozone concentration, and much H_2O_2 accumulation. As the BrO_3^- formation in O_3/UV is restrained under the coexistence of organic contaminants, the O_3/UV is applicable, if water contains sufficient quantity of organic contaminants. Figure 10 shows the relationship between BrO_3^- concentration and cumulative ozone consumption. Figure 10 demonstrates that O_3/UV and O_3/H_2O_2 at the H_2O_2 doses of 100 and 1,000 μM were significantly decreased BrO_3^- formation per ozone consumption in comparison with ozonation and O_3/H_2O_2 at the H_2O_2 dose of 10 μM. Kim et al. [17] reported that BrO_3^- concentration in O_3/H_2O_2 remained less than 10 μg/L, when the molar ratio of H_2O_2 to ozone was above 0.5. In our study of O_3/H_2O_2 at the H_2O_2 dose of 100 μM, BrO_3^- was not produced at ozone consumption less than 250 μM. Thus, our result approximately accorded with the research by Kim et al. [17].

Consequently, if the concentration of organic contaminants is low, the UV irradiation and/or UV/H_2O_2 are applicable to organic contaminants removal without BrO_3^- formation, though it is necessary to beware nitrite formation

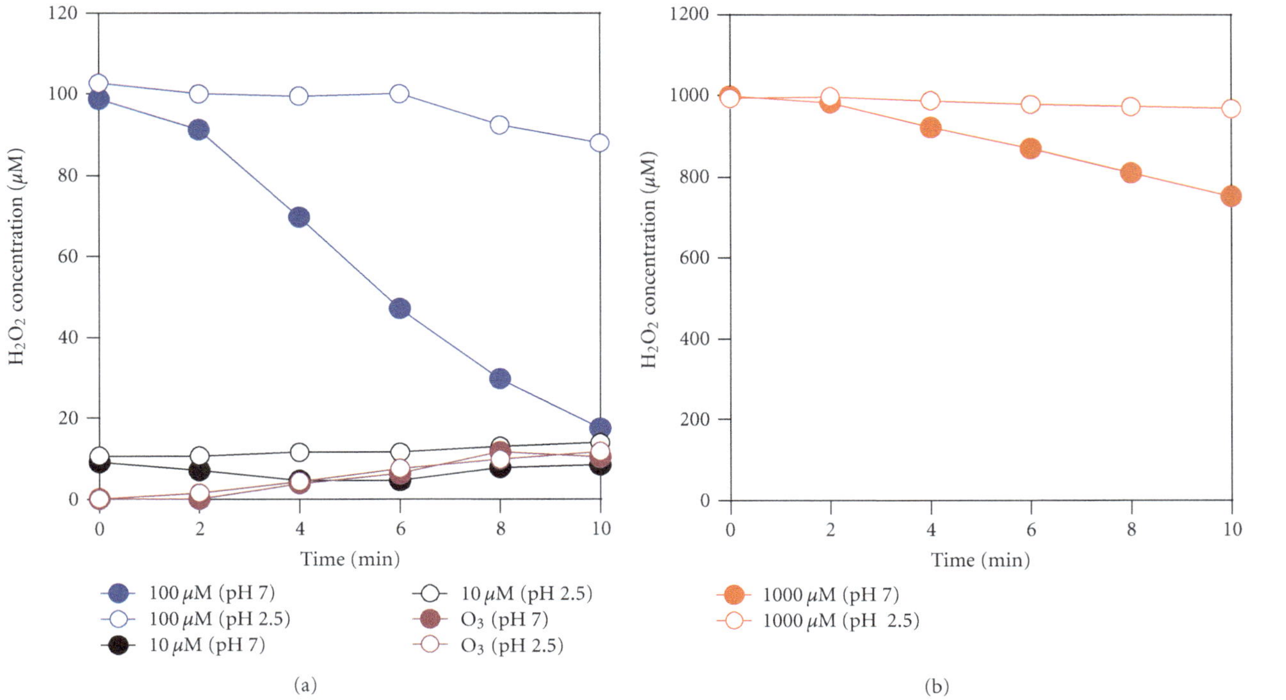

Figure 6: Changes in H_2O_2 concentrations during O_3/H_2O_2 and ozonation without 4-CBA. The concentrations in the legend mean the H_2O_2 dose in O_3/H_2O_2.

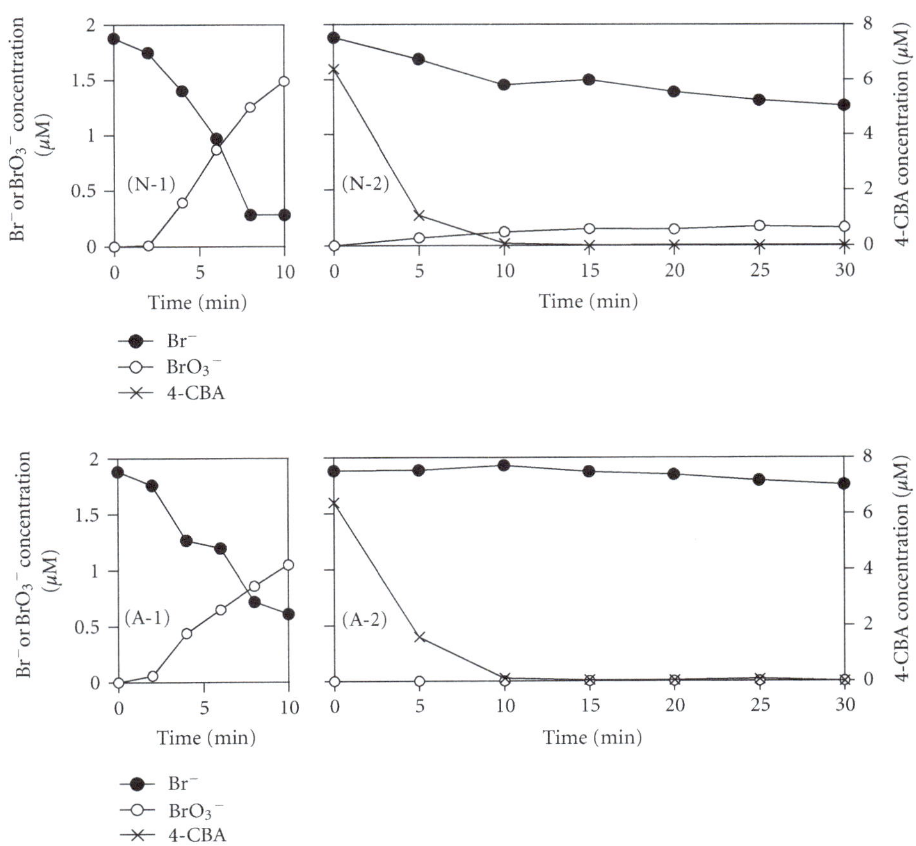

Figure 7: Changes in concentrations of Br^-, BrO_3^-, and 4-CBA during O_3/UV at neutral pH without 4-CBA (N-1), O_3/UV at neutral pH with 4-CBA (N-2), O_3/UV at acidic pH without 4-CBA (A-1), and O_3/UV at acidic pH with 4-CBA (A-2).

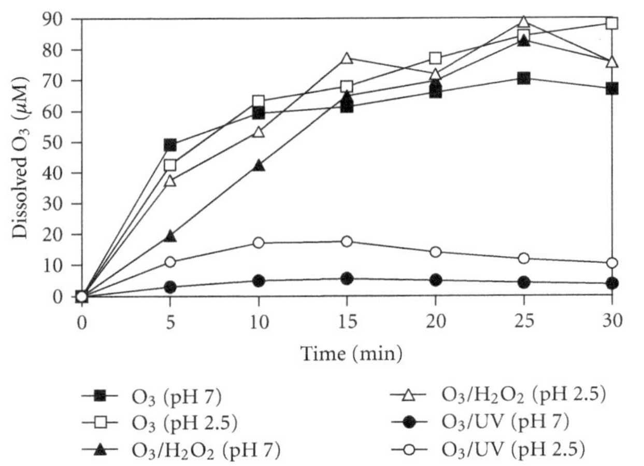

Figure 8: Changes in dissolved ozone concentrations during ozonation, O_3/H_2O_2 (the H_2O_2 dose of 100 μM), and O_3/UV with the addition of 4-CBA.

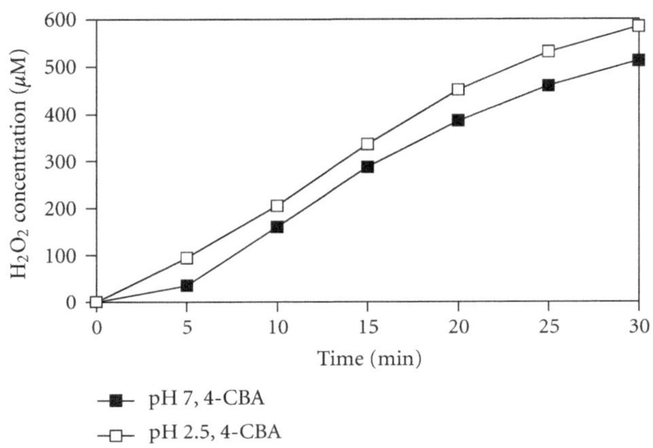

Figure 9: Changes in H_2O_2 concentrations during O_3/UV with 4-CBA.

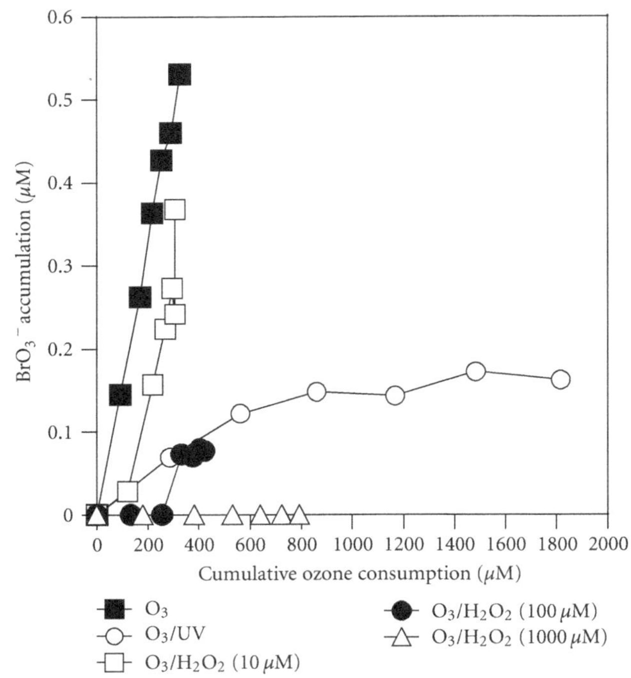

Figure 10: Relationship between cumulative ozone consumption and BrO_3^- accumulation during each treatment with 4-CBA addition at neutral pH. Each treatment continued 30 minutes.

during the treatment [34]. However, if the concentration of organic contaminants is high, O_3/H_2O_2 and O_3/UV should be discussed as advanced oxidation processes. When O_3/H_2O_2 is applied, the H_2O_2 dose should be more than the half of ozone consumption, because low H_2O_2 dose is ineffective for BrO_3^- control. When O_3/UV is applied, the reaction time should be optimized, because extended reaction time increases the BrO_3^- formation potential. The real-time monitoring of UV absorbance of organic contaminants in water [35] may be effective in the optimum control of the reaction time of O_3/UV without BrO_3^- formation.

4. Conclusion

In this research, BrO_3^- formation potential of UV irradiation, H_2O_2 addition, ozonation, UV/H_2O_2, O_3/H_2O_2, and O_3/UV treatment were discussed for organic contaminants removal with restraining BrO_3^- formation using KBr solution with 4-CBA as a model refractory organic contaminant.

The UV irradiation, H_2O_2 addition, and UV/H_2O_2 prevented BrO_3^- formation completely. However, H_2O_2 addition was inapplicable as advanced water treatment because of its weak oxidation ability. The UV irradiation and UV/H_2O_2 could decompose the organic contaminant moderately. Ozonation produced the most BrO_3^- at neutral pH. Although acidification could decrease the BrO_3^- formation, it also deteriorated the oxidation ability of ozonation. Therefore, it was thought to be difficult to apply ozonation to organic contaminants removal with restraining BrO_3^- formation. The O_3/H_2O_2 successfully decreased BrO_3^- formation at the H_2O_2 doses of 100 μM or higher. The degradation rate of 4-CBA was larger than the UV irradiation and UV/H_2O_2. However, the behavior of BrO_3^- formation in the O_3/H_2O_2 at the H_2O_2 dose of 10 μM was similar to that in ozonation because of a deficiency of H_2O_2. The O_3/UV also showed rapid degradation of 4-CBA. Although it produced much BrO_3^- under the absence of 4-CBA, the BrO_3^- formation was strongly inhibited by the coexistence of 4-CBA. Consequently, if the concentration of organic contaminants is low, the UV irradiation and/or UV/H_2O_2 are applicable to organic contaminants removal without BrO_3^- formation. However, if the concentration of organic contaminants is high, O_3/H_2O_2 and O_3/UV should be discussed as advanced oxidation processes because of their higher organic removal efficiency.

References

[1] C. Wu, C. Maurer, Y. Wang, S. Xue, and D. L. Davis, "Water pollution and human health in China," *Environmental Health Perspectives*, vol. 107, no. 4, pp. 251–256, 1999.

[2] A. Gadgil, "Drinking water in developing countries," *Annual Review of Energy and the Environment*, vol. 23, no. 1, pp. 253–286, 1998.

[3] O. A. Jones, J. N. Lester, and N. Voulvoulis, "Pharmaceuticals: a threat to drinking water?" *Trends in Biotechnology*, vol. 23, no. 4, pp. 163–167, 2005.

[4] R. Takanami, H. Ozaki, R. R. Giri, S. Taniguchi, and S. Hayashi, "Antiviral drugs zanamivir and oseltamivir found in wastewater and surface water in Osaka, Japan," *Journal of Water and Environment Technology*, vol. 10, pp. 57–68, 2012.

[5] D. Dotson, C. E. Rodoriquez, and K. G. Linden, "UV disinfection implementation status in US water treatment plants," *Journal AWWA*, vol. 104, pp. 77–78, 2012.

[6] B. L. Loeb, C. M. Thompson, J. Drago, H. Takahara, and S. Baig, "Worldwide ozone capacity for treatment of drinking water and wastewater: a review," *Ozone: Science & Engineering*, vol. 34, pp. 64–77, 2012.

[7] I. Ali and V. K. Gupta, "Advances in water treatment by adsorption technology," *Nature Protocols*, vol. 1, no. 6, pp. 2661–2667, 2007.

[8] U. Von Gunten and Y. Oliveras, "Advanced oxidation of bromide-containig waters: bromate formation mechanisms," *Environmental Science and Technology*, vol. 32, no. 1, pp. 63–70, 1998.

[9] D. Gerrity and S. Snyder, "Review of ozone for water reuse applications: toxicity, regulations, and trace organic contaminant oxidation," *Ozone: Science and Engineering*, vol. 33, no. 4, pp. 253–266, 2011.

[10] B. A. Wols and C. H. M. Hofman-Caris, "Review of photochemical reaction constants of organic micropollutants required for UV advanced oxidation processes in water," *Water Research*, vol. 46, pp. 2815–2827, 2012.

[11] *IARC Monographs on the Evaluation of Carcinogenic Risks to Humans*, vol. 73, International Agency for Research on Cancer, Lyon, France, 1999.

[12] M. S. Siddiqui and G. L. Amy, "Factors affecting DBP formation during ozone-bromide reactions," *Journal of American Water Works Association*, vol. 85, no. 1, pp. 63–72, 1993.

[13] K. Ozekin, P. Westerhoff, G. L. Amy, and M. Siddiqui, "Molecular ozone and radical pathways of bromate formation during ozonation," *Journal of Environmental Engineering*, vol. 124, no. 5, pp. 456–462, 1998.

[14] U. Pinkernell and U. von Gunten, "Bromate minimization during ozonation: mechanistic considerations," *Environmental Science and Technology*, vol. 35, no. 12, pp. 2525–2531, 2001.

[15] U. Von Gunten and Y. Oliveras, "Kinetics of the reaction between hydrogen peroxide and hypobromous acid: implication on water treatment and natural systems," *Water Research*, vol. 31, no. 4, pp. 900–906, 1997.

[16] W. H. Glaze, J. Kang, and D. H. Chapin, "The chemistry of water treatment processes involving ozone, hydrogen peroxide and ultraviolet radiation," *Ozone: Science & Engineering*, vol. 9, pp. 335–352, 1987.

[17] H. S. Kim, H. Yamada, and H. Tsuno, "The control of brominated by-products and the removal of organic pollutants during the ozone/hydrogen peroxide treatment of secondary effluent," *Ozone: Science and Engineering*, vol. 29, no. 1, pp. 23–30, 2007.

[18] M. S. Siddiqui, G. L. Amy, and L. J. McCollum, "Bromate destruction by UV irradiation and electric arc discharge," *Ozone: Science and Engineering*, vol. 18, no. 3, pp. 271–290, 1996.

[19] C. Collivignarelli and S. Sorlini, "AOPs with ozone and UV radiation in drinking water: contaminants removal and effects on disinfection byproducts formation," *Water Science and Technology*, vol. 49, no. 4, pp. 51–56, 2004.

[20] T. Ratpukdi, F. Casey, T. Desutter, and E. Khan, "Bromate formation by ozone-VUV in comparison with ozone and ozone-UV: effects of pH, ozone dose, and VUV power," *Journal of Environmental Engineering*, vol. 137, no. 3, pp. 187–195, 2011.

[21] N. Kishimoto, Y. Morita, H. Tsuno, T. Oomura, and H. Mizutani, "Advanced oxidation effect of ozonation combined with electrolysis," *Water Research*, vol. 39, no. 19, pp. 4661–4672, 2005.

[22] N. Kishimoto, A. Kishimoto, and A. Nakayama, "Rapid removal of bromate ion from water streams with an electrolytic flow cell," *Journal of Water Supply: Research and Technology-AQUA*, vol. 61, pp. 103–110, 2012.

[23] *Standard Methods for the Examination of Water & Wastewater*, American Public Health Association, Washington, DC, USA, 22nd edition, 2012.

[24] A. N. Baga, G. R. A. Johnson, N. B. Nazhat, and R. A. Saadalla-Nazhat, "A simple spectrophotometric determination of hydrogen peroxide at low concentrations in aqueous solution," *Analytica Chimica Acta*, vol. 204, pp. 349–353, 1988.

[25] J. P. Hoare, "4. Oxygen," in *Standard Potentials in Aqueous Solution*, A. J. Bard, R. Parsons, and J. Jordan, Eds., pp. 49–66, Mercel Dekker, New York, NY, USA, 1985.

[26] D. D. Dionysiou, M. T. Suidan, E. Bekou, I. Baudin, and J. M. Laîné, "Effect of ionic strength and hydrogen peroxide on the photocatalytic degradation of 4-chlorobenzoic acid in water," *Applied Catalysis B*, vol. 26, no. 3, pp. 153–171, 2000.

[27] N. Kishimoto, I. Somiya, and Y. Morita, "Dependence of hydroxyl radical production in UV/H_2O_2 photo-oxidation on UV energy absorbed by aqueous solution," *Advances in Asian Environmental Engineering*, vol. 2, no. 2, pp. 8–17, 2002.

[28] J. Staehelin, R. E. Bühler, and J. Hoigné, "Ozone decomposition in water studied by pulse radiolysis. OH and HO_4 as chain intermediates," *Journal of Physical Chemistry*, vol. 88, no. 24, pp. 5999–6004, 1984.

[29] J. C. Kruithof, P. C. Kamp, and M. Belosevic, "UV/H_2O_2-treatment: the ultimate solution for pesticide control and disinfection," *Water Science and Technology: Water Supply*, vol. 2, no. 1, pp. 113–122, 2002.

[30] N. H. Phillip, E. Gürten, and V. Diyamandoğlu, "Transformation of bromine species during decomposition of bromate under UV light from low pressure mercury vapor lamps," *Ozone: Science and Engineering*, vol. 28, no. 4, pp. 217–228, 2006.

[31] M. S. Elovitz, U. Von Gunten, and H. P. Kaiser, "Hydroxyl radical/ozone ratios during ozonation processes. II. The effect of temperature, pH, alkalinity, and DOM properties," *Ozone: Science and Engineering*, vol. 22, no. 2, pp. 123–150, 2000.

[32] W. R. Haag and J. Holgné, "Ozonation of bromide-containing waters: kinetics of formation of hypobromous acid and bromate," *Environmental Science and Technology*, vol. 17, no. 5, pp. 261–267, 1983.

[33] L. Xie and C. Shang, "A review on bromate occurrence and removal strategies in water supply," *Water Science and Technology: Water Supply*, vol. 6, no. 6, pp. 131–136, 2006.

[34] A. J. Martijn and J. C. Kruithof, "UV and UV/H_2O_2 treatment: the silver bullet for by-product and genotoxicity formation in water production," *Ozone: Science & Engineering*, vol. 34, pp. 92–100, 2012.

[35] R. A. Dobbs, R. H. Wise, and R. B. Dean, "The use of ultra-violet absorbance for monitoring the total organic carbon content of water and wastewater," *Water Research*, vol. 6, no. 10, pp. 1173–1180, 1972.

Synthesis of Mesoporous TiO_2-Al_2O_3 Binary Oxides Photocatalyst by Sol-Gel Method Using PEG1000 as Template

Wenjie Zhang,[1] Ruyuan Li,[1] and Hongbo He[2]

[1] *School of Environmental and Chemical Engineering, Shenyang Ligong University, Shenyang 110159, China*
[2] *Institute of Applied Ecology, The Chinese Academy of Sciences, Shenyang 110016, China*

Correspondence should be addressed to Wenjie Zhang, metalzhang@yahoo.com.cn

Academic Editor: Feng Chen

The mesoporous TiO_2-Al_2O_3 binary oxides were prepared by sol-gel method using PEG1000 as template. The effects of PEG1000 concentration on the properties and photocatalytic activity of the materials were investigated. The binary oxides were composed of nanocrystallite anatase TiO_2 and amorphous Al_2O_3. The surfaces of the samples prepared with PEG1000 were rough and had many long grooves on the surfaces. When PEG1000 concentration increased from 0 to 0.048 mol/L, the BET specific surface area of the samples changed from 99 to 140 $m^2 \cdot g^{-1}$, and the average pore size decreased from 22 to 11 nm. The TiO_2-Al_2O_3 binary oxides prepared at PEG1000 concentration of 0.03 mo/L had the maximum photocatalytic activity and high adsorption capacity. Methyl orange photocatalytic degradation and adsorption rates on 600 mg/l of TiO_2-Al_2O_3 binary oxides were 36.1% and 18.6% after 30 min, respectively. Total removal of the dye was achieved after 3 h of reaction.

1. Introduction

TiO_2 is regarded as the most powerful photocatalyst in recent years because of its prospective properties such as high photocatalytic activity, inexpensive, and nontoxic to human life [1–5]. However, photocatalytic activity of TiO_2 cannot meet the demand of large-scale industrial application. Porous photocatalytic materials were used to promote photocatalytic activity and to improve the efficiency of photoelectron conversion because large surface area and porous structure are usually good for strong adsorption ability. Many microporous and mesoporous photocatalytic materials have been prepared accordingly [6–11].

In order to improve photocatalytic activity of porous TiO_2, a popular route is to incorporate other inorganic oxides such as SiO_2 [12], ZrO_2 [13], WO_3 [14], Cu_2O [15], and Ta_2O_5 [16] into the microstructure of porous TiO_2. Although there are reports about TiO_2-Al_2O_3 photocatalytic materials [17–20], these materials are the simple mixtures of TiO_2 and Al_2O_3 particles prepared separately. The mechanical mixture of the two oxides has only limited interaction between external surfaces.

In the present work, mesoporous TiO_2-Al_2O_3 binary oxides were synthesized simultaneously by sol-gel method using a mixed precursor of tetrabutyl titanate and aluminum isopropoxide. Polyethylene glycol (PEG1000) was used as template. The effects of PEG1000 concentration on the physiochemical properties and photocatalytic activity of the materials were investigated. The samples were characterized by X-ray diffraction (XRD), scanning electron microscope (SEM), frontier transform infrared spectrophotometer (FT-IR), X-ray photoelectron spectroscopy (XPS), and N_2-adsorption and desorption techniques. Adsorption capacity and photocatalytic degradation of methyl orange on the materials under UV-light irradiation were investigated to ascertain the enhanced activity of the mesoporous TiO_2-Al_2O_3 binary oxides.

2. Experimental

2.1. Material Preparation. A given mass of tetrabutyl titanate, aluminum isopropoxide [n(Ti)/n(Al) = 12], polyethylene glycol (PEG1000), and concentrated hydrochloric acid were added into absolute ethanol under magnetic stirring.

Another mixture of deionized water and absolute ethanol was dropped into the former solution at a constant speed. The volume ratio of constituents in the final solution was $V[Ti(OC_4H_9)_4] : V(C_2H_5OH) : V(HCl) : V(H_2O) = 1 : 5 : 0.05 : 0.5$. The solution was kept under stirring until the sol and gel formed. The gel was aged for 12 h at ambient temperature and dried at 80°C for 8 h. After that, the samples were grinded and calcinated at 500°C for 3 h. The prepared materials were grinded again before use.

2.2. Characterization. The crystalline phase of the photocatalyst was measured by X-ray diffraction (XRD) with D/max-rB diffractometer using a Cu $K\alpha$ radiation in 2θ range of 10°–80°. The surface morphology was observed by scanning electron microscope (SEM). FT-IR spectrum of the sample was recorded using a FT-IR spectrometer in the wavenumber range of 4000–400 cm^{-1}. The elemental composition of the sample was investigated by X-ray photoelectron spectroscopy (XPS, Multilab 2000). The N_2 adsorption and desorption of the sample was measured by F-Sorb 3400 specific surface area and pore size distribution analytical instrument. Specific surface areas and pore size distribution were calculated from the N_2 desorption isotherms, according to Brunauer-Emmett-Teller (BET) equation and Barrett-Joyner-Halenda (BJH) method, respectively.

2.3. Adsorption and Photocatalytic Activity. The adsorption and photocatalytic activity of the samples were investigated by measuring the decoloration rate of methyl orange (MO) in a lab-scale photocatalytic reactor. The reactor consisted of a 250 mL beaker and a UV lamp hanging over the beaker. The 20 W UV-light lamp with main irradiation wavelength at 253.7 nm was used as light source to trigger the photocatalytic reaction, and the average light intensity was about 1100 $\mu W/cm^2$. The concentration of MO was calculated by Lambert-Beer law using 721E spectrophotometer and TU-1901 UV-Vis double-beam spectrophotometer. The absorbency of MO was measured at 468 nm corresponding to the maximum absorption wavelength. The decoloration rate of methyl orange was calculated by the following equation [21],

$$\omega(\text{Adsorption}) = \frac{(C_0 - C_1)}{C_0} \times 100\%,$$
$$\omega(\text{Degradation}) = \frac{(C_1 - C_2)}{C_0} \times 100\%, \quad (1)$$

where C_0 is the initial dye concentration, C_1 is the concentration after adsorption-desorption equilibrium, and C_2 is the concentration after certain time of irradiation.

The solution of MO and photocatalyst ($V = 50$ mL, $C_{MO} = 10$ mg/L) was mixed in the reactor. The adsorption of MO on photocatalyst was measured in the suspension before photocatalytic reaction. The solution was allowed to reach an adsorption-desorption equilibrium under 30 min of magnetic stirring in the dark. After that, 5 mL of suspension was taken out of the reactor and was filtered through a Millipore filter (pore size 0.45 μm) to remove the catalyst.

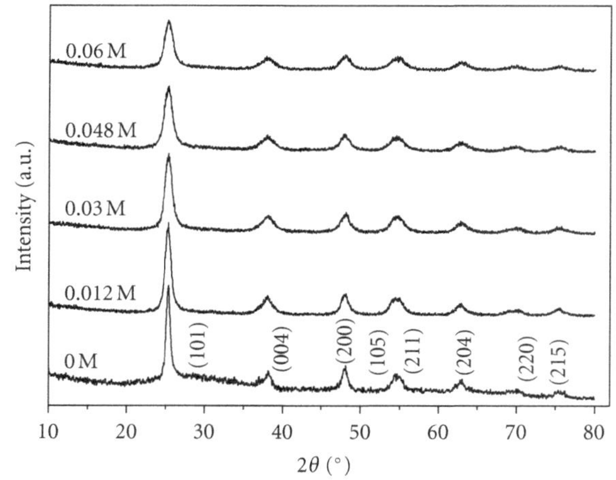

FIGURE 1: XRD patterns of mesoporous TiO_2-Al_2O_3 binary oxides prepared using different concentration of PEG1000.

The filtrated solution was measured for its absorbency. After that, the remaining MO and photocatalyst suspension was put under UV light irradiation. After a distinct time, 5 mL of suspension solution was removed and filtrated to measure the change of MO concentration. If not indicated, the irradiation time in the subsequent experiments was set to 30 min.

3. Results and Discussion

Sol-gel method with the advantages of reacting in low temperature and easily controlled homogeneous multicomposition system has become a popular method to prepare microporous and mesoporous photocatalytic materials in a relatively large scale. In order to obtain porous structure, suitable template needs to be used to bond with precursors in sol-gel system. There are some chemical compounds that have been used as templates, such as surfactants and copolymers. Polyethylene glycol $[H\text{–}(O\text{–}CH_2\text{–}CH_2)_n\text{–}OH]$ is one of nonionic templates and dissolves easily in polar solvents. In sol-gel system, three-dimensional structure of precursor is formed preferably by crosslink action among hydrolysates and PEG. Porous materials with arranged structure can be prepared by removing PEG template consequently.

XRD patterns of mesoporous TiO_2-Al_2O_3 binary oxides prepared using different concentration of PEG1000 are shown in Figure 1. All the diffraction peaks in the patterns indicate the formation of anatase TiO_2 crystallite. The diffraction peaks turned from sharp and high to broad and low when PEG1000 concentration increased from 0 to 0.060 mol/L. The peaks relating to the characteristic peaks of Al_2O_3 were not found. It declares that Al_2O_3 was in its amorphous state in the materials. The anatase TiO_2 crystalline sizes calculated based on Scherrer's formula were 16.1, 12.8, 10.6, 8.8, and 7.6 nm along with increasing PEG1000 concentration. It indicates that the growth of TiO_2 nanocrystals was hindered by PEG1000 addition during sol-gel preparation of the materials.

Template PEG1000 acted as pore forming reagent in sol-gel preparation of TiO_2-Al_2O_3 binary oxides. The precursor molecules were attached to PEG1000 molecules in the sol and formed a three-dimensional structure. After the PEG template was burned out during thermal treatment at high temperature, the resulting material was composed of porous TiO_2-Al_2O_3 binary oxides. The pores inside the oxides prohibited crystallite aggregation during calcination process. When high concentration of PEG1000 was used in the sols, more pores could form in the binary oxides, leading to small crystalline size of anatase TiO_2 in the binary oxides.

The SEM images of the samples are shown in Figure 2. The surface morphology of the TiO_2-Al_2O_3 oxides prepared without PEG1000 was quite smooth. The main surface was fairly flat with very slight roughness. Some particles in the size less than 1 μm scattered on the surface. These particles might be the residues of small particles produced during grinding. The surfaces of the samples prepared with PEG1000 were comparatively rough, and there were many long grooves on the surfaces. Many small particles formed during grinding also scattered over the surfaces.

Surface morphology is quite essential for photocatalytic activity of the materials. It is believed that rough and structured surface with high surface area is capable of providing more active centers. The rough surface of porous TiO_2-Al_2O_3 binary oxides clearly indicates the function of PEG template. There were no large pores existing on the surfaces of the binary oxides prepared with PEG. In other words, the surfaces were quite rough with obvious steps and grooves.

The FT-IR spectra of mesoporous TiO_2-Al_2O_3 binary oxides prepared using different concentration of PEG1000 are shown in Figure 3. The low-frequency absorption band at 499 cm^{-1} is attributed to the Ti–O–Ti vibration of the samples [22, 23]. The band at 1628 cm^{-1} corresponds to the bending vibration of O–H, and the band around 3392 cm^{-1} is assigned to the surface adsorbed water and hydroxyl groups [24, 25]. The broad band between 500 cm^{-1} and 800 cm^{-1} is attributed to Al–O–Al vibration of the samples [26]. No evidence of Ti–O–Al vibration is found, indicating that the samples were composed of TiO_2 and Al_2O_3.

The purpose of this paper was to prepare TiO_2-Al_2O_3 binary oxides. Since tetrabutyl titanate and aluminum isopropoxide were mixed together as precursor in order to make the sol containing both titanium and aluminum, the oxides of titanium and aluminum were thoroughly distributed in the prepared binary oxides. The FT-IR spectra reveal that there was no chemical reaction between titanium dioxide and aluminum oxide.

Figure 4 shows XPS spectra of mesoporous TiO_2-Al_2O_3 binary oxides prepared at PEG1000 concentration of 0.030 mol/L in the sols. The binding energy peaks of $Ti2p_{3/2}$ and $Ti2p_{1/2}$ are approximately at 458.5 eV and 464.2 eV, which are attributed to the typical XPS spectra of Ti^{4+} [27]. The symmetrical binding energy peaks of $Ti2p_{3/2}$ and $Ti2p_{1/2}$ indicate that the titanium was in its full oxidation state and there was no reduced titanium existing in TiO_2-Al_2O_3 oxides. The Al2p peak of the TiO_2-Al_2O_3 composite occurring at the binding energy of 75.1 eV is expected for Al^{3+} in aluminum oxide [28, 29]. The O1s spectrum observed at 529.5 eV is agreed with the electronic binding of O^{2-} in Ti-O and Al-O [30], and the binding energy of 531.8 eV is assigned to O^{2-} of OH adsorbed on the surface of the sample [31]. Therefore, it can be concluded that TiO_2 and Al_2O_3 existed in the binary oxides. As indicated before in XRD patterns, there was no other species formed in the materials.

Table 1: BET specific surface area, pore volume, and pore size of mesoporous TiO_2-Al_2O_3 binary oxides.

$C_{PEG1000}$/mol·L^{-1}	BET specific surface area/$m^2 \cdot g^{-1}$	Pore volume/$cm^3 \cdot g^{-1}$	Average pore size/nm
0	99.2	0.45	22.4
0.012	101.5	0.37	14.7
0.030	128.4	0.39	14.6
0.048	140.1	0.38	10.9
0.060	122.9	0.39	12.6

BET specific surface area, pore volume, and pore size of mesoporous TiO_2-Al_2O_3 binary oxides are listed in Table 1. When PEG1000 concentration increased from 0 to 0.048 mol/L, the BET specific surface area of the samples grew from 99.2 to 140.1 $m^2 \cdot g^{-1}$, and the average pore size decreased from 22.4 to 10.9 nm. Further increase of PEG1000 concentration led to the reverse trend. The addition of PEG1000 template during sol-gel process resulted in small pore size and large specific surface area in the materials.

The PEG1000 molecules which were captured in the gel were consequently burnt out during calcination process, leaving pores inside the binary oxides. The pore skeleton was composed of titanium and aluminum oxides. The pores shrank at high temperature to some extent depending on stability of the pores structure during thermal treatment. As calculated from XRD results, high concentration of PEG1000 in the sols was responsible for more pores forming in the prepared binary oxides, leading to small crystalline size of anatase TiO_2. The aggregation of smaller crystallites is responsible for the formation of smaller pores in the materials [32]. The sample prepared using 0.060 mol/L PEG1000 had smaller surface area than the sample prepared using 0.048 mol/L PEG1000. The PEG1000 molecules might combine and form large micelles at high concentration. As a result, the prepared sample can therefore have larger pore size and smaller surface area than the optimized one.

Figure 5 shows N_2 desorption isotherms (A) and pore size distribution (B) of mesoporous TiO_2-Al_2O_3 binary oxides prepared at $C_{PEG1000}$ of 0.03 mol/L. With the increase of relative pressure P/P_o in Figure 5(a), the adsorption of N_2 on the materials amplified relevantly. When the relative pressure P/P_o was about 0.6 to 0.8, the increment of adsorption was significant because of capillary condensation in the pores of the material. Figure 5(b) clarifies the mesoporous structures of the material. The mesopore size inside the material was below 50 nm, and pore diameter mainly distributed in the range of 5–10 nm. As calculated by BET equation and BJH method, BET specific surface area, pore

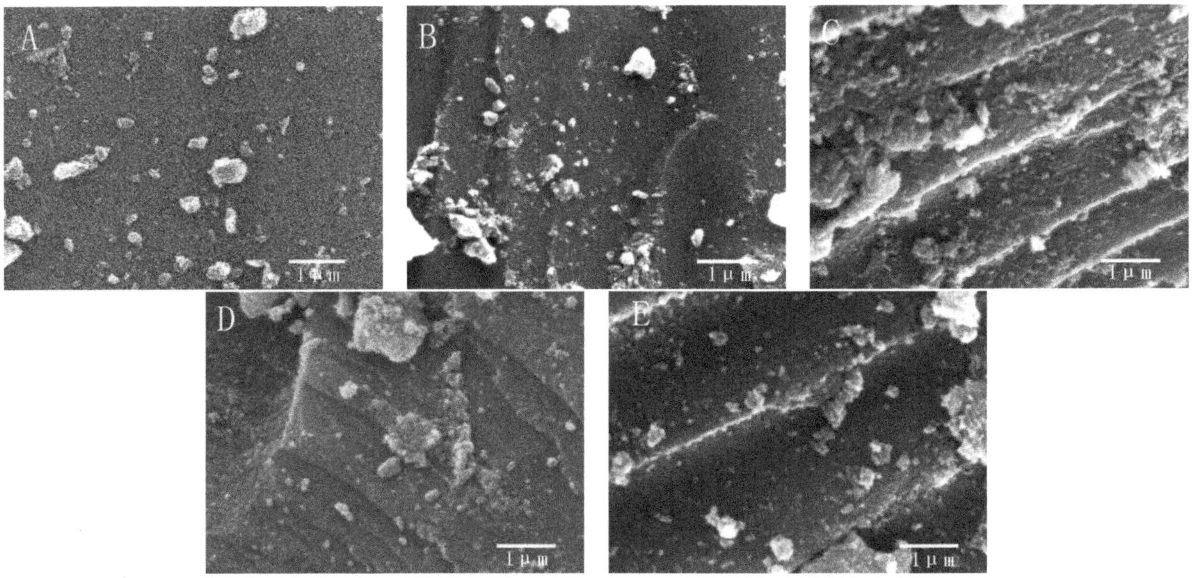

Figure 2: SEM images of mesoporous TiO_2-Al_2O_3 binary oxides prepared using different concentration of PEG1000. (a) 0 mol/L, (b) 0.012 mol/L, (c) 0.030 mol/L, (d) 0.048 mol/L, and (e) 0.060 mol/L.

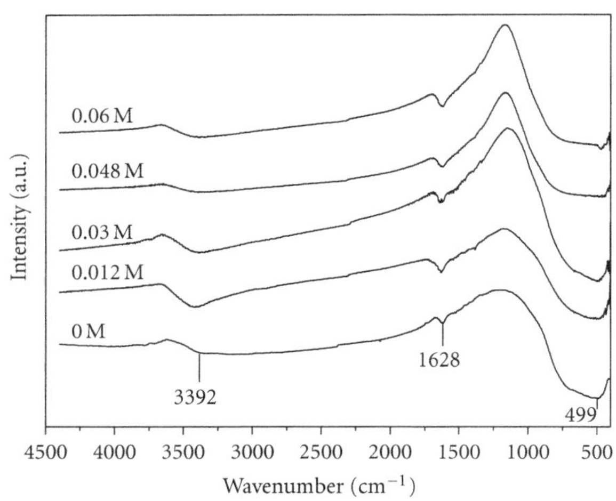

Figure 3: FT-IR spectra of mesoporous TiO_2-Al_2O_3 binary oxides prepared using different concentration of PEG1000.

volume, and average pore size of the material prepared at $C_{PEG1000}$ of 0.03 mol/L were 128.4 $m^2 \cdot g^{-1}$, 0.39 $cm^3 \cdot g^{-1}$, and 14.6 nm, respectively.

The effects of $C_{PEG1000}$ on the adsorption and photocatalytic activity of mesoporous TiO_2-Al_2O_3 binary oxides are shown in Figure 6. Methyl orange adsorption and photocatalytic degradation rates on the material prepared without using PEG1000 were 9.9% and 35.0%, respectively. With increasing PEG1000 concentration, methyl orange adsorption on the samples climbed up to the maximum value of 24.2% at $C_{PEG1000}$ of 0.048 mol/L. Further increase of PEG1000 concentration led to dropping off of methyl orange adsorption on the material. It indicates that adsorption ability of the samples was in accordance with the BET specific surface area. The larger the specific surface area is, the more adsorption capacity the sample has.

Photocatalytic degradation of methyl orange was improved slightly with the increase of PEG1000 concentration up to the maximum rate of 36.1% at $C_{PEG1000}$ of 0.03 mol/L. After that, methyl orange degradation rate declined obviously after using excessive amount of PEG1000 template in the sols. Although the specific surface area and adsorption capacity of the sample prepared at $C_{PEG1000}$ = 0.048 mol/L were the maximum, photocatalytic activity of the sample was only 27.0%. The sample prepared at $C_{PEG1000}$ of 0.03 mol/L had the optimal photocatalytic activity of 36.1% after 30 min of irradiation, and methyl orange adsorption rate on this material was 18.6%. Photocatalytic activity of the samples is related to several main factors. The samples can all adsorb a certain amount of methyl orange molecules, which is essential for the subsequent degradation process. The difference in surface area does not have as much effect on photocatalytic activity as that on adsorption capacity. On the other hand, the growth of TiO_2 nanocrystals was hindered by PEG1000 addition, leading to lowered degradation activity for the samples prepared at high $C_{PEG1000}$.

Although adsorption and photocatalytic degradation of methyl orange can all be regarded as decoloration of the dye in the water, the mechanisms are quite different. After physical adsorption on the materials, the dye remains on the surface of the materials in its initial form. On the other hand, photogenerated electrons and holes and the subsequently produced oxidizing species such as hydroxyl radicals contact with the adsorbed organic substances and initiate deep oxidation of the dye. The TiO_2-Al_2O_3 binary oxides prepared in this paper had high adsorption abilities due to their porous structure and the addition of Al_2O_3 into

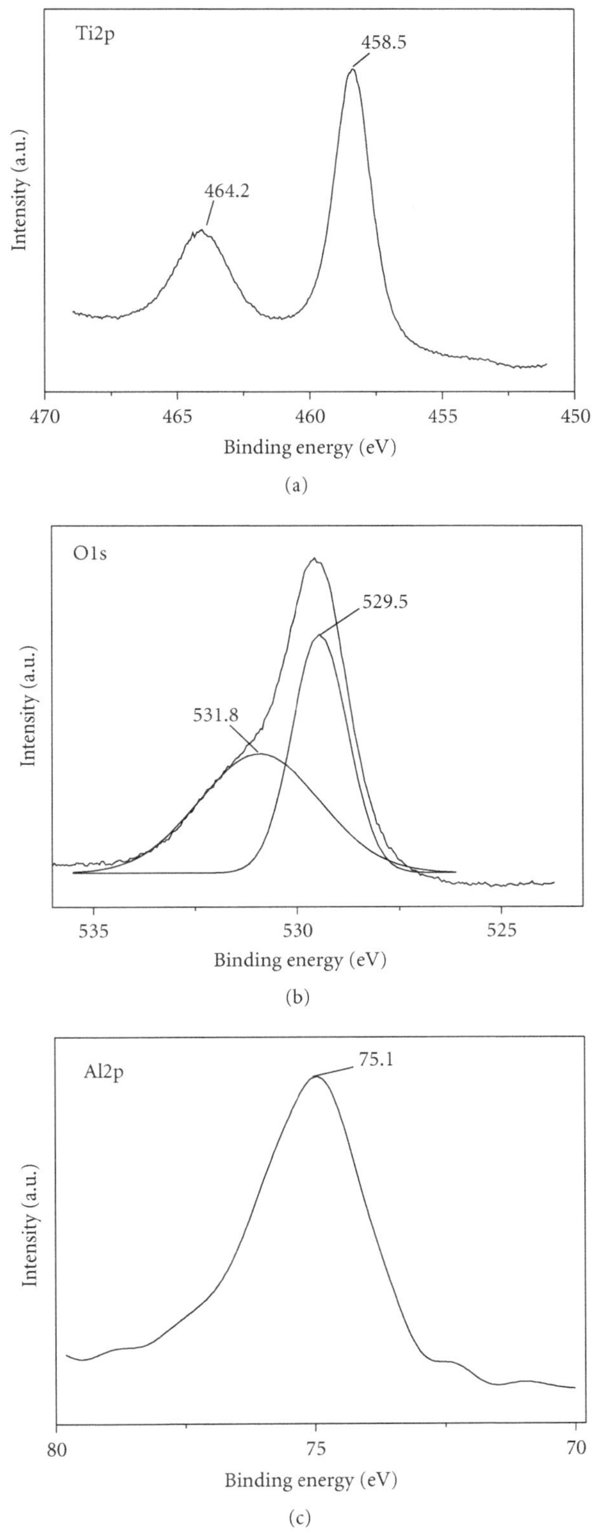

FIGURE 4: XPS spectra of mesoporous TiO_2-Al_2O_3 binary oxides prepared at $C_{PEG1000}$ = 0.030 mol/L.

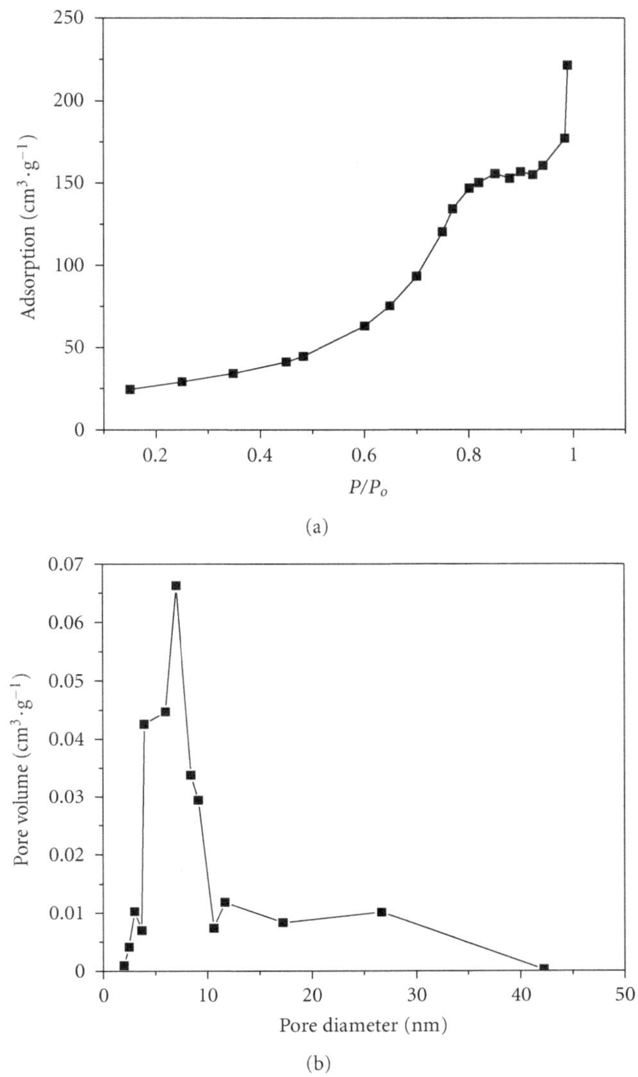

FIGURE 5: N_2 desorption isotherms (a) and pore size distribution (b) of mesoporous TiO_2-Al_2O_3 binary oxides prepared at $C_{PEG1000}$ = 0.03 mol/L.

TiO_2 could also contribute significantly. However, what is more valuable is the photocatalytic activity of the materials because the adsorbed dye can be fully oxidized only through photocatalytic degradation.

The effect of the concentration of mesoporous TiO_2-Al_2O_3 binary oxides (prepared at $C_{PEG1000}$ = 0.03 mol/L) used for methyl orange decoloration was investigated, as shown in Figure 7. Both of adsorption and photocatalytic degradations of methyl orange on TiO_2-Al_2O_3 increased sharply along with increasing photocatalyst concentration. The degradation rate increased much faster at low photocatalyst concentration. The photocatalyst particles might aggregate at high concentration. In spite of the constant increase of adsorption at higher TiO_2-Al_2O_3 oxides concentration, the enhancement rate of degradation was slowed down when photocatalyst concentration was higher than 600 mg/L. When using 600 mg/L of mesoporous TiO_2-Al_2O_3 binary oxides, methyl orange photocatalytic degradation and

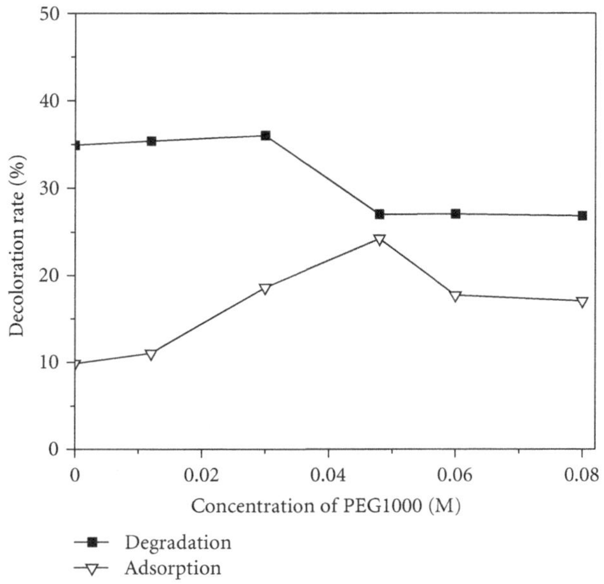

FIGURE 6: Effect of $C_{PEG1000}$ on the photocatalytic activity and adsorption of mesoporous TiO_2-Al_2O_3 binary oxides.

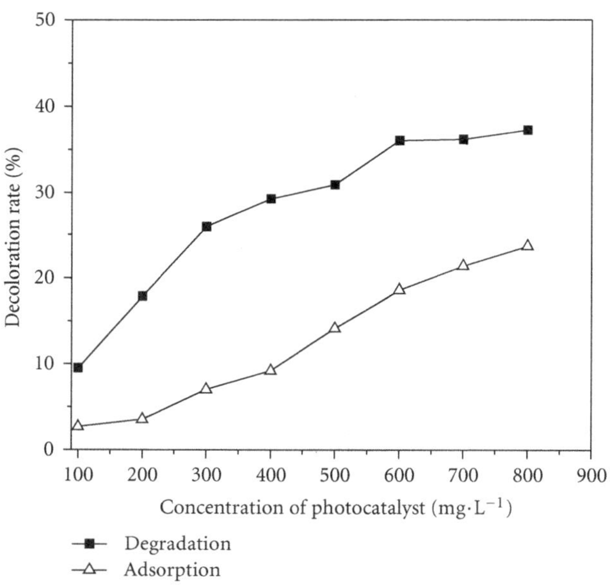

FIGURE 7: The effect of concentration of mesoporous TiO_2-Al_2O_3 binary oxides (prepared at $C_{PEG1000}$ = 0.03 mol/L) on methyl orange decoloration.

adsorption rates were 36.1% and 18.6%, respectively. As indicated before, the degradation rate was achieved in 30 min, and deep oxidation of the dye can be expected with prolonged irradiation time.

Figure 8 shows the UV-Vis spectra of methyl orange solution during photocatalytic degradation on TiO_2-Al_2O_3 oxides. The binary oxides were prepared at $C_{PEG1000}$ of 0.03 mol/L, and photocatalyst concentration was 600 mg/L for decoloration of the dye. The main absorption peaks near

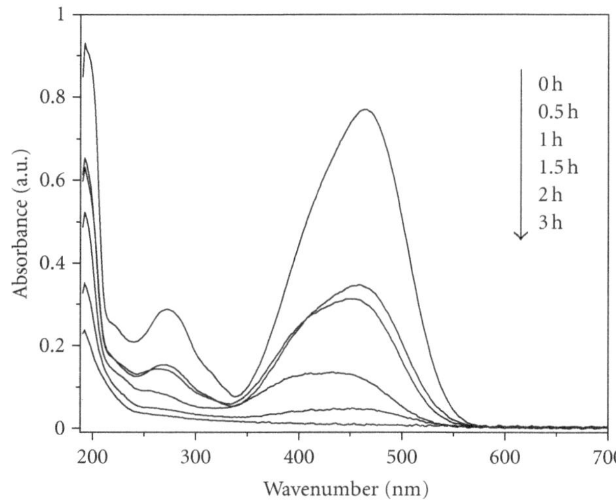

FIGURE 8: The UV-Vis spectra of MO during photocatalytic degradation on TiO_2-Al_2O_3 oxides prepared at $C_{PEG1000}$ = 0.03 mol/L. Photocatalyst concentration was 600 mg/L.

468 nm and 270 nm are assigned to benzene ring and azo-structure of methyl orange molecule. The absorption peaks shrank sharply under UV-light irradiation. Fully degradation of the dye was achieved after 3 h of reaction. The main absorption peaks of methyl orange disappeared after methyl orange molecules broke into small parts during photocatalytic degradation. The high adsorption and photocatalytic activity of the material are attributed to large surface area, mesoporous structure, and small crystallite size.

4. Conclusions

Mesoporous TiO_2-Al_2O_3 binary oxides were synthesized by sol-gel method using PEG1000 as template agent. The effects of PEG1000 concentration on the physicochemical properties of the materials were investigated. The specific surface area and adsorption capacity of the TiO_2-Al_2O_3 oxides were larger than the sample prepared without PEG1000. The TiO_2-Al_2O_3 oxides prepared with $C_{PEG1000}$ of 0.03 mol/L had optimal photocatalytic activity and high adsorption capacity, which are contributed to large surface area, mesoporous structure, and small crystallite size.

Acknowledgments

This work was supported by the National Natural Science Foundation of China (no. 41071161 and 41130524), National Key Basic Research Foundation of China (2011CB403202), and Liaoning Science and Technology Project (2010229002).

References

[1] M. R. Hoffmann, S. T. Martin, W. Choi, and D. W. Bahnemann, "Environmental applications of semiconductor photocatalysis," *Chemical Reviews*, vol. 95, no. 1, pp. 69–96, 1995.

[2] A. Fujishima, T. N. Rao, and D. A. Tryk, "Titanium dioxide photocatalysis," *Journal of Photochemistry and Photobiology C*, vol. 1, no. 1, pp. 1–21, 2000.

[3] M. I. Litter, "Heterogeneous photocatalysis: transition metal ions in photocatalytic systems," *Applied Catalysis B*, vol. 23, no. 2-3, pp. 89–114, 1999.

[4] K. Rajeshwar, M. E. Osugi, W. Chanmanee et al., "Heterogeneous photocatalytic treatment of organic dyes in air and aqueous media," *Journal of Photochemistry and Photobiology C*, vol. 9, no. 4, pp. 171–192, 2008.

[5] G. Plantard, T. Janin, V. Goetz, and S. Brosillon, "Solar photocatalysis treatment of phytosanitary refuses: efficiency of industrial photocatalysts," *Applied Catalysis B*, vol. 115-116, pp. 38–44, 2012.

[6] D. M. Antonelli and J. Y. Ying, "Synthesis of hexagonally packed mesoporous TiO_2 by a modified sol-gel method," *Angewandte Chemie*, vol. 34, no. 18, pp. 2014–2017, 1995.

[7] U. Bach, D. Lupo, P. Comte et al., "Solid-state dye-sensitized mesoporous TiO_2 solar cells with high photon-to-electron conversion efficiencies," *Nature*, vol. 395, no. 6702, pp. 583–585, 1998.

[8] H. Yoshitake, T. Sugihara, and T. Tatsumi, "Preparation of wormhole-like mesoporous TiO_2 with an extremely large surface area and stabilization of its surface by chemical vapor deposition," *Chemistry of Materials*, vol. 14, no. 3, pp. 1023–1029, 2002.

[9] Q. Sheng, S. Yuan, J. Zhang, and F. Chen, "Synthesis of mesoporous titania with high photocatalytic activity by nanocrystalline particle assembly," *Microporous and Mesoporous Materials*, vol. 87, no. 3, pp. 177–184, 2006.

[10] X. Chen, T. Yu, X. Fan et al., "Enhanced activity of mesoporous Nb_2O_5 for photocatalytic hydrogen production," *Applied Surface Science*, vol. 253, no. 20, pp. 8500–8506, 2007.

[11] Z. Wang, Z. Li, H. Zhang, and C. Wang, "Improved photocatalytic activity of mesoporous ZnO-SiO_2 coupled nanofibers," *Catalysis Communications*, vol. 11, no. 4, pp. 257–260, 2009.

[12] J. Jiao, Q. Xu, and L. Li, "Porous TiO_2/SiO_2 composite prepared using PEG as template direction reagent with assistance of supercritical CO_2," *Journal of Colloid and Interface Science*, vol. 316, no. 2, pp. 596–603, 2007.

[13] B. Neppolian, Q. Wang, H. Yamashita, and H. Choi, "Synthesis and characterization of ZrO_2-TiO_2 binary oxide semiconductor nanoparticles: application and interparticle electron transfer process," *Applied Catalysis A*, vol. 333, no. 2, pp. 264–271, 2007.

[14] M. R. Bayati, A. Z. Moshfegh, F. Golestani-Fard, and R. Molaei, "$(WO_3)_x$-$(TiO_2)_{1-x}$ nano-structured porous catalysts grown by micro-arc oxidation method: characterization and formation mechanism," *Materials Chemistry and Physics*, vol. 124, no. 1, pp. 203–207, 2010.

[15] X. Su, J. Zhao, Y. Li et al., "Solution synthesis of Cu_2O/TiO_2 core-shell nanocomposites," *Colloids and Surfaces A*, vol. 349, no. 1-3, pp. 151–155, 2009.

[16] M. Stodolny and M. Laniecki, "Synthesis and characterization of mesoporous Ta_2O_5-TiO_2 photocatalysts for water splitting," *Catalysis Today*, vol. 142, no. 3-4, pp. 314–319, 2009.

[17] T. Miwa, S. Kaneco, H. Katsumata et al., "Photocatalytic hydrogen production from aqueous methanol solution with CuO/Al_2O_3/TiO_2 nanocomposite," *International Journal of Hydrogen Energy*, vol. 35, no. 13, pp. 6554–6560, 2010.

[18] H. Choi, E. Stathatos, and D. D. Dionysiou, "Sol-gel preparation of mesoporous photocatalytic TiO_2 films and TiO_2/Al_2O_3 composite membranes for environmental applications," *Applied Catalysis B*, vol. 63, no. 1-2, pp. 60–67, 2006.

[19] C. Anderson and A. J. Bard, "Improved photocatalytic activity and characterization of mixed TiO_2/SiO_2 and TiO_2/Al_2O_3 materials," *Journal of Physical Chemistry B*, vol. 101, no. 14, pp. 2611–2616, 1997.

[20] W. Li, S. Li, M. Zhang, and K. Tao, "The effect of phosphorus on nano-sized TiO_2 particles dispersed on Al_2O_3," *Colloids and Surfaces A*, vol. 272, no. 3, pp. 189–193, 2006.

[21] Y. Kim and M. Yoon, "TiO_2/Y-zeolite encapsulating intramolecular charge transfer molecules: a new photocatalyst for photoreduction of methyl orange in aqueous medium," *Journal of Molecular Catalysis A*, vol. 168, no. 1-2, pp. 257–263, 2001.

[22] Q. Li, H. Su, and T. Tan, "Synthesis of ion-imprinted chitosan-TiO_2 adsorbent and its multi-functional performances," *Biochemical Engineering Journal*, vol. 38, no. 2, pp. 212–218, 2008.

[23] G. Martra, V. Augugliaro, S. Coluccia et al., "Photocatalytic oxidation of gaseous toluene on polycrystalline TiO_2: FT-IR investigation of surface reactivity of different types of catalysts," *Studies in Surface Science and Catalysis*, vol. 130, pp. 665–670, 2000.

[24] N. Venkatachalam, M. Palanichamy, and V. Murugesan, "Sol-gel preparation and characterization of alkaline earth metal doped nano TiO_2: efficient photocatalytic degradation of 4-chlorophenol," *Journal of Molecular Catalysis A*, vol. 273, no. 1-2, pp. 177–185, 2007.

[25] M. Huang, C. Xu, Z. Wu, Y. Huang, J. Lin, and J. Wu, "Photocatalytic discolorization of methyl orange solution by Pt modified TiO_2 loaded on natural zeolite," *Dyes and Pigments*, vol. 77, no. 2, pp. 327–334, 2008.

[26] C. J. Doss and R. Zallen, "Raman studies of sol-gel alumina: finite-size effects in nanocrystalline AlO(OH)," *Physical Review B*, vol. 48, no. 21, pp. 15626–15637, 1993.

[27] X. Zhang, M. Zhou, and L. Lei, "Preparation of anatase TiO_2 supported on alumina by different metal organic chemical vapor deposition methods," *Applied Catalysis A*, vol. 282, no. 1-2, pp. 285–293, 2005.

[28] M. R. Bayati, H. R. Zargar, A. Talimian, A. Ziaee, and R. Molaei, "Characterization of Al_2O_3-TiO_2 nano porous solar absorbers derived via MAO/sol-gel hybrid process," *Surface and Coatings Technology*, vol. 205, no. 7, pp. 2483–2489, 2010.

[29] Y. Xin, K. Huo, T. Hu, G. Tang, and P. K. Chu, "Mechanical properties of Al_2O_3/Al bi-layer coated AZ91 magnesium alloy," *Thin Solid Films*, vol. 517, no. 17, pp. 5357–5360, 2009.

[30] Y. Xu, B. Lei, L. Guo, W. Zhou, and Y. Liu, "Preparation, characterization and photocatalytic activity of manganese doped TiO_2 immobilized on silica gel," *Journal of Hazardous Materials*, vol. 160, no. 1, pp. 78–82, 2008.

[31] Y. Xie, X. Zhao, Y. Li, Q. Zhao, X. Zhou, and Q. Yuan, "CTAB-assisted synthesis of mesoporous F-N-codoped TiO_2 powders with high visible-light-driven catalytic activity and adsorption capacity," *Journal of Solid State Chemistry*, vol. 181, no. 8, pp. 1936–1942, 2008.

[32] M. Zhou, J. Yu, and H. Yu, "Effects of urea on the microstructure and photocatalytic activity of bimodal mesoporous titania microspheres," *Journal of Molecular Catalysis A*, vol. 313, no. 1-2, pp. 107–113, 2009.

7

A Novel Extension Decision-Making Method for Selecting Solar Power Systems

Meng-Hui Wang

Department of Electrical Engineering, National Chin-Yi University of Technology, Taichung 41110, Taiwan

Correspondence should be addressed to Meng-Hui Wang; wangmh@ncut.edu.tw

Academic Editor: Tapas Mallick

Due to the complex parameters of a solar power system, the designer not only must think about the load demand but also needs to consider the price, weight, and annual power generating capacity (APGC) and maximum power of the solar system. It is an important task to find the optimal solar power system with many parameters. Therefore, this paper presents a novel decision-making method based on the extension theory; we call it extension decision-making method (EDMM). Using the EDMM can make it quick to select the optimal solar power system. The paper proposed this method not only to provide a useful estimated tool for the solar system engineers but also to supply the important reference with the installation of solar systems to the consumer.

1. Introduction

In recent years, global warming caused the sea levels rising. As a result, the loss of coastal wetlands causes the erosion of the coastline and the continued flooding in the low-lying area. Regional climate changes also caused serious striking against the environmental ecology and agriculture. The greenhouse effect and global warming are becoming increasingly apparent. In the past two decades, the economy has grown rapidly in Taiwan, and it made the energy requirement increase greatly. However, there is an extreme shortage of energy resources in Taiwan. Also, 99 percent of the energy must rely on import from the foreign countries. Facing the fossil-fuel shortage crisis, Taiwan is actively opening up the renewable energy at present. The solar energy is one of the renewable energy resources with the highest acceptance in Taiwan.

Because Taiwan is located in subtropics, the temperature and the sunlight suit to develop the solar power system. Solar power system has high potential. When the solar power system operates, it neither produces the air pollution nor generates a large number of CO_2. Therefore, this system is very helpful to reduce the global warming and the demand of fossil energy. The solar power system is also a clean and safe power supply system for businesses and families [1, 2].

The semiconductor industry fairly matures in Taiwan. The electronic power technology development is stepping into a new stage. There is a quite good basis on manufacturing technology and R&D manpower. By the global photovoltaic (PV) industry growth trends, it is high competition to develop the solar power system in Taiwan. According to the industry statistics until 2007, there are more than 70 solar power system companies and solar power system output is 430 MW which is the fourth in the world. The PVs industry production is worth 53.5 billion dollars. It is beneficial to save the energy and reduce CO_2 and the industrial development. In part of the solar development, the single crystal silicon solar cell efficiency has reached 21%, and the polysilicon solar cell efficiency is up to 18.5%. The high efficient design and the packaging technologies in the PV module technology have been established. In 2012, we expect to be completed to install 20,000 families, and the capacity of 60 MW installations. The solar energy system can generate 7,200 million kWh per year. It promotes the development of the domestic PV industry, creates the domestic markets, and develops the industries.

When the solar power system is installed, it needs to consider the loading demand and the installing area. For the user, it is important to consider with the recovered cost, APGC, and the total weight of solar power system [3]. Confronting the variety of solar power system in the market, the solar system is an important task of how to select the best option from the specifications provided. Therefore, this paper presents a novel decision-making method based on the extension theory [4–6]; we call it extension decision-making method (EDMM). First, according to the specifications of PV cell provided by the manufacturer, this paper uses the commercial software to simulate every solar power system and to compute the APGC of solar system. Then, this paper uses the proposed EDMM to count the evaluation index of every system. Using the evaluation index can make it easy to select the optimal system. This proposed method not only provides the system engineers with as a very useful assessment tool, but also supplies the users with the important reference when setting a solar power system. In addition, this paper proposes the method which can be added or subtracted to the characteristic and weight by the user's requirement. We hope that the proposed method will lead to further investigation for other renewable energy systems.

2. Review of Extension Theory

In the standard set, an element either belongs to or not belongs; so, the range of the standard set is $\{0, 1\}$, which can be used to solve a two-valued problem. In contrast to the standard set, the fuzzy set allows the description of concepts in which the boundary is not explicit. It concerns not only whether an element belongs to the set but also to what degree it belongs to the set. The range of a fuzzy set is $[0, 1]$. The extension set extends the fuzzy set from $[0, 1]$ to $(-\infty, \infty)$. As a result, it allows us to define a set that includes any data in the domain. Extension theory tries to solve the incompatibility or contradiction problems by the transformation of the matter element [7, 8]. The comparisons of the standard sets, fuzzy sets, and extension sets are shown in Table 1. Some definitions of extension theory are introduced in the next section.

2.1. Matter-Element Theory

2.1.1. Definition of Matter-Element. Defining the name of matter as N, the characteristics of the matter as c, and the value of c as v, a matter-element in extension theory can be described as follows:

$$R = (N, c, v), \qquad (1)$$

where N, c, and v were called the three fundamental elements of the matter-element. If the value of the characteristic has a classical domain or a range, then we define the matter-element for the classical domain as follows:

$$R = (N, c, v) = \left(N, c, \left\langle v^L, v^U \right\rangle\right), \qquad (2)$$

where v^L and v^U are the lower bound and upper bound of a classical domain.

2.1.2. Multidimensional Matter-Element. Assuming $R = (N, C, V)$ to be a multidimensional matter-element, $C = [c_1, c_2, \ldots, c_n]$ a characteristic vector, and $V = [v_1, v_2, \ldots, v_n]$ a value vector of C, then a multidimensional matter-element is defined as

$$R = \begin{bmatrix} N, c_1, v_1 \\ c_2, v_2 \\ \vdots \\ c_n, v_n \end{bmatrix} = \begin{bmatrix} R_1 \\ R_2 \\ \vdots \\ R_n \end{bmatrix}, \qquad (3)$$

where $R_i = (N, c_i, v_i)$ $(i = 1, 2, \ldots, n)$ is defined as the sub-matter-element of R.

2.2. Extension Set Theory

2.2.1. Definition of Extension Set. Let U be the universe of discourse; then, an extension set $\widetilde{A}$ on U is defined as a set of ordered pairs as follows:

$$\widetilde{A} = \{(v, y) \mid v \in U, y = K(v) \in (-\infty, \infty)\}, \qquad (4)$$

where $y = K(x)$ is called the correlation function for extension set $\widetilde{A}$. The $K(x)$ maps each element of U to a membership grade between $-\infty$ and ∞. The higher degree and the more elements belong to the set. In a special condition, when $0 \le K(x) \le 1$, it corresponds to a normal fuzzy set. $K(x) \le -1$ implies that the element x has no chance to belong to the set. When $-1 < K(x) < 0$, it is called an extension domain, which means that the element x still has a chance to become part of the set.

2.2.2. Definition of Correlation Function. The correlation functions have many forms dependent on application. If we set $V_o = \langle a, b \rangle$ and $V = \langle c, d \rangle$ being two intervals in the real number field, and $V_o \in V$, then the correlation function in the extension theory can be defined as follows:

$$K(v) = \frac{\rho(v, V_o)}{D(v, V_o, V)}, \qquad (5)$$

where

$$D(v, V_o, V) = \begin{cases} \rho(v, V) - \rho(v, V_o), & v \notin V_o, \\ -1, & v \in V_o, \end{cases}$$

$$\rho(v, V_o) = \left| x - \frac{a+b}{2} \right| - \frac{b-a}{2}, \qquad (6)$$

$$\rho(v, V) = \left| x - \frac{c+d}{2} \right| - \frac{d-c}{2}.$$

The correlation function can be used to calculate the membership grade between x and V_o as shown in Figure 1.

2.3. Decision-Making Method for Solar Power System. Using the decision-making method for the solar power systems, it must have a software suit to simulate APGC, the maximum power, and the current of the solar power systems. Then, we can use the characteristics of solar power system to build the matter-element model of solar systems and to calculate the estimated index for selecting rank.

Table 1: Three different sorts of mathematical sets.

Compared item	Standard set	Fuzzy set	Extension set
Research objects	Data variables	Linguistic variables	Contradictory problems
Model	Mathematics model	Fuzzy mathematics model	Matter-element model
Descriptive function	Transfer function	Membership function	Correlation function
Descriptive property	Precision	Ambiguity	Extension
Range of set	$C_A(x) \in (0,1)$	$\mu_A(x) \in [0,1]$	$K_A(x) \in (-\infty,\infty)$

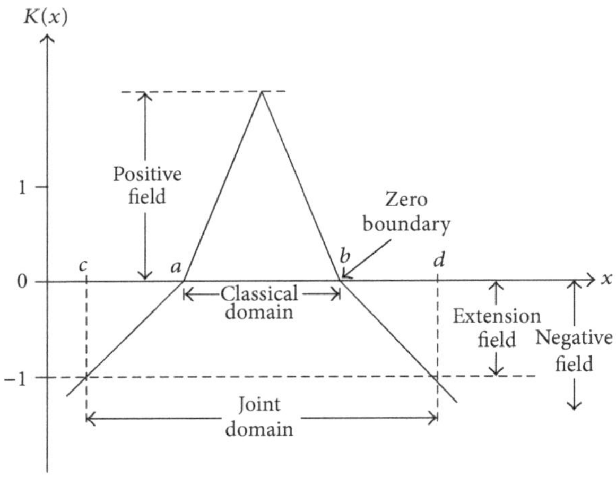

Figure 1: The proposed extended relation function.

2.4. Test Conditions. The paper is based on the simulated software to unify the location, area, the direction of harmonization, and the horizontal angle for the solar power system. It simulates all the conditions of each solar power system installed in the location of the latitude and longitude, respectively, (24, 9) and (120, 40), in Taichung. Then, it simulates the solar power system and the horizontal included angle is 23 degrees and the panel toward the south. The simulated total area of the solar power system is 50 square meters. There are established five groups of the matter-element model according to the characteristics which are the APGC, the maximum power, the maximum current, the total cost, and the total weight of the solar power systems.

The first three data statistics calculate from the commercial simulation software, the typical simulated results of the solar power system can be shown in Figures 2 and 3, there are easy to calculate the APCG, maximum power, and current of the tested solar system.

The two other data are provided from the producers of solar power systems. Table 2 shows the specification of different solar systems. Then, the paper uses the earlier 5 characteristics to create the matter-element model as shown in Table 3. Here, the A-class represents the best solar module. The opposite E-class represents the worst solar module. Table 3 shows that the $c1$ is APGC, $c2$ is the maximum power, $c3$ is the total cost, $c4$ is the maximum current, and $c5$ is the total weight.

Table 2: Specifications of the different PV modules.

PV samples	Characteristics				
	APGC (kwh/year)	Max. power (kw)	Cost (10^3 USD)	Max. current (A)	Weight (kg)
Module 1	4962	4.2	23.6	240	380
Module 2	4575	3.8	23.6	238	395
Module 3	11969	10.3	42.1	253	800
Module 4	3890	3.7	51.5	119	1030
Module 5	6998	6.2	30.1	174	644
Module 6	6629	5.6	24.8	330	608
Module 7	6685	6.0	27.7	341	608
Module 8	5251	4.3	44.1	288	662
Module 9	4528	4.1	40.8	131	836
Module 10	6833	6.2	26.8	363	886
Module 11	6286	5.1	52.3	220	420
Module 12	7255	6.3	29.7	178	570
Module 13	7858	6.7	29.3	186	532
Module 14	7812	6.3	30.9	356	600
Module 15	7915	6.2	32.7	356	601
Module 16	6543	5.7	32.2	324	672
Module 17	7284	6.4	23.5	127	812
Module 18	7415	5.5	30.7	332	550
Module 19	12614	10.8	47.3	257	800
Module 20	12297	10.5	42.4	255	800
Module 21	11094	9.5	49.0	172	1150
Module 22	5824	4.6	29.7	278	814
Module 23	4335	4.1	51.3	199	624
Module 24	7303	6.0	33.1	338	589
Module 25	3903	3.5	49.3	221	516
Module 26	8209	7.1	24.9	236	480
Module 27	4307	3.9	47.0	125	941
Module 28	6943	5.9	30.5	338	476
Module 29	10200	8.8	40.1	219	660
Module 30	2134	2.1	38.6	121	620

2.5. The Proposed Extension Decision-Making Method. This paper has been successfully using computer software to implement the proposed method for the selection of solar power systems; the details of the proposed method are shown as follows.

TABLE 3: Matter-element models of different classes.

Classes	Matter-element models
A-Class	$R1 = \begin{Bmatrix} A & c1 & \langle 80, 100\rangle \\ & c2 & \langle 80, 100\rangle \\ & c3 & \langle 0, 20\rangle \\ & c4 & \langle 80, 100\rangle \\ & c5 & \langle 0, 20\rangle \end{Bmatrix}$
B-Class	$R2 = \begin{Bmatrix} B & c1 & \langle 60, 80\rangle \\ & c2 & \langle 60, 80\rangle \\ & c3 & \langle 20, 40\rangle \\ & c4 & \langle 60, 80\rangle \\ & c5 & \langle 20, 40\rangle \end{Bmatrix}$
C-Class	$R3 = \begin{Bmatrix} C & c1 & \langle 40, 60\rangle \\ & c2 & \langle 40, 60\rangle \\ & c3 & \langle 40, 60\rangle \\ & c4 & \langle 40, 60\rangle \\ & c5 & \langle 40, 60\rangle \end{Bmatrix}$
D-Class	$R4 = \begin{Bmatrix} D & c1 & \langle 20, 40\rangle \\ & c2 & \langle 20, 40\rangle \\ & c3 & \langle 60, 80\rangle \\ & c4 & \langle 20, 40\rangle \\ & c5 & \langle 60, 80\rangle \end{Bmatrix}$
E-Class	$R5 = \begin{Bmatrix} E & c1 & \langle 0, 20\rangle \\ & c2 & \langle 0, 20\rangle \\ & c3 & \langle 80, 100\rangle \\ & c4 & \langle 0, 20\rangle \\ & c5 & \langle 80, 100\rangle \end{Bmatrix}$

Step 1. Normalize the values of the characteristics into an interval between 0 and 100 as in (7). This process will be beneficial for estimation of solar systems. Consider

$$V_j^{\text{new}} = \left(\frac{V_j^{\text{old}} - V_{\min}}{V_{\max} - V_{\min}}\right) \times 100, \quad (7)$$

$$V_{\max} = \max\{V_j\}, \qquad V_{\min} = \min\{V_j\}. \quad (8)$$

Step 2. Establish the matter-element model of every rank according to Table 3, where

$$R_i = \begin{Bmatrix} \text{class} & c1 & v_{i1} \\ & c2 & v_{i2} \\ & c3 & v_{i3} \\ & c4 & v_{i4} \\ & c5 & v_{i5} \end{Bmatrix}. \quad (9)$$

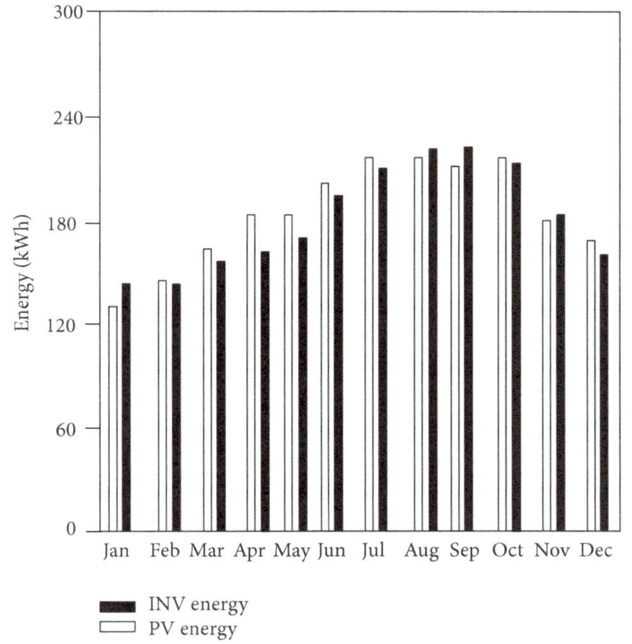

FIGURE 2: The typical APGC curves of the solar power system.

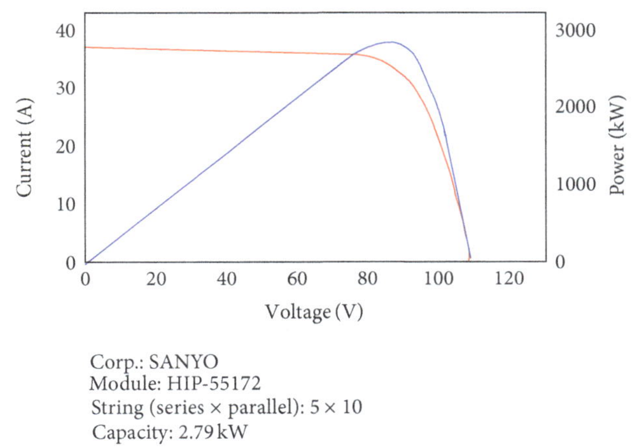

FIGURE 3: The typical maximum power and current curves of the solar power systems.

Step 3. Calculate the correlation index of every rank with the proposed extended correlation function as follows:

$$K_{ij} = \begin{cases} -\rho(v_i, V_{ij}), & v_i \in V_{ij}, \\ \dfrac{\rho(v_i, V_{ij})}{\rho(V_i, V_{pi}) - \rho(V_i, V_{ij})}, & v_i \notin V_{ij}, \end{cases} \quad (10)$$

$$i = 1, 2, \ldots, 5; \quad j = 1, 2, \ldots, 5.$$

Step 4. Set the weights of the estimated characteristics, W_{i1}, W_{i2}, W_{i3}, W_{i4}, and W_{i5}, depending on the importance of every characteristic in the estimated process of the user. In this paper, the five weights are set at 0.3, 0.15, 0.3, 0.15, and 0.1, which means that the APGC and cost are more important than the other characteristics.

Step 5. Calculate the relation indexes for every rank as follows:

$$\lambda_i = \sum_{j=1}^{5} W_{ij} K_{ij}, \quad i = 1, 2, \ldots, 5. \tag{11}$$

Step 6. Normalize the values of the relation indexes into an interval between −100 and 100 as in (12). This process will be beneficial for the estimated process. Consider

$$\lambda_i' = \begin{cases} \dfrac{\lambda_i}{\lambda_{\max}} \times 100, & \lambda_i > 0, \\ \dfrac{\lambda_i}{|\lambda_{\min}|} \times 100, & \lambda_i < 0, \end{cases} \tag{12}$$

where

$$\lambda_{\min} = \min\{\lambda_i\}; \qquad \lambda_{\max} = \max\{\lambda_i\}. \tag{13}$$

Step 7. Rank the normalized correlation indexes to estimate the rank of the tested solar system. Generally, the high correlation index implies that the solar power system has the high probability to belong to this class. Therefore, using the correlation index can be easy to determine the class of system that is most useful for the user and engineer.

3. Test Results and Discussion

The paper provides the decision-making method of solar power system that uses the MATLAB software to write the computing program. The testing results of the proposed method are shown in Table 4. It can observe the correlated index of various solar systems with the different rank, it should be noticed that the correlated index is based on the demanded conditions of the user. It can observe the correlated index of various solar power systems from Table 4. For example, module 1 can be a D-class solar power system due to large correlated index with the D-class, and module 19 is an A-class or better solar power system due to large correlated index with the A-class. On the contrary, the module 25 is an E-class or bad solar power system in the demanded conditions of the user. Table 5 compiles statistics of the top 5 according to A and E classes with correlated index the top 5 of the A-class are modules 19, 10, 20, 3, and 7, respectively, and the top 5 of the E-class include modules 4, 25, 9, 27, and 30, respectively. Therefore, the proposed EDMM is useful to determine the optimal solar power system based on the predetermined specifications of the user or solar system engineer.

Table 4: The correlation results of solar modules.

Sample	Rank				
	A	B	D	C	E
Module 1	−40	−76	38	14	−56
Module 2	−20	−79	10	12	−53
Module 3	46	−52	14	15	−80
Module 4	−100	−94	−54	−85	26
Module 5	−37	18	3	44	−29
Module 6	25	9	−53	10	−68
Module 7	37	11	−50	24	−40
Module 8	−60	22	100	−29	−46
Module 9	−76	−62	20	−28	9
Module 10	62	48	2	36	−36
Module 11	−8	−67	8	−19	−61
Module 12	−33	10	6	60	−31
Module 13	−17	−15	13	37	−36
Module 14	18	36	−53	43	−51
Module 15	17	50	−49	39	−53
Module 16	1	48	−36	12	−42
Module 17	−54	−62	−57	66	1
Module 18	5	24	−45	42	−46
Module 19	66	−77	−62	7	−91
Module 20	56	−60	19	11	−82
Module 21	12	−64	−27	−75	−72
Module 22	−39	17	42	2	−26
Module 23	−82	3	26	−59	−44
Module 24	8	42	−42	54	−49
Module 25	−53	−72	−32	−27	16
Module 26	31	−34	−58	30	−72
Module 27	−89	−80	17	−62	3
Module 28	20	17	−48	33	−46
Module 29	−29	24	−38	10	−66
Module 30	−84	4	−74	27	2

Table 5: The top 5 of A-class and E-class.

Ranks	A-class	E-class
1st	Module 19	Module 4
2nd	Module 10	Module 25
3rd	Module 20	Module 9
4th	Module 3	Module 27
5th	Module 7	Module 30

4. Conclusion

The paper combines the commercial solar simulated software with the proposed EDMM to determine the optimal solar power system based on the predetermined specifications of the user or the engineer. The calculation of the proposed decision-making algorithm is also fast and very simple. It can be easily implemented by PC software. Test results show that the proposed method cannot only estimate the class of solar system, but it can also be applied to the selecting tools of other related renewable power systems such as wind power and full cells systems. The proposed method is also a new decision-making scheme for other engineering problems.

References

[1] H. Wang, Y. Li, J. Wang, and Z. Wang, "Analysis of photovoltaic charging system based on MPPT," in *Proceedings of the Pacific-Asia Workshop on Computational Intelligence and Industrial Application (PACIIA '08)*, pp. 498–501, December 2008.

[2] M. Taherbaneh and K. Faez, "Maximum power point estimation for photovoltaic systems using neural networks," in *Proceedings of the IEEE International Conference on Control and Automation ThC4-5 (ICCA '07)*, pp. 1614–1619, Guangzhou, China, June 2007.

[3] M. Taherbaneh, H. G. Fard, A. H. Rezaie, and S. Karbasian, "Combination of fuzzy-based maximum power point tracker and sun tracker for deployable solar panels in photovoltaic systems," in *Proceedings of the IEEE International Conference on Fuzzy Systems (FUZZ '07)*, July 2007.

[4] M. H. Wang, "A novel extension method for transformer fault diagnosis," *IEEE Transactions on Power Delivery*, vol. 22, no. 8, pp. 63–64, 2002.

[5] M. H. Wang, "Application of extension theory to vibration fault diagnosis of generator sets," *IEE Proceedings*, vol. 151, no. 4, pp. 503–508, 2004.

[6] W. Cai, "Extension managementengineering and applications," *International Journal of Operations and Quantitative Management*. In press.

[7] M. H. Wang and C. P. Hung, "Extension neural network and its applications," *Neural Networks*, vol. 16, no. 5-6, pp. 779–784, 2003.

[8] M. H. Wang, Y. F. Tseng, H. C. Chen, and K. H. Chao, "A novel clustering algorithm based on the extension theory and genetic algorithm," *Expert Systems with Applications*, vol. 36, no. 4, pp. 8269–8276, 2009.

8

A Comprehensive Review and Analysis of Solar Photovoltaic Array Configurations under Partial Shaded Conditions

R. Ramaprabha and B. L. Mathur

Department of EEE, SSN College of Engineering, Kalavakkam-603 110, Chennai, India

Correspondence should be addressed to R. Ramaprabha, ramaprabhasuresh@gmail.com

Academic Editor: Songyuan Dai

The aim of this paper is to investigate the effects of partial shading on energy output of different Solar Photovoltaic Array (SPVA) configurations and to mitigate the losses faced in Solar Photovoltaic (SPV) systems by incorporating bypass diodes. Owing to the practical difficulty of conducting experiments on varied array sizes, a generalized MATLAB M-code has been developed for any required array size, configuration, shading patterns, and number of bypass diodes. The proposed model which also includes the insolation-dependent shunt resistance can provide sufficient degree of precision without increasing the computational effort. All the configurations have been analyzed and comparative study is made for different random shading patterns to determine the configuration less susceptible to power losses under partial shading. Inferences have been drawn by testing several shading scenarios.

1. Introduction

Solar photovoltaic array is formed by series/parallel combination of SPV modules to attain a desired voltage and current level. The major challenge in using a SPV source containing a number of cells in series is to deal with its nonlinear internal resistance. The problem gets complex when the array receives nonuniform irradiance or partially shaded. In a larger SPVA, the occurrence of partial shading is common due to tree leaves falling over it, birds or bird litters on the array, shade of a neighboring construction, and so forth. In a series connected string of cells, all the cells carry the same current. Even though a few cells under shade produce less photon current, these cells are also forced to carry the same current as the other fully illuminated cells. The shaded cells may get reverse biased, acting as loads, draining power from fully illuminated cells. If the system is not appropriately protected, hot-spot problem [1] can arise and in several cases, the system can be irreversibly damaged. Nowadays there is an increasing trend to integrate the SPV arrays at the design level in the building itself. In such cases it is difficult to avoid partial shading of array due to neighboring buildings throughout the day in all the seasons. In conventional SPV systems, these shadows lower the overall generation power to a larger degree than what is expected. Hence the SPV installation cost is increased, because the number of SPV modules must be increased [2] and as a result, SPV power generation will be less attractive. This makes the study of partial shading of SPV modules a key issue. Moreover it is very important to understand the characteristics of SPVA under partial shaded conditions to use SPV installations effectively under all conditions.

In recent years, the impact of partial shading on the SPV array performance has been widely discussed [3–6]. With a physical SPV module it is difficult to study the effects of partial shading since the field testing is costly, time consuming and depends heavily on the prevailing weather conditions. Moreover, it is difficult to maintain the same shade under varying numbers of shaded and fully illuminated cells throughout the experiment. However it is convenient to carry out the simulation study with the help of a computer model. In most of the studies [7–10], the effect of partial shading in reducing the output power of the SPVA has been discussed. But little attention has been paid to the power dissipated by the shaded cells affecting the array life and utilization of the array for the worst shaded case. The

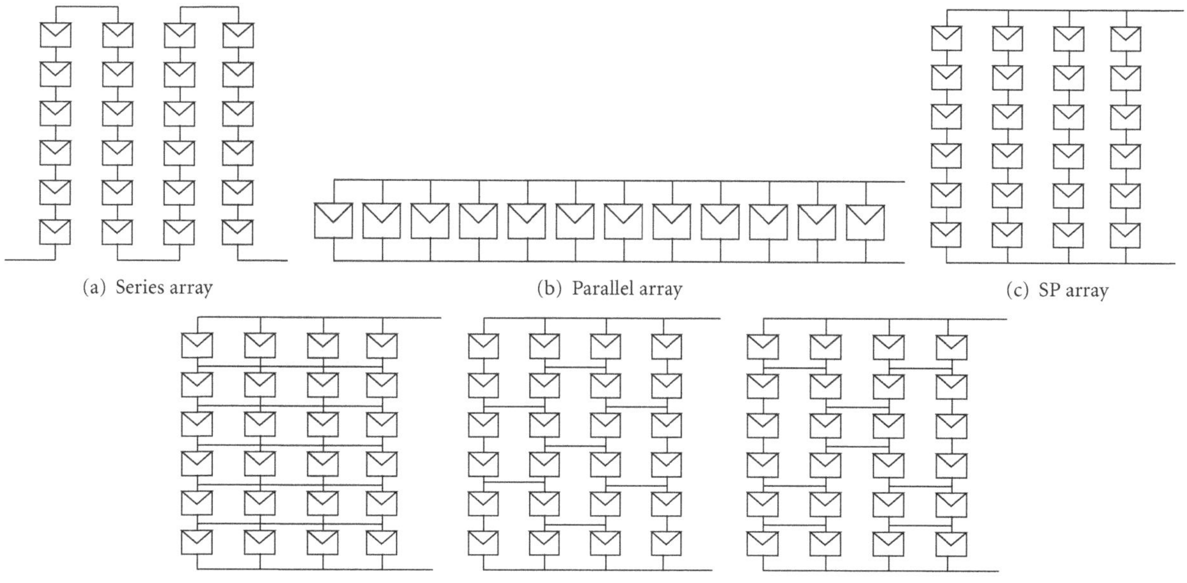

FIGURE 1: Schematic diagrams of SPVA configurations.

harmful effects in basic configurations and their comparison have been discussed in [11]. Common use of bypass diodes in antiparallel with the series-connected SPV modules can partially mitigate the power reduction due to partial shadow [11]. In such cases a more sophisticated Maximum Power Point Tracking (MPPT) algorithms capable to disregard local power maximums is required [12–16]. Alternatively, the maximum available DC power can be improved if the connection of the SPV modules can be reconfigured such that panels with similar operating conditions are connected in the same series string. Moreover the parallel configuration should be dominant under partial shaded conditions [11, 16–18]. However high output current at low voltage in parallel configuration will have to be properly conditioned to the required level by using suitable DC-DC converter. Hence it is required to opt for derived configurations. In this paper, for different configuration types, the generalized MATLAB programs have been developed which are capable of simulating any number of modules connected in series, parallel or combined for any type of shading patterns and any number of bypass diodes. The comparative study is made among the configurations and conclusions have been presented.

2. Review of Different SPVA Configurations

Several SPVA configurations have been proposed in the literature as shown in Figures 1(a) to 1(f) [1, 4, 19, 20]. They are series, parallel, series-parallel (SP) total cross-tied (TCT) bridge-linked (BL) and Honey-comb (HC) configurations [21, 22]. Series and parallel configurations are the basic configurations (Figures 1(a) and 1(b)) and the performance of these configurations has been discussed in detail by [11]. The major drawbacks of using the series or parallel configuration are that the current and voltage are less respectively. In SP configuration, shown in Figure 1(c) the modules are first connected in series to get the requisite voltage and then series-connected modules are paralleled. TCT configuration is derived from the SP configuration by connecting ties across rows of the junctions. In TCT configuration (Figure 1(d)), the voltages across the ties are equal. The sum of currents across the various ties is equal. The power is obtained as SP configuration. In BL configuration the modules are connected in a bridge rectifier fashion as shown in Figure 1(e). From the diagram it is seen that four modules constitute a bridge. Here two modules in the bridge are connected in series and then they are connected in parallel. Ties are present between the bridges. Hence the voltage and current values are obtained by appropriately adding voltages in series and currents in parallel.

The modifications have been made in BL configuration to arrive at a new configuration called HC configuration [21, 22]. The advantages of TCT and BL configurations have been combined together in HC configuration. Sometimes, insolation pattern on an array may be such that consecutive modules in a column of array receive equal insolation and other modules in a same column receive different insolation. In this case, it is not necessary to select TCT as it has so many ties. BL may also cause power loss as it has fewer ties in this case. So we have to select ties properly. This is obtained by connecting ties across variants of two, four, and six modules. This is done in HC configuration as shown in Figure 1(f) [21, 22].

3. Simulation of Configurations under Partial Shaded Conditions

Quaschning and Hanitsch [1] proposed a numerical algorithm to simulate the mismatch in individual SPV cells and their shading levels. But it requires each element to be represented by a mathematical expression. Even though this produces accurate results, the model is complex and requires more computation time and higher memory requirement. Kaushika and Gautam [4] developed a computational network analysis approach to compare the configurations. Karatepe et al. [10] proposed a module-based and cell-based model for analyzing the array configurations. Giraud and Salameh [20] proposed a neural network-based model to investigate the effects of passing clouds on a grid-connected SPV system using battery storage. The importance of selecting the proper size of the SPV array and batteries in such systems has been discussed by [23]. It is required for the stable operation of SPV system with a sudden and large change in SPV power because of irradiance variation, caused by shading, and so forth. Shading caused due to passing clouds also has a financial claim on the utility. Jewell and Unruh [24] have carried out an economic analysis to estimate the cost of the fluctuations in power generation from a SPV source. Based on the literature it is understood that not only the size of the SPVA but also its configuration that significantly affects its power output, and therefore, the performance of the system under partially shaded conditions. From the above discussion, it may be concluded that, while it is very important to model, study, and understand the effects of shading on SPV arrays, a simple tool is not available for the purpose. Therefore, it is felt that there is a need for a flexible, interactive, and comprehensive simulation model capable to predict the SPV characteristics (including multiple peaks) and output power under partially shaded conditions. Patel and Agarwal [25, 26] have proposed a MATLAB based-simulator cum learning tool to understand the characteristics of a large SPV array by considering the model in I quadrant given in Figure 2. They have developed a model for SP configuration with bypass diodes. The model used by [25] neglects the effect of shunt resistance. Swaleh and Green [27] discussed the impact of R_{sh} under partial shaded conditions. In order to obtain the realistic model which provides the practical maximum power point values, it is mandatory to include the effect of varying R_{sh} with respect to environmental parameters particularly for crystalline type SPV modules. Hence the proposed model includes the insolation-dependent shunt resistance and the basic model equations used by [25] have been replaced by the improved model equations used by Villalva et al. [28]. The model equations (A.1)–(A.9) given in the appendix are used for modelling the SPV system. Equations (A.1) to (A.9) relating the SPV parameters with irradiance and temperature have been taken from [28] excluding (A.6). The parameters of (A.6) have been experimentally determined. The dependence of R_{sh} is found to be negligible and hence neglected to reduce the complexity of the model. The equations given in the appendix are for single SPV module.

Modeling of a large array with shading patterns is very complex. In this work, software has been developed for all the configurations having any number of assemblies, strings, substrings, and so forth. The software is capable of considering/ignoring the effect of varying insolation on R_{sh}. This software gives the output power, voltage, and current values for any irradiance and temperature patterns. Before going in detail about the software some of the terminologies are introduced with the help of Figure 3. Most of the SPV arrays in real time are large in size. It is cumbersome to enter the individual irradiance and temperature values for each module [25, 26]. Therefore groups of modules have been considered based on shading pattern. The representation of the terminologies has been explained with 6 × 4 array shown in Figure 3. The terminologies used in the proposed software are as following.

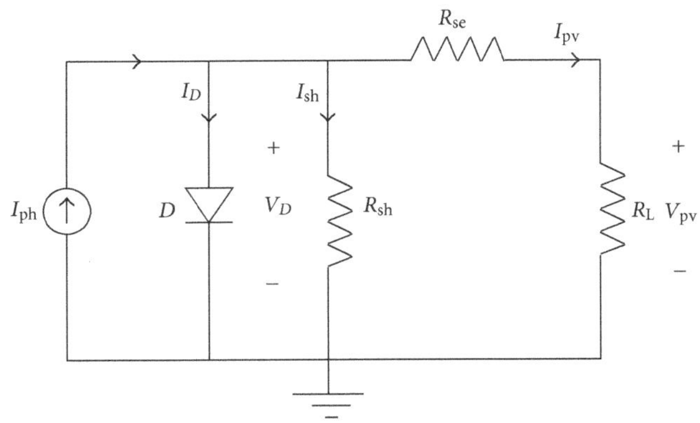

FIGURE 2: Electrical equivalent circuit model of a SPVA in I quadrant.

(i) Modules that always refer to a typical SPV panel consisting of a group of 36 cells connected in series. An antiparallel diode shunting 36/18 cells connected/ignored can be programmed.

(ii) Modules that are receiving the same irradiance connected in series form a "substring."

(iii) Several substrings that are receiving different irradiance but connected in series form a "string."

(iv) Identical strings that are connected in parallel form an "assembly."

(v) Assemblies that are connected in parallel form an "array."

As the importance of bypass diodes is well known, a bypass diode has been included as a part of every module in the M-file code. This section considers that each module is connected with a bypass diode. To include the effect of bypass diode, negative voltages caused by shading is taken as diode forward drop (~0.7 V) in M-file coding.

The architecture of the developed software is shown in Figure 4.

The individual block of Figure 4 is presented in the form of flow chart from Figure 5(a) to Figure 5(f). Figure 5(a) is common for all the configurations, after which there are subtle differences in the calculations of the various configurations. These are depicted from Figure 5(b) to Figure 5(f). The *V-I* and *V-P* characteristics for all the configurations including insolation-dependent R_{sh} is as shown in Figure 6 for a shading pattern shown in Figure 3. In the series configuration it is seen that the number of peaks correspond

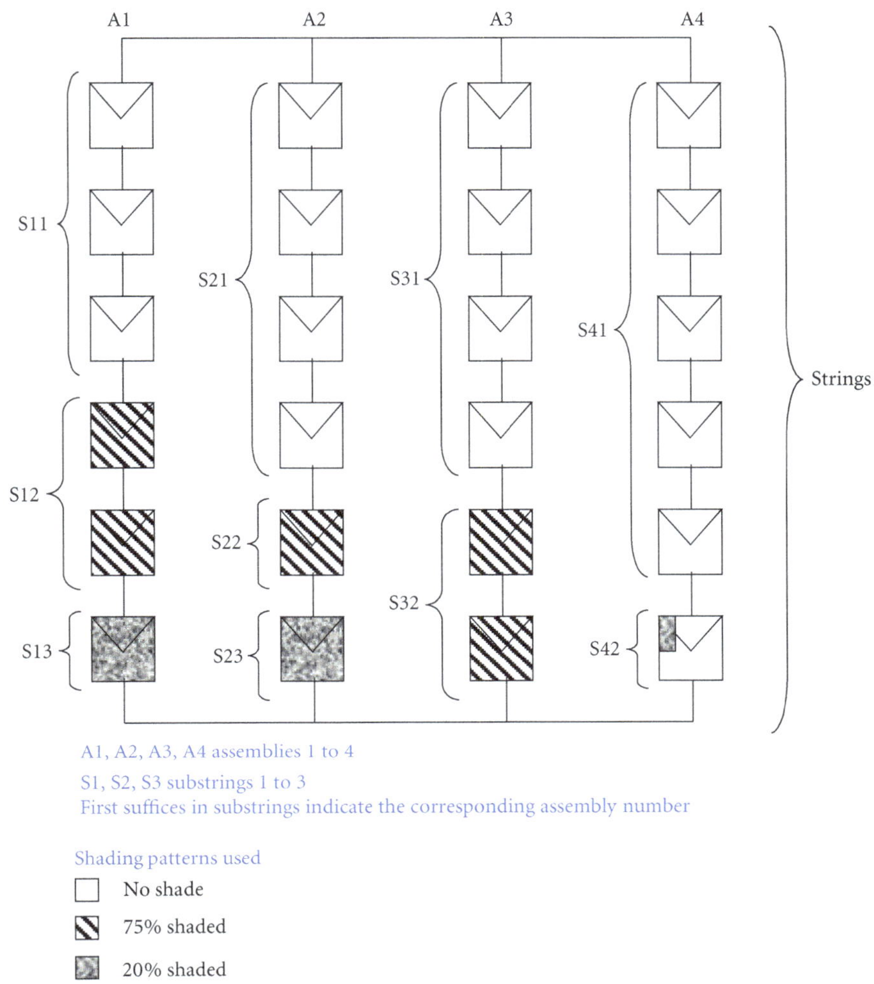

FIGURE 3: Illustration of 6 × 4 array with a particular shading pattern.

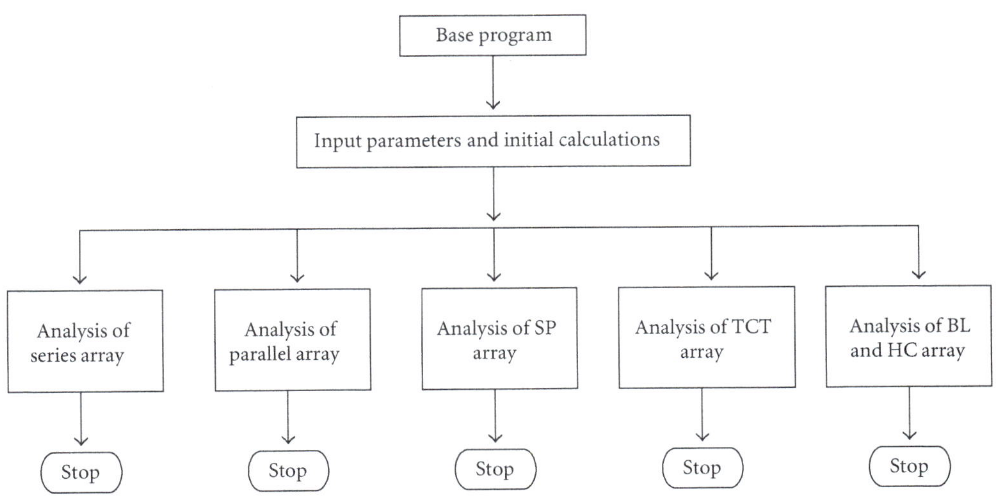

FIGURE 4: Architecture of the developed software.

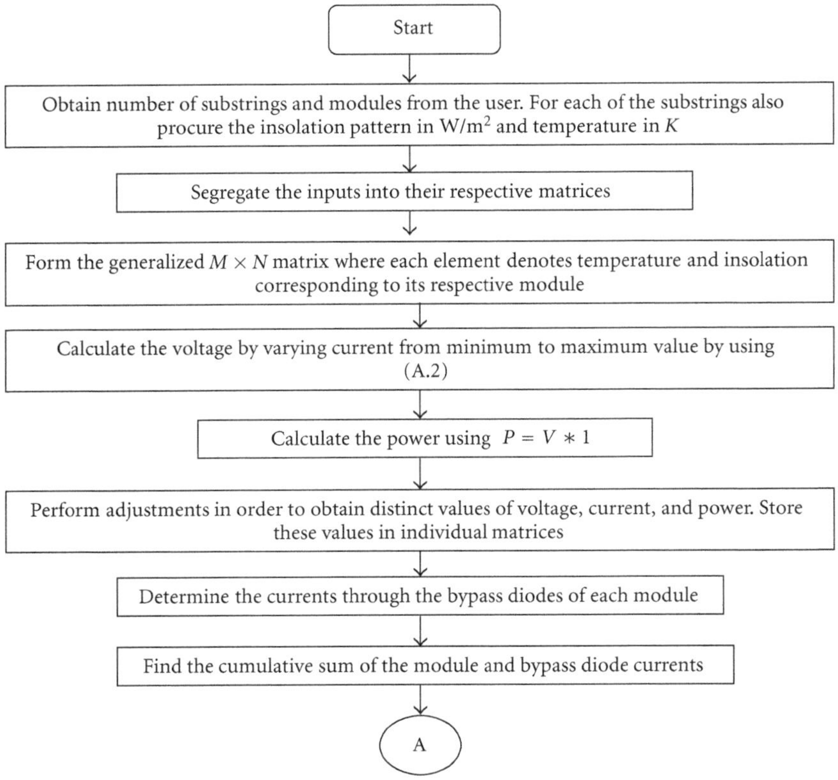

(a) Flow chart for initial part of all the configurations

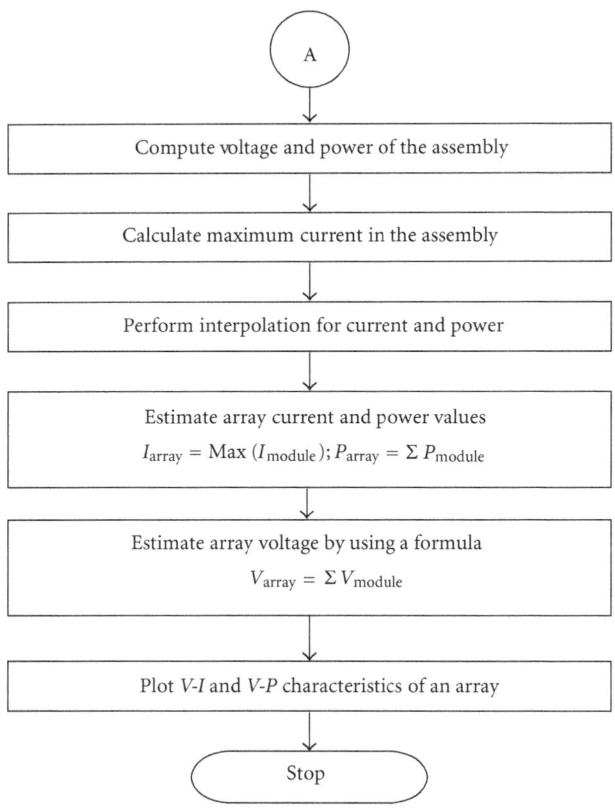

(b) Flow chart for Series configuration

FIGURE 5: Continued.

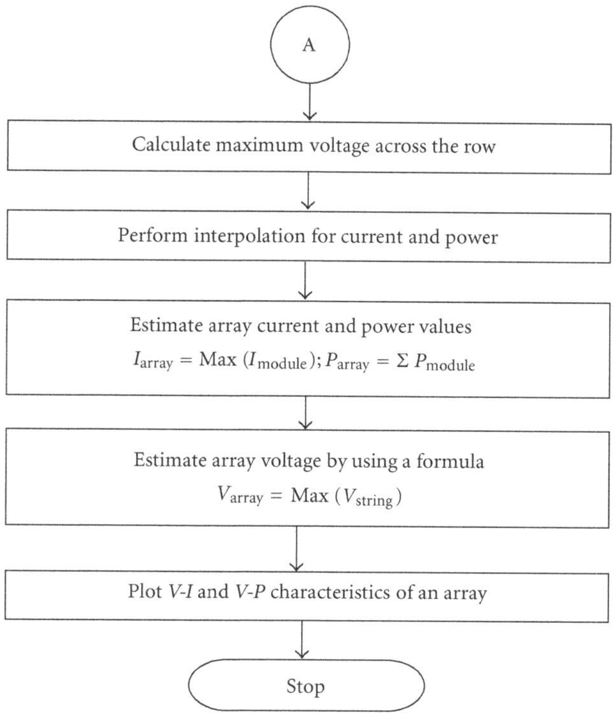

(c) Flow chart for parallel configuration

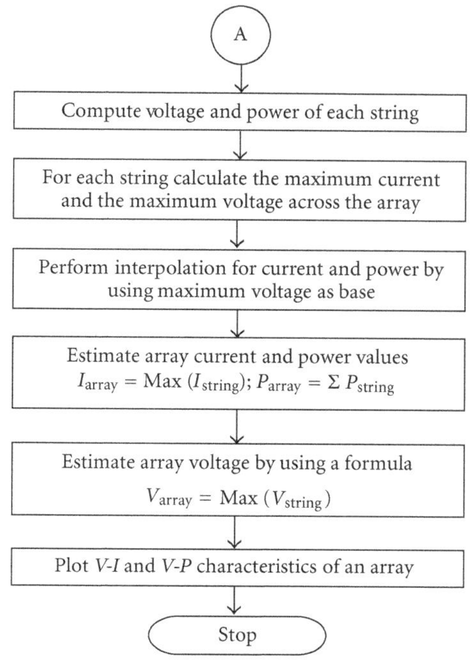

(d) Flow chart for SP configuration

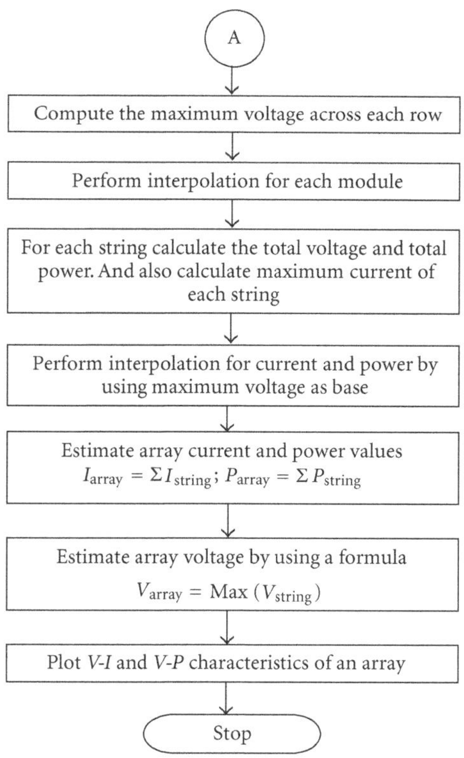

(e) Flow chart for TCT configuration

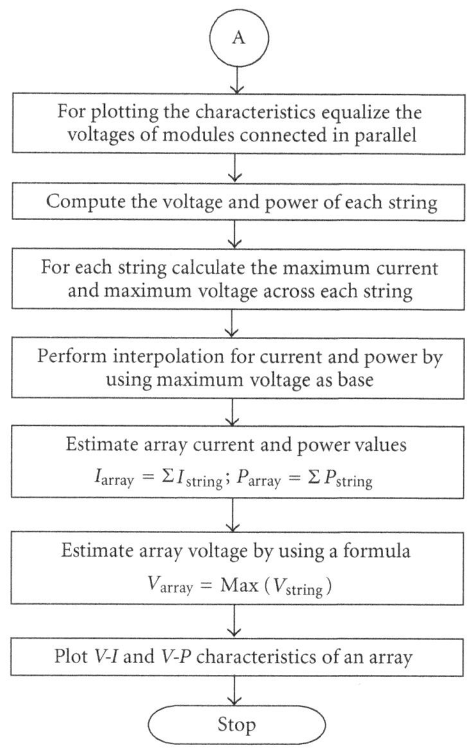

(f) Flow chart for BL and HC configurations (flow charts for BL and HC configurations are same but the ties have to be changed)

Figure 5: Flowchart for coding all the configurations.

TABLE 1: Comparison of power with and without the effect of insolation-dependent R_{sh}.

Configuration	P_m(W) constant R_{sh} (R_{sh} = 145.62 Ω)	P_m(W) with insolation-dependent R_{sh}
Series	**520.7**, 408.3, 193.1 (three peaks)	**488.2**, 387.8, 185.4 (three peaks)
Parallel	**564.8** (one peak)	**565.3** (one peak)
SP	421.6, 455.5, **458.9**, 370.5 (four peaks)	410.7, **435.8**, 434.2, 346.1 (four peaks)
TCT	341.4, 480.2, **551.6**, 416.4 (four peaks)	443.5, **468.4**, 463.3, 372.7 (four peaks)
BL	362.7, 479.6, **483.1**, 447.8 (four peaks)	410.7, **435.8**, 433.5, 393.1 (four peaks)
HC	428.1,445.2, **448.7**, 418.3 (four peaks)	394.3, 436.1, **442.8**, 385.6 (four peaks)

TABLE 2: Comparisons of Configurations under Uniform Irradiance Conditions.

Configuration	P_m(W)	V_m(V)	I_m (A)
Series	676.8	292.1	2.32
Parallel	676.8	12.26	55.2
SP	676.8	73.05	9.27
TCT	676.8	73.05	9.27
BL	676.8	73.05	9.27
HC	676.8	73.05	9.27

to the number of shading patterns and current is less compared with other configurations. In this configuration it is understood that if even one module is shaded it affects the output power considerably. In the parallel configuration it is seen that there are no multiple peaks. This is because all the modules are connected in parallel; therefore no module can be forced to carry more than its share of current. In parallel configuration the voltage is less. SP configuration provides higher power at considerable voltage and current values. Hence it can be inferred that SP configuration negates the defects of series and parallel configurations. In TCT configuration due to the inclusion of ties, the flaws of the series configuration have been avoided. This is because none of the modules are connected in series. Hence stress on modules is reduced. In BL configuration few modules in a string are connected in series and these are connected in parallel. Therefore it subjected to lesser stress than SP configuration. The generalized MATLAB program has been extended for HC with modifications. The flowchart for HC configuration is similar to BL. While writing the program the difference in the tie connections has been taken care of (Figure 5(f)).

4. Impact of Including the Effect of Varying Shunt Resistance in the Model

Table 1 shows the comparison between power values with and without the varying shunt resistance. The input pattern is given as shown in Figure 3. It is seen that the power values change when varying shunt resistance is included. The power values in the third column of Table 1 matches very closely with practical values. Hence shunt resistance should be included in order to obtain the realistic modeling of SPV array.

5. Comparison of Array Configurations with and without Bypass Diode

For the analysis of array configurations without bypass diode, two quadrant characteristics have to be taken care of. [29, 30]. Hence the additional term is included in the mathematical model as shown in Figure 7 [31] and the same set of programs has been modified with the model represented by (A.10). Table 2 shows the power, voltage, and current values under uniform irradiance conditions. This corresponds to an irradiance of 1000 W/m^2 and a temperature of 25°C (298.15 K). It is seen that almost all the configurations provide the same power under uniform irradiance conditions.

Table 3 shows the comparison of power with and without a bypass diode for a 6 × 4 array. The input pattern is as in Figure 3. Even though the use of bypass diode introduces multiple peaks, it is seen from Table 3 that a higher power is obtained by using a bypass diode.

6. Comparison of Different Array Configurations for Different Shading Scenarios

Here the case where one bypass diode across a group of 36 cells (one bypass diode per module) has been considered. The array sizes are 2 × 4, 4 × 2, 2 × 6, 6 × 2, 3 × 4, 4 × 3, 4 × 6, 6 × 4, 3 × 3 and 4 × 4. An array size can be designated by $M \times N$, where M indicates number of modules connected in series and N indicates number of strings in parallel. Fifteen different random shading patterns are generated for each of the ten different array sizes. One of the 15 random patterns of irradiance is shown in Figure 8 and corresponding shading matrix for different array sizes are shown in Figure 9. In Figure 8, the shading patterns E, L, R, and X are very low values which replicate the bird litters or single leaves closing completely the cell in a larger array. Practically it is found that some of the bird litters are difficult to remove from the array which causes permanent shade on the cell so that particular cell receives very low insolation at all times. The maximum power obtainable from each configuration is computed for each of these shading patterns. The mean value of this power and its maximum and minimum values for different shading patterns have been tabulated vide Table 4 in which the values highlighted with bold letters indicate the global peak values whereas other values are local peak values.

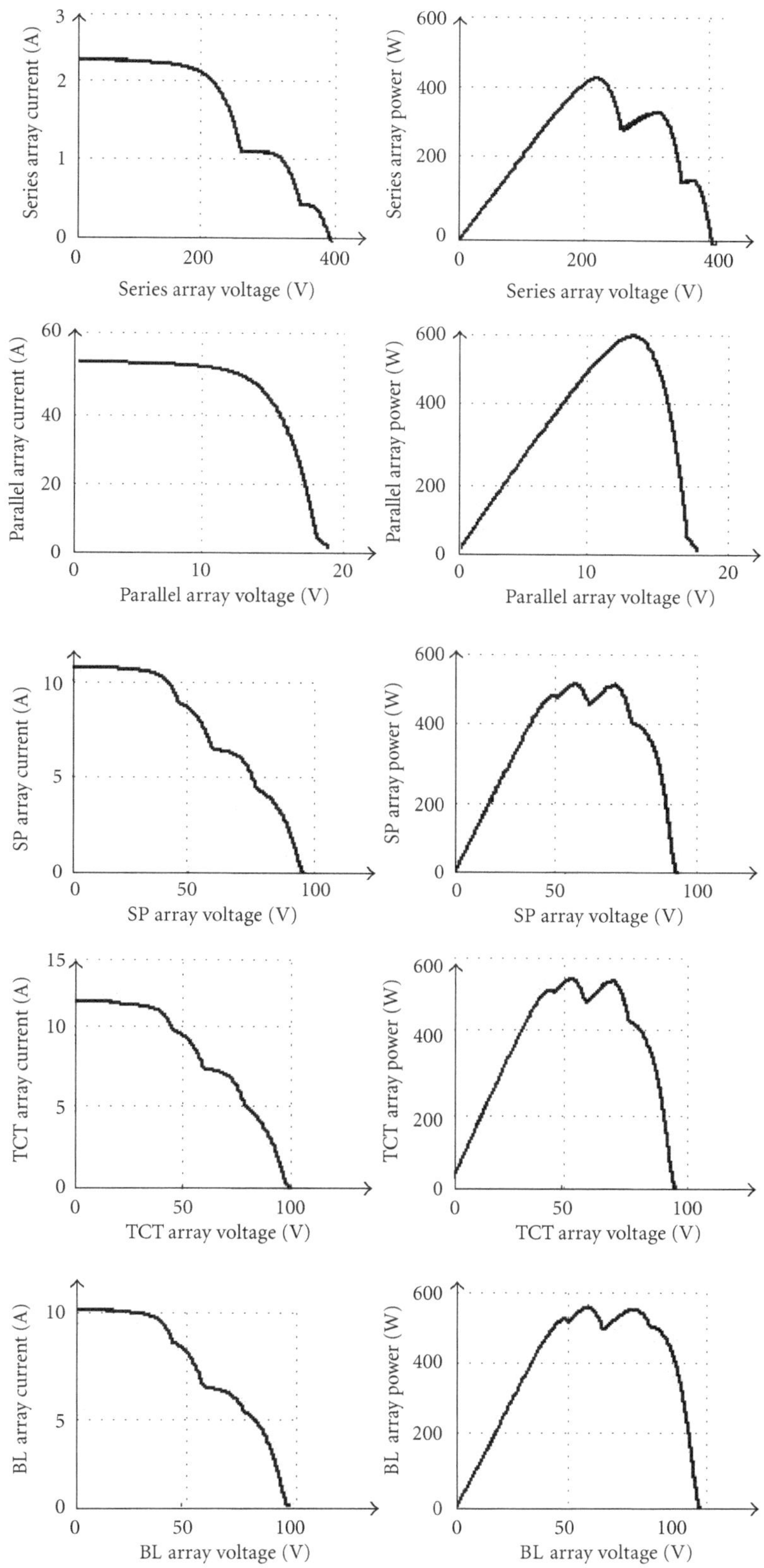

Figure 6: Continued.

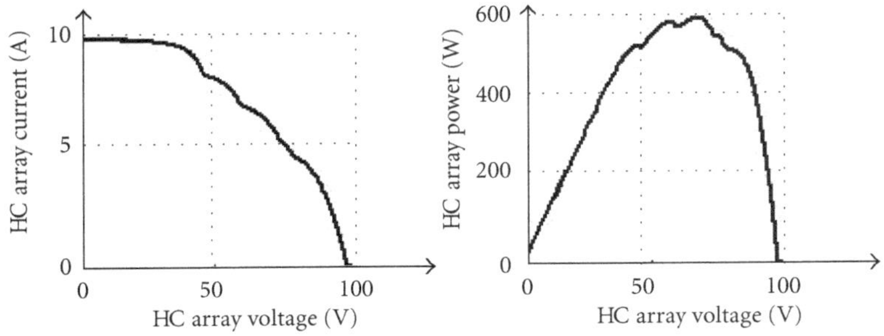

FIGURE 6: Simulated *V-I* and *V-P* characteristics of SPVA configurations for a shading pattern shown in Figure 3.

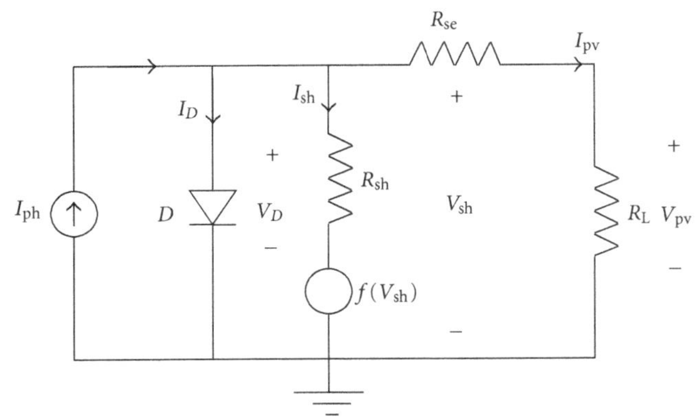

FIGURE 7: Bishop's model to represent the SPVA under partial shaded condition.

From Table 4 it can be inferred that depending on the size of array and type of shading pattern different configurations are preferred. But in most of the cases TCT closely followed by HC are the preferred configurations. It is observed that wherever the modules with similar shade are grouped in a string, HC is better in which less ties are there as compared to TCT.

7. Practical Verification

A few results obtained from the software were verified. Figure 10 shows a set up of 3 × 3 SPV array. SOLKAR (Model No. 3712/0507) solar module is used to setup the array.

The electronic load [30] was used to verify the characteristics. GWINSTEK GDS-1022 DSO was used to trace the practical characteristics. It is calibrated using Fluke 5500 A Multi-Product Calibrator. For different irradiances and temperatures the practical characteristics are easily traced out using electronic load method and the relevant data traced by DSO are stored in Excel spreadsheet to calculate *V-P* characteristics and for comparison of model parameters. Solar irradiance level/insolation of 1000 W/m^2 corresponds to a short circuit current of 2.55 A as per the datasheet of SOLKAR modules. In all the experiments the solar insolation has been measured as proportional to short circuit current. Outputs were verified for uniform as well as partial shaded conditions. The sample snapshot of digital storage oscilloscope has been shown in Figure 10 for the four types of configurations (SP, TCT, BL, and HC) for a particular shading pattern. The calculated *P-V* characteristics for Figure 11 are shown in Figure 12. The practical verification was done for several artificially introduced input shading patterns. The outputs obtained were closer to the outputs obtained from simulation which took into consideration the effect of varying R_{sh}. Irradiance level of a module was assumed proportional to the short circuit current and different shadows were introduced by tilting the module of the stand.

8. Effect of Using More Bypass Diodes

The concept of using bypass diode is extended in this section. One diode is connected across a group of 18 cells in a module (2 bypass diodes per module) is considered. Table 5 gives the comparison between mean value of the power for 6 × 4 configurations with one bypass per module and two bypass diodes per module for fifteen random shading patterns.

From Tables 5 and 6, it is observed that the improvement in the power when two bypass diodes are used in the single module. This study can be extended to select the optimum number of diodes used in a module to get the maximum power under partial shaded conditions. If the number of bypass diodes used in a module is increased or in other words the number of cells grouped is minimized, the maximum output can be obtained.

The generalized program developed has been used to choose the optimum array configuration for the 10.5 kW array installed in the SSN research center (14 × 10 array)

TABLE 3: Comparison of Configurations Power with and without Bypass Diode.

Configuration	P_m(W) (without bypass diode)	P_m(W) (with bypass diode)
Series	325.52	**488.2**, 387.8, 185.4 (Three Peaks)
Parallel	557.2	**565.3** (one peak)
SP	408.08	410.7, **435.8**, 434.2, 346.1 (Four Peaks)
TCT	448.97	443.5, **468.4**, 463.3, 372.7 (Four Peaks)
BL	428.98	410.7, **435.8**, 433.5, 393.1 (Four peaks)
HC	440.78	394.3, 436.1, **442.8**, 385.6 (Four Peaks)

TABLE 4: Mean and Range of the maximum power for different configurations with different sizes under random shading patterns (*Readings practically verified vide Section 7).

Array Size	Configuration	Mean Value of Maximum Power (W)	Range of Maximum Power (W)	
			Maximum Value	Minimum Value
2 × 4*	SP	93.32	130.8	45.68
2 × 4*	TCT	94.67	131.40	45.43
2 × 4*	BL	105.58	161.90	47.61
2 × 4*	**HC**	**121.98**	**191.30**	**52.45**
4 × 2*	SP	103.96	154.30	54.80
4 × 2*	**TCT**	**117.70**	171.50	65.97
4 × 2*	BL	104.00	154.30	54.80
4 × 2*	HC	114.02	173.80	56.15
2 × 6	SP	145.51	217.40	83.43
2 × 6	TCT	149.78	239.40	95.20
2 × 6	BL	175.42	269.90	126.50
2 × 6	**HC**	**187.23**	**279.01**	**131.90**
6 × 2	SP	143.13	187.60	82.37
6 × 2	TCT	160.09	209.20	96.73
6 × 2	BL	143.11	187.80	82.27
6 × 2	**HC**	**165.34**	**215.90**	**84.11**
3 × 4	SP	128.14	190.50	62.96
3 × 4	TCT	142.58	207.00	84.04
3 × 4	**BL**	**146.70**	**205.60**	**67.86**
3 × 4	HC	144.93	226.80	66.93
4 × 3	SP	125.79	201.20	43.28
4 × 3	TCT	132.37	209.08	61.02
4 × 3	BL	132.89	226.40	73.70
4 × 3	**HC**	**137.95**	**225.10**	**55.31**
3 × 3*	SP	85.72	135.20	51.33
3 × 3*	**TCT**	**92.79**	145.70	54.58
3 × 3*	BL	84.50	135.20	46.94
3 × 3*	HC	83.22	128.60	53.73
4 × 4	SP	145.31	250.10	93.72
4 × 4	**TCT**	**164.62**	272.80	91.33
4 × 4	BL	145.03	225.80	93.72
4 × 4	HC	144.14	225.40	85.17
4 × 6	SP	186.66	302.10	131.97
4 × 6	TCT	211.54	375.47	146.56
4 × 6	BL	186.06	317.32	104.47
4 × 6	**HC**	**227.06**	**392.47**	**117.29**
6 × 4	SP	197.86	324.12	110.47
6 × 4	**TCT**	**234.57**	**383.12**	**113.29**
6 × 4	BL	184.48	334.36	98.65
6 × 4	HC	219.15	387.93	98.97

Pattern label	A	B	C	D	E	F	G	H	I	J	K	L	M	N	O	P	Q	R	S	T	U	V	W	X
Irradiance in W/m^2	355	259	657	780	29	667	352	627	618	310	994	42	196	420	902	468	263	97	602	643	814	829	492	79

FIGURE 8: One of the 15 random patterns of irradiance.

F	C	R	Q
G	O	V	R

2 × 4

O	O	R	Q	T	E
N	K	V	U	B	R

2 × 6

P	Q	I
R	J	C
D	S	O

3 × 3

M	H	T	R
A	M	J	H
J	M	K	P

3 × 4

E	N	B	L
C	J	K	Q
E	T	X	Q
S	V	O	B

4 × 4

G	B	I	P	B	J
H	G	J	D	J	M
U	F	R	P	M	P
F	Q	N	O	E	H

4 × 6

K	E
Q	I
Q	F
Q	J

4 × 2

K	I	B
U	D	A
I	C	T
S	I	N

4 × 3

A	R
A	F
L	H
T	G
F	H
K	O

6 × 2

A	A	A	B
D	W	W	G
V	O	O	V
C	F	I	H
D	D	V	T
D	D	E	S

6 × 4

FIGURE 9: Shading matrixes for different array sizes with shading pattern of Figure 8.

TABLE 5: Comparison between mean value of the maximum power for 6 × 4 configurations with one bypass diode and two bypass diodes per module under random shading patterns.

Array size	Configuration	Mean value of maximum power (W)		Difference in mean value of power (W)
		One diode per module	Two diodes per module	
6 × 4	SP	197.86	218.87	21.01
6 × 4	**TCT**	**234.57**	**262.20**	**27.63**
6 × 4	BL	184.48	206.20	21.72
6 × 4	HC	219.15	245.60	26.45

TABLE 6: Deviation of RMSD and mean value for 6 × 4 array when one bypass diode across the module and two bypass diodes across two groups of 18 cells in the module.

Array size	Configuration	Maximum power (W)		Voltage at MPP (V)	
		Mean value	RMSD value	Mean value	RMSD value
6 × 4	SP	21.01	59.38	6.35	2.25
6 × 4	TCT	28.28	51.83	2.61	5.70
6 × 4	BL	22.68	57.71	4.77	2.91
6 × 4	HC	26.72	53.83	4.55	4.01

Table 7: Mean value and range of maximum power for 14 × 10 configuration shown in Figure 12.

Array size	Configuration	Mean value of maximum power (W)	Range of maximum power (W)	
			Maximum value	Minimum value
14 × 10	SP	2462.51	3289.00	1777.12
14 × 10	TCT	2628.83	3396.14	2054.37
14 × 10	BL	2561.17	3309.34	1876.46
14 × 10	**HC**	**2675.17**	**3562**	**2023**

Figure 10: Practical setup of a 3×3 array employing tilting modules for different shades.

Table 8

S. no	Parameters	Values
1	Rated power (P)	37.08 W
2	Voltage at maximum power (V_m)	16.56 V
3	Current at maximum power (I_m)	2.25 A
4	Open circuit voltage (V_{oc})	21.24 V
5	Short circuit current (I_{sc})	2.55 A
6	No. of series cells (N_s)	36
7	Type	Monocrystalline

which is shown in Figure 13. In this array for each module 18 cells are grouped together and bypass diode is connected in anti-parallel with that. In the generalized program the specifications according to the datasheet of BEL laboratories and the array size have been altered for the study. For this case, the maximum power is about 10.5 kW at maximum voltage and about 65.69 V under uniform irradiance condition (at $G = 1000\,\mathrm{W/m^2}$ and $T = 25°\mathrm{C}$). The rated power per module is 75 W.

Table 7 gives the comparison of different array configurations under 20 random shading patterns. It is found that the HC configuration is dominated under partial shaded condition for the existing array.

9. Conclusion

Analysis of various SPVA configurations with respect to environmental parameters by developing a more realistic model using MATLAB M-file has been presented. In analysis a recent configuration, HC configuration, has also been taken for comparison. In order to obtain the maximum possible power under partial shaded conditions it is mandatory to connect a bypass diode in anti-parallel with a module or group of cells to avoid the stress on the shaded cells. This setup would reduce the problems of hot-spot as well as provide a higher power when compared to a SPVA without bypass diode. After analyzing the various configurations for different random shading patterns for varied sizes, it is also observed that in most cases TCT gave a higher amount of power when compared to the other configurations but in some cases where the array was asymmetrical or where the number of columns receiving same insolation was more when compared to the number of rows, HC configuration provided a higher power when compared to the other configurations because of less ties. Hence we can conclude that TCT is the best configuration closely followed by HC. After analyzing various configurations, it can be concluded that TCT is the best configuration for symmetrical array size and HC configuration for asymmetrical array sizes. The generalized program developed here can be used for any array size, any number of bypass diodes across group of cells, and for any module by simply changing the specifications of the module used in the program. Moreover, the results confirm that this approach often allows attaining a higher electrical energy production compared to that attainable with SPV arrays with a proper layout.

Appendix

The equations used to develop a simulation model of a SPV cell are: [28–31]

$$I_{PV} = I_{ph} - I_r\left[\exp\left\{\frac{V_{PV} + I_{PV}R_{se}}{V_t}\right\} - 1\right] - \frac{(V_{PV} + I_{PV}R_{se})}{R_{sh}}. \quad (A.1)$$

For coding purpose (A.1) has been rearranged as

$$V_{PV} = \ln\left\{\frac{I_{ph} - I_{PV} - (V_{PV} + I_{PV}R_{se})/R_{sh} + I_r}{I_r}\right\} \times V_t - I_{PV}R_{se}, \quad (A.2)$$

where

$$I_{ph} = \left\{I_{ph,ref}[1 + \alpha(T - T_{ref})]\right\}\frac{G}{G_{ref}},$$
$$I_{ph,ref} = I_{sc,ref}, \quad I_{ph,ref} = \frac{R_{sh} + R_{se}}{R_{sh}} \times I_{sc,ref}, \quad (A.3)$$

$$I_r = \frac{I_{sc,ref} + \alpha(T - T_{ref})}{\exp((V_{oc,ref} + \beta(T - T_{ref}))/nV_T) - 1},$$
$$I_{r,ref} = \frac{I_{sc,ref}}{\exp(V_{oc,ref}/V_{t,ref}) - 1}, \quad (A.4)$$

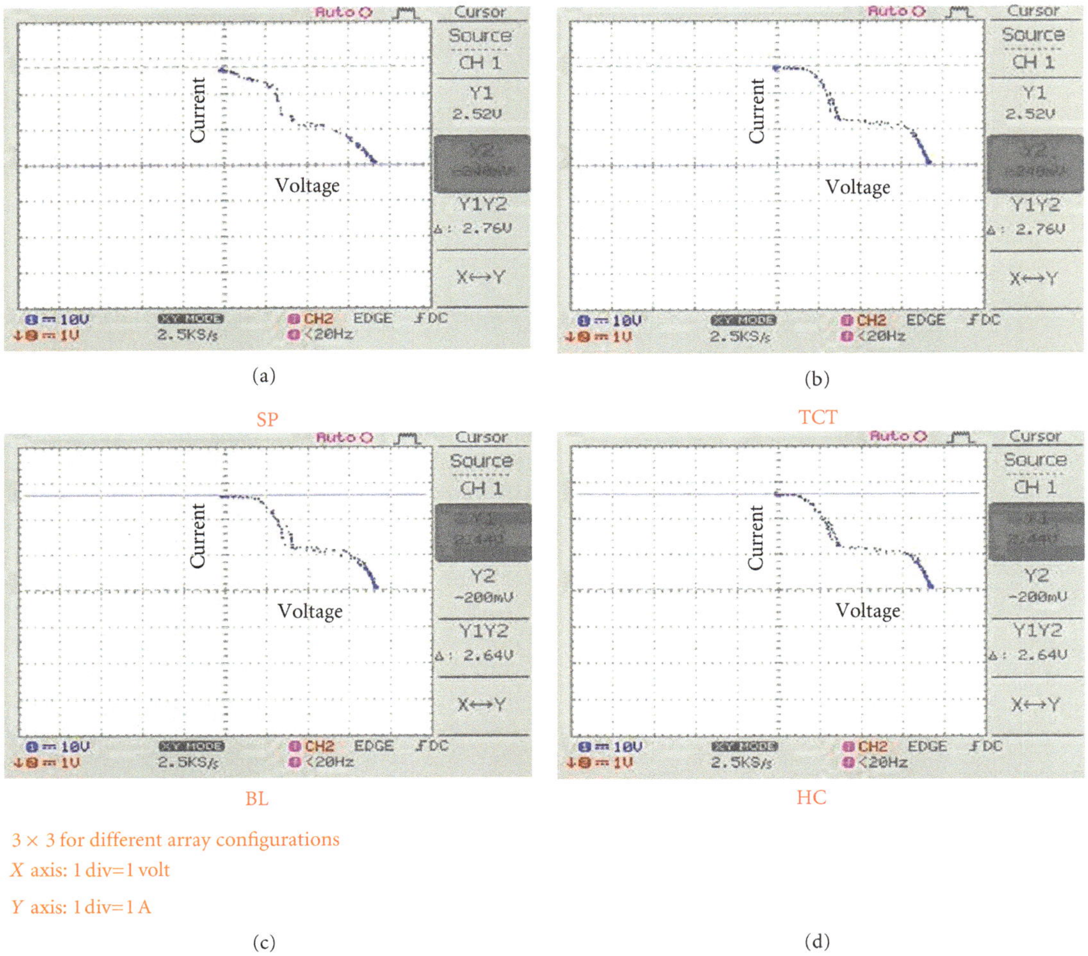

FIGURE 11: Snapshot of I-V characteristics for a 3 × 3 SPVA for different configurations.

$$V_t = V_{t,ref}\frac{T}{T_{ref}}, \qquad V_{t,ref} = \frac{n_{ref}kT_{ref}}{q} \tag{A.5}$$

$$R_{sh} = \frac{3.6}{G - 0.086}, \tag{A.6}$$

$$I_m = I_{m,ref} \times G, \qquad V_m = V_{m,ref} + \{\beta(T - T_{ref})\}, \tag{A.7}$$

$$\begin{aligned} R_{se}\frac{G}{G_{ref}} &= \frac{V_{t,ref}}{I_{r,ref}} e^{-(V_{m,ref}+I_{m,ref}R_{se,ref})/V_{t,ref}} \\ &\quad + R_{se,ref} - \frac{G}{G_{ref}}\left(\frac{V_t}{I_r} e^{-(V_m+I_mR_s)/V_t} + R_{se}\right), \end{aligned} \tag{A.8}$$

$$n = n_{ref}\frac{T}{T_{ref}}, \tag{A.9}$$

The electrical behavior of the solar cell can be described by

$$\begin{aligned} I_{PV} &= I_{ph} - I_r\left[\exp\left\{\frac{V_{PV} + I_{PV}R_{se}}{V_t}\right\} - 1\right] - \frac{(V_{PV} + I_{PV}R_{se})}{R_{sh}} \\ &\quad - a\frac{(V_{PV} + I_{PV}R_{se})}{R_{sh}}\left(1 - \frac{V_{PV} + I_{PV}R_{se}}{V_{br}}\right)^{-m}, \end{aligned} \tag{A.10}$$

where V_{br}: junction break down voltage, a: fraction of ohmic current involved in avalanche breakdown, and m: avalanche break down exponent.

The electrical behavior of the solar cell can be described by (A.10) over the whole voltage range. The unknown parameters are "a", "V_{br}", and "m". These parameters are calculated by extracting parameters in those areas of practical V-I characteristic which are more significant.

The V-I characteristics of the solar cell under reverse biased conditions for dark condition have been measured. Breakdown voltage is calculated by linear regression of the straight line of voltage against the inverse of current near breakdown region from the dark characteristics [32]. The breakdown voltage V_{br} of the cell is found to be 13.5 V. The other two parameters are found by tuning them in model by

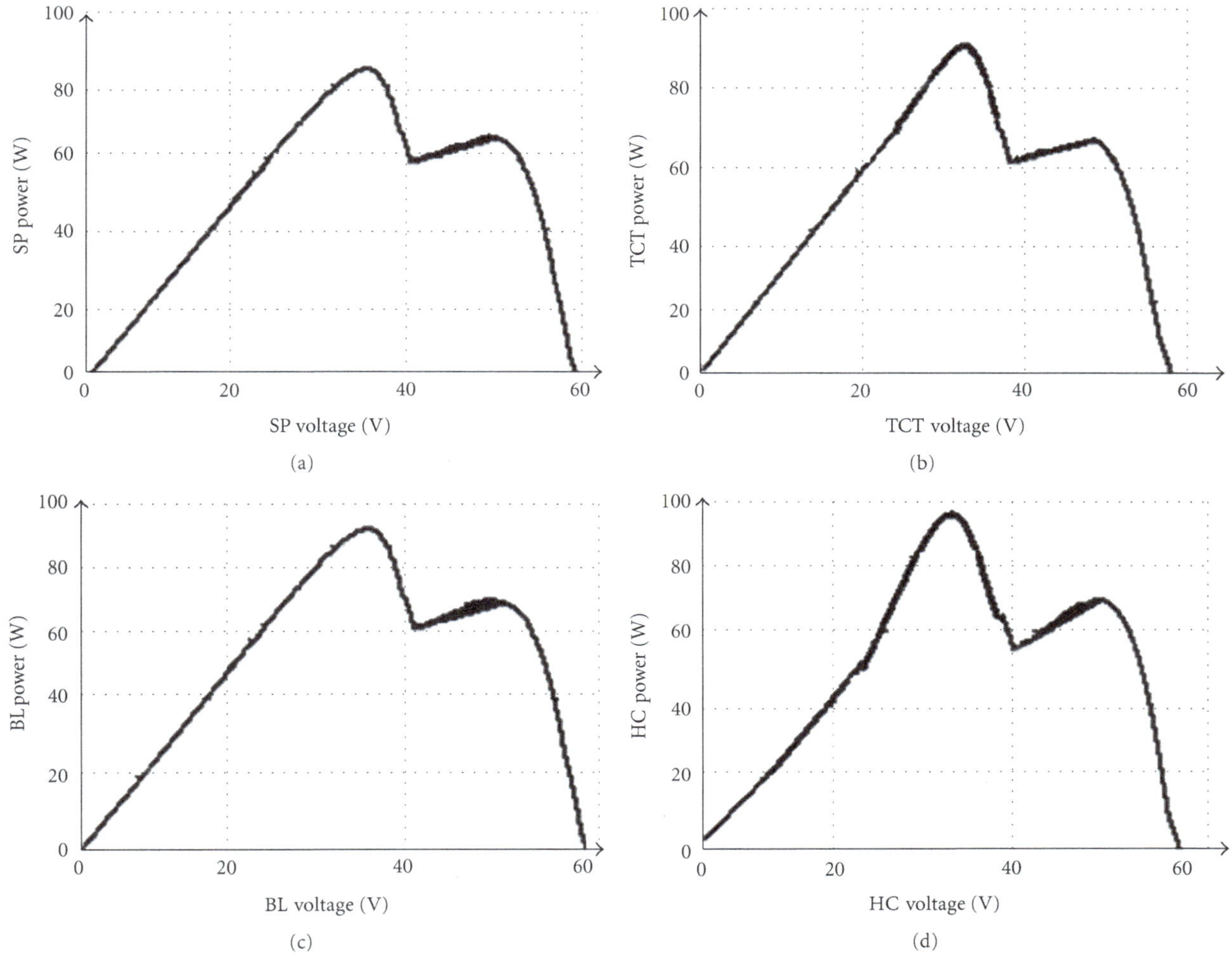

FIGURE 12: Calculated P-V characteristics for Figure 11.

FIGURE 13: 10 kW SPVA (14 × 10) installed in open terrace of EEE department block by SSNRC (BELL Laboratories).

trial and error method so as to match with the experimental characteristics. The values of "a" and "m" were found as 0.10 and −3.70, respectively. These values are used by proper multiplication factor for modelling SPV modules/arrays.

The parameters of the SOLKAR solar module used for practical verification at STC ($G = 1000\,\mathrm{W/m^2}$ and $T = 25°\mathrm{C}$) are given in Table 8.

Nomenclature

I_{PV}: Solar module output current (A)
V_{PV}: Solar module output voltage (V)
I_{ph}: Photo current of the SPV module (A)
I_r: Diode reverse saturation current in the equivalent circuit (μA)
R_{se}: Series resistance in the equivalent circuit of the module (mΩ)
R_{sh}: Parallel resistance in the equivalent circuit of the module (Ω)
n: Diode ideality factor
q: Electron charge ($=1.602 \times 10^{-19}$ C)
k: Boltzman's constant ($= 1.381 \times 10^{-23}$ J/K)
T: Temperature (Kelvin)
V_t: Thermal voltage (= nkT/q)
G: Irradiance level (at reference condition $G = 1000\,\mathrm{W/m^2}$)
α: Short-circuit current temperature coefficient
β: Open-circuit voltage temperature coefficient
I_{sc}: Short-circuit current of the module

V_{oc}: Open-circuit voltage of the module
V_m: Maximum power point voltage
I_m: Maximum power point current
P_m: Maximum power
ref: Additions subscripts indicate the parameters at reference conditions.

Index

SPVA: Solar photovoltaic array
SP: Series-parallel configuration
TCT: Total cross-tied configuration
BL: Bridge linked configuration
HC: Honey-Comb configuration
MPP: Maximum power point.

Acknowledgment

The authors wish to thank the management of the college for providing all the experimental and computational facilities to carry out this work.

References

[1] V. Quaschning and R. Hanitsch, "Numerical simulation of current-voltage characteristics of photovoltaic systems with shaded solar cells," *Solar Energy*, vol. 56, no. 6, pp. 513–520, 1996.
[2] J. H. R. Enslin, M. S. Wolf, D. B. Snyman, and W. Swiegers, "Integrated photovoltaic maximum power point tracking converter," *IEEE Transactions on Industrial Electronics*, vol. 44, no. 6, pp. 769–773, 1997.
[3] W. Herrmann, W. Wiesner, and W. Vaassen, "Hot spot investigations on PV modules - new concepts for a test standard and consequences for module design with respect to bypass diodes," in *Proceedings of the 26th IEEE Photovoltaic Specialists Conference*, pp. 1129–1132, October 1997.
[4] N. D. Kaushika and N. K. Gautam, "Energy yield simulations of interconnected solar PV arrays," *IEEE Transactions on Energy Conversion*, vol. 18, no. 1, pp. 127–134, 2003.
[5] M. Klenk, S. Keller, L. Weber et al., "Investigation of the hot-spot behaviour and formation in crystalline silicon Power cells, PV in Europe, From PV technology to energy solutions," in *Proceedings of the International Conference*, pp. 272–275, 2002.
[6] A. Woyte, J. Nijs, and R. Belmans, "Partial shadowing of photovoltaic arrays with different system configurations: literature review and field test results," *Solar Energy*, vol. 74, no. 3, pp. 217–233, 2003.
[7] H. S. Rauschenbach, "Electrical output of shadowed solar arrays," *IEEE Transactions on Electron Devices*, vol. 18, no. 8, pp. 483–490, 1971.
[8] M. C. Alonso-García, J. M. Ruiz, and W. Herrmann, "Computer simulation of shading effects in photovoltaic arrays," *Renewable Energy*, vol. 31, no. 12, pp. 1986–1993, 2006.
[9] M. C. Alonso-García, J. M. Ruiz, and F. Chenlo, "Experimental study of mismatch and shading effects in the I-V characteristic of a photovoltaic module," *Solar Energy Materials and Solar Cells*, vol. 90, no. 3, pp. 329–340, 2006.
[10] E. Karatepe, M. Boztepe, and M. Colak, "Development of suitable model for characterizing photovoltaic arrays with shaded solar cells," *Solar Energy*, pp. 329–340, 2007.
[11] R. Ramaprabha and B. L. Mathur, "Effect of shading on series and parallel connected solar PV modules," *Journal of Modern Applied Science*, vol. 3, no. 10, pp. 32–41, 2009.
[12] W. Xiao, N. Ozog, and W. G. Dunford, "Topology study of photovoltaic interface for maximum power point tracking," *IEEE Transactions on Industrial Electronics*, vol. 54, no. 3, pp. 1696–1704, 2007.
[13] T. Noguchi, S. Togashi, and R. Nakamoto, "Short-current pulse-based maximum-power-point tracking method for multiple photovoltaic-and-converter module system," *IEEE Transactions on Industrial Electronics*, vol. 49, no. 1, pp. 217–223, 2002.
[14] E. Román, R. Alonso, P. Ibañez, S. Elorduizapatarietxe, and D. Goitia, "Intelligent PV module for grid-connected PV systems," *IEEE Transactions on Industrial Electronics*, vol. 53, no. 4, pp. 1066–1073, 2006.
[15] N. Femia, G. Lisi, G. Petrone, G. Spagnuolo, and M. Vitelli, "Distributed maximum power point tracking of photovoltaic arrays: novel approach and system analysis," *IEEE Transactions on Industrial Electronics*, vol. 55, no. 7, pp. 2610–2621, 2008.
[16] L. Gao, R. A. Dougal, S. Liu, and A. P. Iotova, "Parallel-connected solar PV system to address partial and rapidly fluctuating shadow conditions," *IEEE Transactions on Industrial Electronics*, vol. 56, no. 5, pp. 1548–1556, 2009.
[17] R. Gules, J. D. P. Pacheco, H. L. Hey, and J. Imhoff, "A maximum power point tracking system with parallel connection for PV stand-alone applications," *IEEE Transactions on Industrial Electronics*, vol. 55, no. 7, pp. 2674–2683, 2008.
[18] C. Gonzalez and R. Weaver, "Circuit design considerations for photovoltaic modules and systems," in *Proceedings of the 14th IEEE Photovoltaic Specialist Conference*, pp. 528–535, January 1980.
[19] N. F. Shepard and R. S. Sugimura, "The integration of bypass diode with terrestrial photovoltaic modules and arrays," in *Proceedings of the 17th IEEE Photovoltaic Specialist Conference*, pp. 676–681, 1984.
[20] F. Giraud and Z. Salameh, "Analysis of the effects of a passing cloud on a grid-interactive photovoltaic system with battery storage using neural networks," *IEEE Transactions on Energy Conversion*, vol. 14, no. 4, pp. 1572–1577, 1999.
[21] Y. J. Wang and P. C. Hsu, "Analysis of partially shaded PV modules using piecewise linear parallel branches model," *World Academy of Science, Engineering and Technology*, vol. 60, pp. 783–789, 2009.
[22] Y. J. Wang and P. C. Hsu, "An investigation on partial shading of PV modules with different connection configurations of PV cells," *International Journal on Energy*, vol. 36, no. 5, pp. 3069–3078, 2011.
[23] M. G. Jaboori, M. M. Saied, and A. R. Hanafy, "A contribution to the simulation and design optimization of photovoltaic systems," *IEEE Transactions on Energy Conversion*, vol. 6, no. 3, pp. 401–406, 1991.
[24] W. T. Jewell and T. D. Unruh, "Limits on cloud-induced fluctuation in photovoltaic generation," *IEEE Transactions on Energy Conversion*, vol. 5, no. 1, pp. 8–14, 1990.
[25] H. Patel and V. Agarwal, "MATLAB-based modeling to study the effects of partial shading on PV array characteristics," *IEEE Transactions on Energy Conversion*, vol. 23, no. 1, pp. 302–310, 2008.
[26] H. Patel and V. Agarwal, "Maximum power point tracking scheme for PV systems operating under partially shaded conditions," *IEEE Transactions on Industrial Electronics*, vol. 55, no. 4, pp. 1689–1698, 2008.

[27] M. S. Swaleh and M. A. Green, "Effect of shunt resistance and bypass diodes on the shadow tolerance of solar cell modules," *Solar Cells*, vol. 5, no. 2, pp. 183–198, 1982.

[28] M. G. Villalva, J. R. Gazoli, and E. R. Filho, "Comprehensive approach to modeling and simulation of photovoltaic arrays," *IEEE Transactions on Power Electronics*, vol. 24, no. 5, pp. 1198–1208, 2009.

[29] S. Silvestre, A. Boronat, and A. Chouder, "Study of bypass diodes configuration on PV modules," *Applied Energy*, vol. 86, no. 9, pp. 1632–1640, 2009.

[30] R. Ramaprabha and B. L. Mathur, "Modelling and simulation of solar PV array under partial shaded conditions," in *Proceedings of the IEEE-International Conference on Sustainable Energy Technologies (ICSET '08)*, pp. 7–11, SMU Conference Centre, Singapore, November 2008.

[31] J. W. Bishop, "Computer simulation of the effects of electrical mismatches in photovoltaic cell interconnection circuits," *Solar Cells*, vol. 25, no. 1, pp. 73–89, 1988.

[32] M. C. Alonso-García and J. M. Ruíz, "Analysis and modelling the reverse characteristic of photovoltaic cells," *Solar Energy Materials and Solar Cells*, vol. 90, no. 7-8, pp. 1105–1120, 2006.

Preparation of a Modified PTFE Fibrous Photo-Fenton Catalyst and Its Optimization towards the Degradation of Organic Dye

Zhizhong Ding,[1] Yongchun Dong,[1,2] and Bing Li[1]

[1] *Division of Textile Chemistry & Ecology, School of Textiles, Tianjin Polytechnic University, 399 Binshui Western Road, Xiqing District, Tianjin 300387, China*
[2] *State Key Laboratory Breeding Base of Photocatalysis, Fuzhou University, Fuzhou 350002, China*

Correspondence should be addressed to Yongchun Dong, dyefib@yahoo.com.cn

Academic Editor: Meenakshisundaram Swaminathan

Polytetrafluoroethylene (PTFE) fiber was grafted with acrylic acid to impart the carboxyl groups onto the fiber surface, which were used to coordinate with both transition metal ions Fe(III) and Cu(II) and a rare metal ion Ce(III) to prepare the metal grafted PTFE fiber complexes as the novel heterogeneous Fenton catalysts for the degradation of the azo dye in water under visible irradiation. Some factors affecting the preparation process, such as nature and concentration of metal ions in the coordination solution, grafting degree of PTFE and reaction temperature were optimized with respect to the content and strength of metal fixation on the fiber and dye degradation efficiency. The results indicated that increasing metal ion concentrations in solution and grafting degree of PTFE fiber as well as higher coordination temperature led to a significant increase in metal content, especially Fe(III) and Cu(II) content of the complexes. Fe(III) ions fixed on the fiber showed the better catalytic performance than Cu(II) and Ce(III) ions fixed when three different complexes with similar metal content being employed, respectively. Moreover, Increasing Fe content or incorporation of Cu(II) ions could significantly improve the catalytic activity of the complexes.

1. Introduction

Dye-containing wastewater from different industrial fields is a principal source of environmental contamination. And these dyes are known to be largely nonbiodegradable and toxic to aquatic plants and animals. In recent years, heterogeneous Fenton process has become a research focus in the treatment of these dyes in water because of its application over a wide pH range and easy separation of the catalyst after the reaction. The modified polyacrylonitrile (PAN) fiber Fe complexes have been regarded as the attractive heterogeneous Fenton catalysts with low cost and high effectiveness [1–5]. However, it was reported that the modifications, especially amidoximation can lead to a significant decrease in the mechanical performance of PAN fiber. Moreover, the tensile strength of the amidoximated PAN fiber was found to be further lower after coordinating with Fe ions in our previous work [2, 6]. This may limit its commercial applications in the future because good mechanical performance is necessary for effective using of the fibers as the catalysts for Fenton system.

Polytetrafluoroethylene (PTFE) fiber has received more and more attention in the past decades because of their outstanding thermal, chemical stability, and mechanical properties even under very harsh conditions [7–9]. In recent years, there have been many reports of using modified PTFE fibers and films containing functional groups for the concentration and separation of metal ions from aqueous media [10–12]. However, little attention has been paid to the heterogeneous Fenton catalyst produced by PTFE modification and metal coordination and its catalytic property for degradation of organic dye. Therefore, our investigations have been directed to the preparation of a fibrous PTFE catalyst with both good catalytic activity and mechanical strength. In this work, the influence of grafting degree of PTFE fiber, nature, and concentration of metal ion in the coordination solution and reaction temperature was optimized with respect to the content and strength of metal

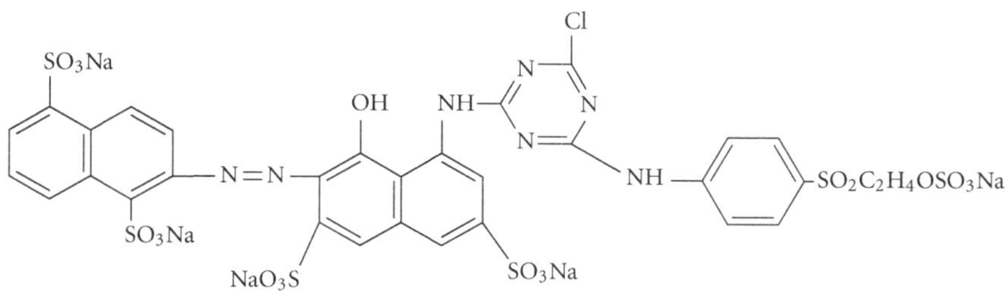

Scheme 1: Chemical structure of RR 195.

fixation on the fiber and decomposition efficiency of organic dye in aqueous solutions.

2. Experimental

2.1. Materials and Reagents. PTFE fibers were commercially available and provided by Fuxin Sengfuli Polymer Materials Co., China. Acrylic acid, hydrogen peroxide (30%, w/w), and ammonium ferrous sulphate were of analytical grade. $FeCl_3 \cdot 7H_2O$, $CuSO_4 \cdot 5H_2O$ and $CeCl_3 \cdot 7H_2O$ were used as the metal ions sources, respectively. Horseradish peroxidase (POD), N,N-diethyl-p-phenylene-deamine (DPD) and Reactive Red 195 (abbr. RR 195) was of laboratory agent grade and used without further purification. RR 195 was a typical azo dye and its molecular structure is presented in Scheme 1. Double distilled and deionized water was used throughout the study.

2.2. Preparation of Metallic Modified PTFE Fiber Complexes

2.2.1. Graft Polymerization of PTFE Fibers. Before modification, PTFE fibers were treated with acetone at room temperature for 30 min, and then dried in a vacuum oven at 50°C for 48 h. 10.0 g of the dried fibers was immersed into the monomer solution containing 100 mL acrylic acid and 200 mL deionized water in a 500 mL flask. 3.0 wt% ammonium ferrous sulphate was also used in the monomer solution to minimize the homopolymerization of acrylic acid during the reaction. The flask containing the monomer solution and fibers was bubbled with nitrogen for 20 min to remove oxygen and sealed. The mixture in the flask was directly subjected to gamma-rays of irradiation from a ^{60}Co source at a dose rate of 0.5 $kGyh^{-1}$ for a given time. After irradiation, the grafted PTFE fibers (denoted as PAA-g-PTFE) were took out and washed with hot water, and acetone repeatedly to remove the residual monomer, homopolymers adhered to the fiber surface. Finally, the obtained grafted fibers were dried in a vacuum oven at 50°C for 24 h and weighed. The grafting degree was obtained using the following formula: $G_f\% = (W_g/W_o - 1) \times 100\%$, where W_o and W_g are the weights of the original and grafted fibers, respectively. It was noticed that the breaking strength of the original PTFE fiber was measured to be 93.35 CN, and little decrease in its breaking strength was observed after the polymerization.

2.2.2. Transition Metal PAA-g-PTFE Complexes. 5.0 g of PAA-g-PTFE was immersed in 150 mL of $FeCl_3$ or $CuSO_4$ aqueous solution, and then the mixture was treated at 50°C and pH 2.0-3.0 for a given time under continuous agitation. The resulting fibrous metal complexes were filtered, washed with deionized water and dried under vacuum at 60°C for 4.0 h to obtain Fe(III-) grafted PTFE fiber complexes (denoted as Fe-PAA-g-PTFE) or Cu(II-) grafted PTFE fiber complexes (denoted as Cu-PAA-g-PTFE). The residual concentrations of Fe(III) or Cu(II) ions in the solutions after coordination were determined by using a WXF120 atomic absorption spectrometry (Beijing Rayleigh Analytical Instrument Corp., China) for calculating the Fe content ($Q_{\text{Fe-PTFE}}$) or Cu content ($Q_{\text{Cu-PTFE}}$) of the complexes.

2.2.3. Rare Metal PAA-g-PTFE Complexes. The complexes were prepared similarly to the transition metal PAA-g-PTFE complexes by using $CeCl_3$ in the coordination solution. And the final complexes were denoted as Ce-PAA-g-PTFE. The Ce content ($Q_{\text{Ce-PTFE}}$) of the complexes was measured and calculated by determining the residual Ce(III) ions in solution after coordination through a Na_2-EDTA-based titrimetric method.

2.2.4. Cu-Fe Bimetallic PAA-g-PTFE Complexes. These complexes were also synthesized similarly to monometallic PAA-g-PTFE complexes by using the mixed coordination solutions containing different concentration of Fe(III) and Cu(II) ions during the coordination processes. Cu-Fe bimetallic PAA-g-PTFE complexes (denoted as Cu-Fe-PAA-g-PTFE) with different molar ratios of Fe(III) to Cu(II) ion (abbr. Cu/Fe molar ratio) were produced by adjusting the concentrations of the metal ions in coordination solution, and in which the concentrations of total metal ions were kept constant at 0.10 mol L^{-1}.

2.3. Catalytic Activity of the Complex towards H_2O_2 Decomposition. The decomposition of H_2O_2 was carried out at the 150 mL Pyrex vessels inside the photoreaction reaction system. A 400 W high pressure mercury lamp (Foshan Osram Illumination Co., China) was used as the illuminating source for photocatalytic reaction. A cutoff filter was used to ensure irradiation only by visible light ($\lambda > 420$ nm) for the high pressure mercury lamp. The intensity of visible light irradiation over the surface of test solution inside photoreaction system was measured to be 9.65 Wcm^{-2} using

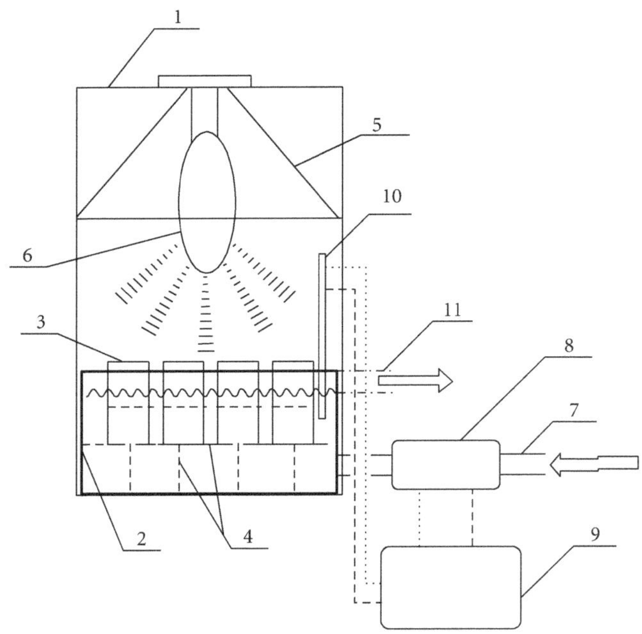

(1) Chamber
(2) Water bath
(3) Glass vessel
(4) Support
(5) Lamp-chimney
(6) Mercury lamp
(7) Water in
(8) Electromagnetic valve
(9) Relay
(10) Thermometer
(11) Water out

FIGURE 1: The schematic diagram of photoreaction system.

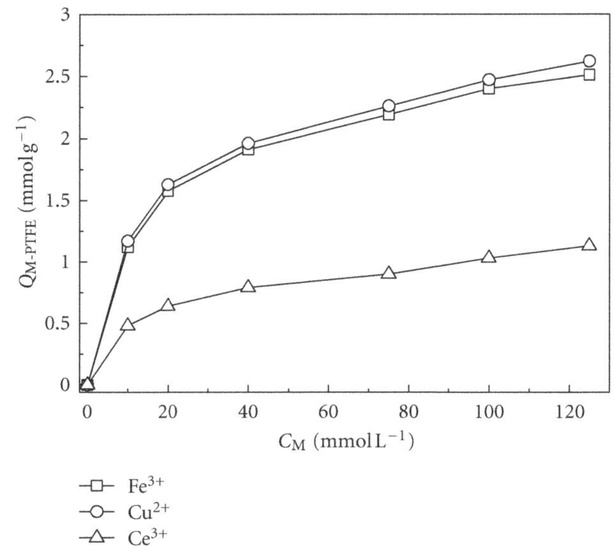

FIGURE 2: Effect of C_M on $Q_{M\text{-}PTFE}$ of the complexes during the preparation.

FZ-A radiometer (BNU Light and Electronic Instrumental Co., China), respectively. The diagram of the photoreaction system was presented in Figure 1.

1.0 g of the complex was placed into the vessels containing 100 mL H_2O_2 solution of concentration 10.0 mmol L^{-1} at 25°C. The residual concentration of H_2O_2 in the solution during the reaction was determined though the spectrophotometric DPD method, in which the DPD was oxidized by H_2O_2 based on the POD catalyzed reaction [13]. The analysis of the residual concentration of H_2O_2 was repeated three times for each sample and the results averaged.

2.4. Dye Degradation Procedure. The dye degradation was performed in the photoreaction reaction system mentioned above under visible irradiation. 0.40 g of metal PAA-g-PTFE complex was added into 100 mL of test solution containing 0.04 mmol L^{-1} RR 195 and 6.0 mmol L^{-1} H_2O_2. The reaction temperature was kept at 25 ± 1°C, and the pH of the solution was adjusted with 0.10 mmol L^{-1} HNO_3 and/or NaOH aqueous solution and measured using DHS-25C digital pH meter (Shanghai Jingmi Instrumental Co., China). The degradation of the dye was initialized after the adsorption/desorption equilibrium of the dye on the complex was reached in the dark for 2.0 h. At given irradiation time intervals, 1-2 mL of the test solution was sampled from the reaction vessel and analyzed immediately by recording the UV-vis spectrum of the dye at the λ_{max} (523 nm) with a UV-2401 Shimadzu spectrophotometer. The decoloration percentage of the dye was expressed as $D\% = (1 - C_d/C_{d,0}) \times 100\%$, where $C_{d,0}$ and C_d are the initial and residual concentrations of the dye (mmol L^{-1}), respectively. Moreover, total organic carbon (TOC) was assayed in the dye degradation process by using a Phoenix 8000 TOC analyzer (Tekmar-Dehrmann Inc., USA) and TOC removal percentage of the dye was calculated as follows: TOC removal % $= (1 - TOC_t/TOC_0) \times 100\%$, where TOC_0 and TOC_t are the TOC values (mg L^{-1}) at reaction times 0 and t, respectively.

2.5. Metal Ions Leaching from the Complex. 0.40 g of metal PTFE fiber complex was placed into in 100 mL solution containing 6.0 mmol L^{-1} H_2O_2. The leaching of Fe(III) or/and Cu(II) ions from the complex in solution at room temperature and under visible light irradiation was carried out for 300 min. And the concentrations of Fe(III) or Cu(II) ions in solution at intervals of time were measured using atomic absorption spectrometry method, respectively.

3. Results and Discussion

3.1. Optimization of Coordination Parameters with respect to Metal Contents on PAA-g-PTFE

3.1.1. Metal Ion Concentration in Coordination Solution. In order to investigate the effect of metal ion concentration in coordination solution (C_M) on metal content on PAA-g-PTFE ($Q_{M\text{-}PTFE}$), three metal ions including Fe(III), Cu(II), and Ce(III) ions were selected to react with PAA-g-PTFE with G_f% of 34.53 at 50°C, and results were shown in Figure 2.

It is seen from Figure 2 that all the $Q_{M\text{-}PTFE}$ values significantly increase with increasing C_M, and the increasing tendency gradually declines when C_M being above

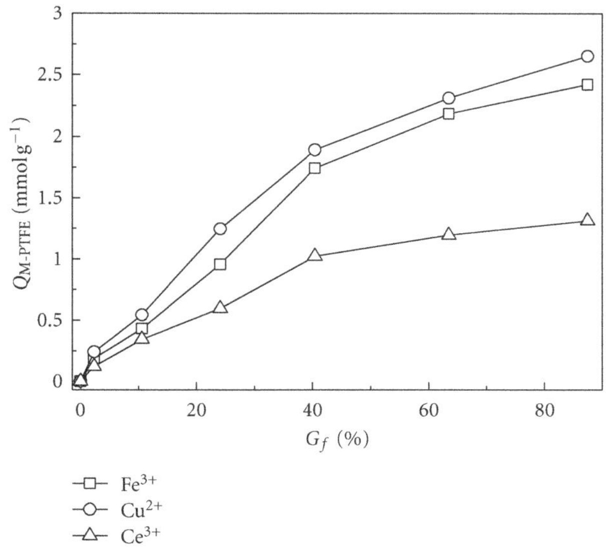

FIGURE 3: $Q_{M\text{-PTFE}}$ of the complexes as a function of graft degree of PAA-g-PTFE.

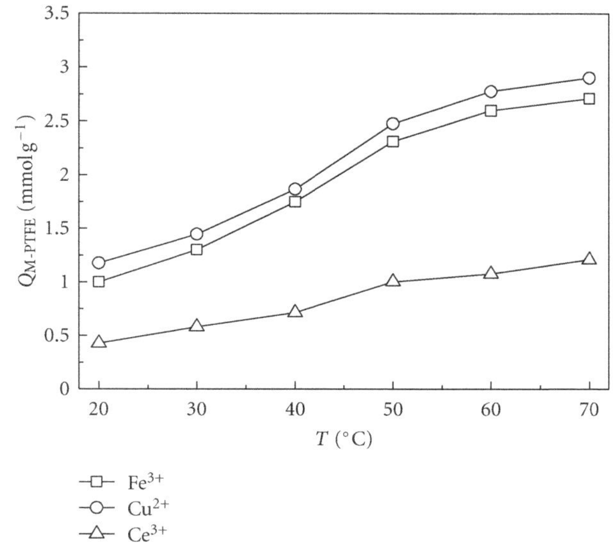

FIGURE 4: $Q_{M\text{-PTFE}}$ of the complexes at different reaction temperatures.

40.0 mmol L^{-1}, especially for Ce(III) ions. This suggests that higher C_M can enhance the coordination of PAA-g-PTFE with metal ions, particularly two transition metal ions. The previous works [11, 12, 14] have found that the carboxyl groups on PAA-g-PTFE are the main coordinating sites with metal ions. Furthermore, it has been proposed that the oxygen of the carboxyl group has a pair of electrons that can add themselves to a proton or a metal ion to form a complex through a coordinated covalent bond. The metal ions with empty orbitals act as a Lewis acid capable of accepting electron pairs. In contrast, the carboxyl groups that have nonshared electron pairs function as Lewis bases donating their electrons pair [15, 16]. Moreover, transition metal contents ($Q_{Fe\text{-PTFE}}$ and $Q_{Cu\text{-PTFE}}$) on the fiber are much more than Ce content ($Q_{Ce\text{-PTFE}}$) on the fiber, especially at higher metal concentration in solution. This is because the 4f electrons of rare earth metal ion are in the inner orbital of the atomic structure, so that the ligand field has less effect on them and the rare earth metal ions have much lower LFSE values (4.18 kJ moL^{-1}) than d-transition metal ions ($\geq$ 418 kJ moL^{-1}) [5, 17].

3.1.2. Graft Degree of PAA-g-PTFE. To determine the dependence of $Q_{M\text{-PTFE}}$ value on graft degree of PAA-g-PTFE, the complexes were produced by the coordination of PAA-g-PTFE with different G_f% in 0.10 mol L^{-1} metal ion aqueous solution at 50°C, and relationship between G_f% and $Q_{M\text{-PTFE}}$ values was presented in Figure 3.

Figure 3 shows that the $Q_{M\text{-PTFE}}$ value of metal PAA-g-PTFE complexes was greatly enhanced with G_f% of PAA-g-PTFE, particularly when Fe(III) and Cu(II) ion being used. This indicates that increasing G_f% value can benefit the fixation of three metal ions on the surface of PAA-g-PTFE and form the metal PAA-g-PTFE complexes with high $Q_{M\text{-PTFE}}$ because PAA-g-PTFE with high G_f% values has much more carboxyl groups on their surface than those with low G_f% values, thus more easily react with metal ions in solution.

3.1.3. Temperature. PAA-g-PTFE with G_f% of 34.53 was employed to coordinate with three metal ions, respectively, in the solution, for investigating the impact of reaction temperature on $Q_{M\text{-PTFE}}$ value of the resulting metal PAA-g-PTFE complexes, and results were shown in Figure 4.

Figure 4 shows that $Q_{M\text{-PTFE}}$ value of the complexes is increased with elevation of reaction temperature from 20 to 70°C. This obviously demonstrates that it is favorable for the coordination with the temperature rising, which implies that the coordination process is an endothermic process [11]. A possible reason is that higher temperature increases mobility of the polymer chains and the swelling of the fiber in the reaction medium, thus improving the interactions between metal ion and the fiber and resulting in more metal ions fixed on them.

3.2. Comparison of Metal Ions toward H_2O_2 Decomposition. The catalytic activity of three metal PAA-g-PTFE complexes with similar $Q_{M\text{-PTFE}}$ value (approximately 1.30 mmol L^{-1}) was examined with respect to the decomposition of H_2O_2, respectively, and the decay ratios of H_2O_2 as the function of irradiation time were shown in Figure 5.

It is observed from Figure 5 that the decay ratios of H_2O_2 gradually decline with the increasing irradiation time and vary with the nature of metal ions. The decay ratio of H_2O_2 is the lowest for Fe-PAA-g-PTFE, followed by Cu-PAA-g-PTFE, and the highest for Ce-PAA-g-PTFE. This indicates that Fe-PAA-g-PTFE has a stronger catalytic activity than the other complexes at the same conditions. Additionally, from the viewpoint of the toxicity and cost of three metal ions,

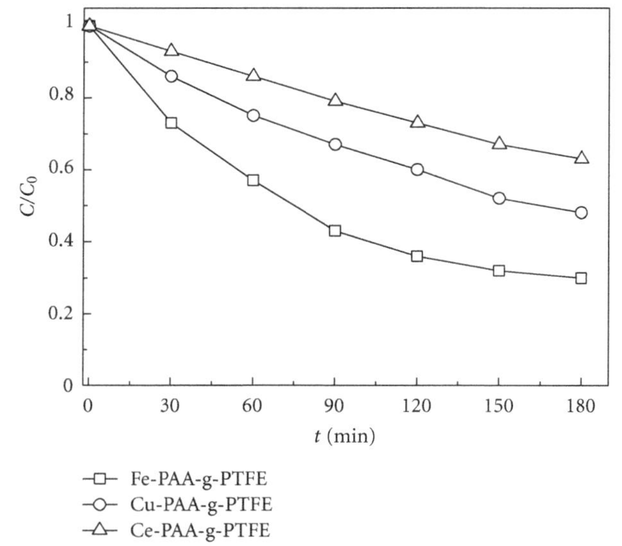

Figure 5: Decomposition of H_2O_2 in the presence of the different catalysts.

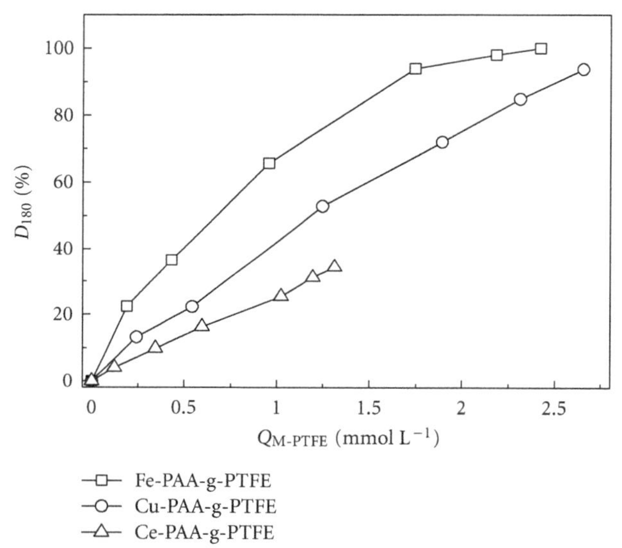

Figure 6: Relationship between $Q_{\text{M-PTFE}}$ and D_{180}%.

Fe(III) ions should been selected to prepare the catalyst for the majority of the future research and practical application.

3.3. Effect of Metal Ion on Dye Degradation. The dye degradations were conducted in the presence of three metal PAA-g-PTFE complexes with different $Q_{\text{M-PTFE}}$ values as the heterogeneous Fenton catalysts, respectively, and the influence of $Q_{\text{M-PTFE}}$ value on decoloration percentage of the dye within 180 min (D_{180}%) was presented in Figure 6.

Figure 6 shows that increasing $Q_{\text{M-PTFE}}$ value can remarkably enhance the dye degradation for all the complexes, particularly Fe-PAA-g-PTFE. Moreover, comparing three metal PAA-g-PTFE complexes with similar $Q_{\text{M-PTFE}}$ values, the highest D_{180}% is achieved when Fe-PAA-g-PTF is used. This means that Fe-PAA-g-PTFE exhibits the strongest catalytic activity for dye degradation, of three complexes. The main reason is that Fe-PAA-g-PTFE leads to the enhanced decomposition of H_2O_2 in solution indicated in Figure 5, thus possibly producing more hydroxyl radicals, which is responsible for the dye degradation. This reaction process is described by (1) and (2). On the other hand, the formed Fe(II) ions on Fe-PAA-g-PTFE may be oxidized to Fe(IV) ions in this work, which are regarded as another reactive intermediate from the photo-Fenton reaction which can decompose dye molecules. A possible formation of Fe(IV) ions was expressed by (3)

$$Fe^{3+}/\text{PTFE} + H_2O_2 \xrightarrow{h\nu} Fe^{2+}/\text{PTFE} + HO_2^{\bullet} + H^{\oplus} \quad (1)$$

$$Fe^{2+}/\text{PTFE} + H_2O_2 \xrightarrow{h\nu} Fe^{3+}/\text{PTFE} + OH^{\ominus} + HO^{\bullet} \quad (2)$$

$$Fe^{2+}/\text{PTFE} + H_2O_2 \xrightarrow{h\nu} Fe^{4+}/\text{PTFE} \xrightarrow{h\nu} Fe^{3+}/\text{PTFE} \quad (3)$$

$$\text{Dye} + Fe^{3+}/\text{PTFE} \xrightarrow{h\nu} \text{Dye}^{*} \cdots Fe^{3+}/\text{PTFE} \longrightarrow \text{Dye}^{+\bullet} + Fe^{2+}/\text{PTFE}. \quad (4)$$

Additionally, the photosensitization of the azo dyes can promote a conversion of Fe(III) to Fe(II) ions during photo-assisted Fenton reaction [12]. In this work, it is believed that RR 195 could act as the sensitizer of Fe-PAA-g-PTFE to lead to a charge transfer with the concomitant quenching of the excited dye* and the formation of dye^{+}. This process was suggested by (4).

It is well known the hydroxyl radicals is a powerful oxidant and can ultimately cause mineralization of textile dyes. Consequently, the mineralization of the dye was examined by measuring the TOC values of test solutions during the degradation when three complexes with the similar $Q_{\text{M-PTFE}}$ values were employed, respectively.

Figure 7 shows that TOC removal % of the dye is as a function of irradiation time. It is found that TOC removal % gradually increases with the prolongation of irradiation time. Specifically, TOC removal % for Fe-PAA-g-PTF is much higher than those for the other complexes at the same time. These results demonstrate that three metal PAA-g-PTFE complexes as the photocatalysts can result in a drastic breakdown of dye molecules, and then convert them into water, carbon dioxide, and inorganic compounds. It should also be pointed out that TOC removal % is mainly determined by nature of metal ions, which is well consistent with the impact of nature of metal ions on the degradation mentioned above.

In order to study the catalytic activity of the complexes in a wide pH range, the dye degradation was carried out at different solution pH within 180 min of reaction time. The catalytic activity of the complexes was determined with respect to the dye degradation with the results presented in Figure 8.

Figure 8 shows that the dye degradation can be operated in the presence of the complex over a wide pH 3–9 range, indicating that a significant breakdown of the dye occur, especially at acidic media. Moreover, the catalytic activity

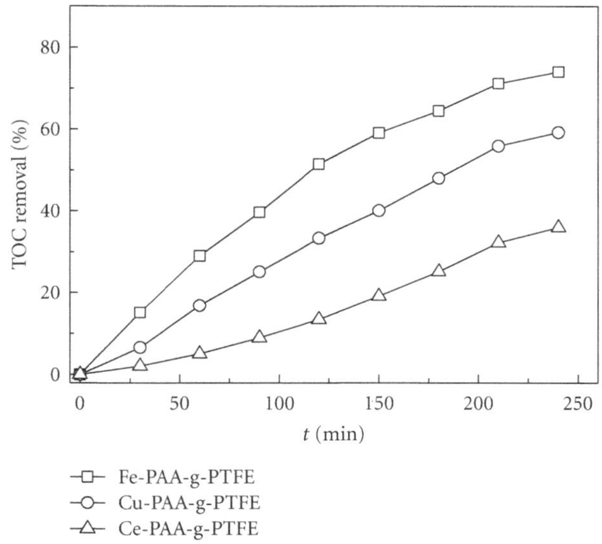

FIGURE 7: Mineralization of the dye with different catalysts.

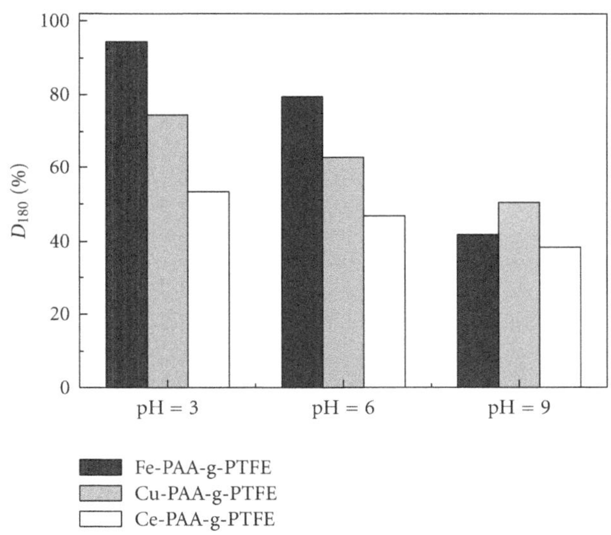

FIGURE 8: Dye degradation at different pH levels for three catalysts.

of three complexes at acid and neutral medium is ranked in this order: Fe-PAA-g-PTFE > Cu-PAA-g-PTFE > Ce-PAA-g-PTFE. However, higher pH range, especially alkaline pH medium dramatically decreases the catalytic activity of three complexes. It is worth noticing that Cu-PAA-g-PTFE shows a stronger activity than Fe-PAA-g-PTFE at pH 9. This is similar to the performance of Cu/MCM-41 as a highly stable and pH-insensitive heterogeneous photo-Fenton-like Fenton catalyst [18]. Some researches [18, 19] suggested that an advantage of Cu^{2+} ion as a Fenton-like catalyst is its characteristics of a wide-working pH range compared to the conventional Fenton catalyst. Therefore it is believed that Cu-PAA-g-PTFE behaves similarly to Fe-PAA-g-PTFE, and it has a higher catalytic activity over a wide pH range.

3.4. Role of Cu(II) Ion as an Assisted Metal Ion. It has been suggested that the performance of supported Fe catalysts is often improved by adding certain metals. And copper ion has a role as an assistant in photocatalysis [5]. Moreover, it has been presented that the outstanding catalytic activity of bimetal catalysts may be generally ascribed to synergistic effects between copper and iron redox couples. In order to meet the demands for rapid and stable dye degradation in future commercial application, a series of Cu-Fe bimetal PAA-g-PTFE complexes with different Cu/Fe molar ratios were prepared and used in the dye degradation at pH 9 in this work. And Cu/Fe molar ratio of metal PAA-g-PTFE complexes was observed to have a significant impact on the decomposition efficiency of the dye, and the results are given in Table 1.

The data in Table 1 indicate that higher concentration of Cu(II) ions in the solutions increased their Cu/Fe molar ratio. When the molar concentration of Cu(II) ions equals to that of Fe(III) ions in the solution, the Cu/Fe molar ratio of Cu-Fe-PAA-g-PTF is 2.87, much more than 1.0, suggesting that PAA-g-PTF has much higher affinity for Cu(II) ions than Fe(III) ions at the same conditions. More importantly, $D_{180}\%$ and TOC removal % of dye degradation in the presence of the bimetal PAA-g-PTFE complexes is much higher than those in the presence of the monometal PAA-g-PTE complexes, and the maximum $D_{180}\%$ and TOC removal % was achieved for Cu-Fe-PAA-g-PTFE (ii). These results confirm that catalytic activity of metal PAA-g-PTFE complexes depends highly upon their Cu/Fe molar ratio, and Cu-Fe-PAA-g-PTFE (ii) with a Cu/Fe molar ratio of 2.87 shows the best catalytic activity at alkaline pH medium. This demonstrates the existence of synergetic effect between loaded Fe(III) and Cu(II) ions in the bimetal PAA-g-PTFE complexes, which improved their catalytic activity to a great extent. The main reason may be that the H_2O_2 molecules absorbed on the surface of Cu-Fe-PAA-g-PTFE can reduce Fe(III) to Fe(II) ions through the Fenton-like process under visible light irradiation, and the generated Fe(II) ions are then oxidized by H_2O_2 to complete the Fe(III)/Fe(II) ions cycle and produce hydroxyl radicals. Meanwhile, the formed Fe(II) ions can also react with the loaded Cu(II) ions to form the Cu(I) and Fe(III) ions (5) to accelerate the transformation of Fe(III)/Fe(II) ions and Cu(II)/Cu(I) ions in Cu-Fe-PAA-g-PTFE, thus giving rise to additional hydroxyl radicals during the degradation:

$$Fe^{3+}/PTFE + Cu^{+}/PTFE \rightleftharpoons Fe^{2+}/PTFE + Cu^{2+}/PTFE. \quad (5)$$

3.5. Metal Ions Leaching from the Catalyst. One problem with the heterogeneous photocatalysts is metal leaching from the catalyst that causes secondary contamination to wastewater in pilot-scale application. The leaching concentrations of metal ions from metal PAA-g-PTFE complexes in solution were measured. The results indicated that metal ions were continuously leached out from the complexes after the reaction started, and the final measured concentrations of both metal ions in the solution within 300 min at different pH levels were shown in Tables 2 and 3.

Table 1: Effect of Cu/Fe molar ratio of the catalysts on dye decomposition.

Samples	C_M (mol L^{-1})		$Q_{M\text{-}PTFE}$ (mmol L^{-1})			Cu/Fe	D_{180}%	TOC Removal %
	Fe(III)	Cu(II)	Fe(III)	Cu(II)	Total			
Fe-PAA-g-PTFE	0.100	0.000	1.89	0.00	1.89	0	47.3	37.3
Cu-Fe-PAA-g-PTFE (i)	0.075	0.025	1.13	0.75	1.88	0.66	61.2	54.6
Cu-Fe-PAA-g-PTFE (ii)	0.050	0.050	0.54	1.55	2.09	2.87	84.6	72.7
Cu-Fe-PAA-g-PTFE (iii)	0.025	0.075	0.31	1.75	2.06	5.65	78.4	56.7
Cu-PAA-g-PTFE	0.000	0.100	0.00	1.98	1.98	/	51.9	45.1

Table 2: Leaching of Fe ions from Fe-PAA-g-PTFE with different Q_{Fe} values.

Fe-PAA-g-PTFE	Fe(III) ions concentration in solution (mg L^{-1})		
	pH = 3	pH = 7	pH = 9
Q_{Fe} = 0.432 mmol L^{-1}	0.49	0.37	0.07
Q_{Fe} = 0.955 mmol L^{-1}	0.72	0.57	0.15
Q_{Fe} = 1.743 mmol L^{-1}	0.95	0.73	0.22
Q_{Fe} = 2.183 mmol L^{-1}	1.22	0.94	0.43
Q_{Fe} = 2.421 mmol L^{-1}	1.53	1.23	0.57

Table 3: Impact of Cu/Fe molar ratio on leaching of metal ions from Cu-Fe bimetallic PTFE fiber complexes.

Samples	Metal ions concentration in solution (mg L^{-1})					
	pH = 3		pH = 7		pH = 9	
	Fe(III)	Cu(II)	Fe(III)	Cu(II)	Fe(III)	Cu(II)
Fe-PAA-g-PTFE	0.93	0.00	0.82	0.00	0.29	0.00
Cu-Fe-PAA-g-PTFE (i)	0.63	0.35	0.53	0.29	0.19	0.12
Cu-Fe-PAA-g-PTFE (ii)	0.21	0.76	0.13	0.49	0.06	0.24
Cu-Fe-PAA-g-PTFE (iii)	0.17	0.92	0.10	0.45	0.08	0.19
Cu-PAA-g-PTFE	0.00	1.25	0.00	0.89	0.00	0.31

It can be noticed from Table 2 that increasing Q_{Fe} value is accompanied by increasing Fe ion concentration in solution at three investigated pH levels. This means that Fe-PAA-g-PTFE with higher Q_{Fe} value can leach more Fe(III) ions than that with lower Q_{Fe} value. It is obvious from the results shown in Table 3 that Fe(III) and Cu(II) ion concentrations in solution are much affected by Cu/Fe molar ratio of the complexes. On the other hand, higher pH level caused a gradual decrease in leached Fe(III) or Cu(II) ion concentration in solution, revealing that metal ions leaching from the complexes is dependent on the solution pH. Moreover, it should be pointed out that the final leached concentration of Fe(III) or Cu(II) ions is lower than 1.0 mg L^{-1} for both complexes, especially at pH 7 and 9. This result implies that both metal ions were attached on the surface of the modified PTFE via the hydroxyl group. Thus, these complexes can be regarded as the effective and stable heterogeneous photocatalysts for the dye degradation over a broad pH range due to a minimal threshold concentration of 3–15 mg L^{-1} Fe which allows the homogeneous Fenton reaction to proceed within a reasonable period of time. Additionally, it is well known that Cu ion is hazardous to both marine animals and humans. The discharge standard was suggested by the Environmental Protection Agency is 1.0 mg L^{-1} [18], which is higher than the final leached concentration of Cu(II) ions for Cu-Fe bimetallic PTFE fiber complexes except Cu-Fe-PAA-g-PTFE (iii) and Cu-PAA-g-PTFE.

4. Conclusions

The novel heterogeneous photo-Fenton catalysts were produced by a modification of PTFE fibers through graft polymerization of acrylic acid to introduce the carboxyl groups onto the fiber surface, which were used to coordinate with transition metal ions (Fe(III) and Cu(II)) or rare metal ion (Ce(III)). Increasing metal ion concentrations in solution and grafting degree of PTFE fiber as well as higher coordination temperature could lead to a significant increase in metal content, especially Fe(III) and Cu(II) content of the catalysts. Moreover, Fe(III) and Cu(II) ions reacted with the modified PTFE fibers more easily than Ce(III) ion at the same conditions to form the complexes with higher metal content. Fe(III) ions fixed on the fiber showed a much powerful catalytic performance for the decomposition of H_2O_2 and the oxidative degradation of the dye than Cu(II) and Ce(III) ions fixed when three different complexes with similar metal content being employed as the catalysts, respectively. Also, taking into the toxicity of Cu (II) and Ce(III) ions and the cheapness of iron salt, thus, Fe(III) modified PTFE fiber complex was proposed to act as an effective catalyst for the dye degradation under visible irradiation. And increasing Fe content could remarkably improve the catalytic activity of the complex. Comparing with the three monometallic complexes, Cu-Fe bimetallic modified PTFE fiber complexes exhibited the excellent catalytic activity because of the existence of synergetic effect between Fe(III) and Cu(II) ions. Their catalytic activity depended highly on Cu/Fe molar ratio. 2.87 is the optimum Cu/Fe molar ratio to obtain the best catalytic activity for the bimetallic complexes. However, increasing the amount of Fe(III) or/and Cu(II) ions fixed on the fibers resulted in the relatively higher concentrations of both ions leaching from them, especially at low pH ranges. To sum up, the highest dye degradation efficiency and the accepted concentrations of metal ions leaching from the complexes were found when it was prepared with a mixture of 0.05 mol L^{-1} Fe(III) and 0.05 mol L^{-1} Cu(II) ions at 50°C by poly acrylic acid grafted PTFE fiber as the supported material using a simple exhausting method.

Acknowledgments

The authors thank the Tianjin Municipal Science and Technology Committee for a Research Program of Application Foundation and Advanced Technology (11JCZDJ24600). This research was also supported in part by a Grant from the Natural Science Foundation of China (20773093).

References

[1] V. V. Ishtchenko, K. D. Huddersman, and R. F. Vitkovskaya, "Part 1. Production of a modified PAN fibrous catalyst and its optimisation towards the decomposition of hydrogen peroxide," *Applied Catalysis A*, vol. 242, no. 1, pp. 123–137, 2003.

[2] V. V. Ishtchenko, R. F. Vitkovskaya, and K. D. Huddersman, "Investigation of the mechanical and physico-chemical properties of a modified PAN fibrous catalyst," *Applied Catalysis A*, vol. 242, no. 2, pp. 221–231, 2003.

[3] R. F. Vitkovskaya, I. G. Rumynskaya, E. P. Romanova, and L. Y. Tereshchenko, "Fibre catalyst from modified polyacrylonitrile fibres," *Fibre Chemistry*, vol. 35, no. 3, pp. 202–207, 2003.

[4] Z. Han, Y. Dong, and S. Dong, "Copper-iron bimetal modified PAN fiber complexes as novel heterogeneous Fenton catalysts for degradation of organic dye under visible light irradiation," *Journal of Hazardous Materials*, vol. 189, no. 1-2, pp. 241–248, 2011.

[5] Y. Dong, Z. Han, S. Dong, J. Wu, and Z. Ding, "Enhanced catalytic activity of Fe bimetallic modified PAN fiber complexes prepared with different assisted metal ions for degradation of organic dye," *Catalysis Today*, vol. 175, no. 1, pp. 299–309, 2011.

[6] Z. Han, Y. Dong, and S. Dong, "Comparative study on the mechanical and thermal properties of two different modified PAN fibers and their Fe complexes," *Materials and Design*, vol. 31, no. 6, pp. 2784–2789, 2010.

[7] S. Parra, I. Guasaquillo, O. Enea et al., "Abatement of an azo dye on structured C-Nafion/Fe-ion surfaces by photo-Fenton reactions leading to carboxylate intermediates with a remarkable biodegradability increase of the treated solution," *The Journal of Physical Chemistry B*, vol. 107, no. 29, pp. 7026–7035, 2003.

[8] S. Wu, E. T. Kang, K. G. Neoh, H. S. Han, and K. L. Tan, "Surface modification of poly(tetrafluoroethylene) films by graft copolymerization for adhesion improvement with evaporated copper," *Macromolecules*, vol. 32, no. 1, pp. 186–193, 1999.

[9] M. C. Zhang, E. T. Kang, and K. G. Neoh, "Consecutive graft copolymerization of glycidyl methacrylate and aniline on poly(tetrafluoroethylene) films," *Langmuir*, vol. 16, no. 24, pp. 9666–9672, 2000.

[10] Q. Zhang, S. Zhang, S. Chen, P. Li, T. Qin, and S. Yuan, "Preparation and characterization of a strong basic anion exchanger by radiation-induced grafting of styrene onto poly(tetrafluoroethylene) fiber," *Journal of Colloid and Interface Science*, vol. 322, no. 2, pp. 421–428, 2008.

[11] C. Xiong and C. Yao, "Preparation and application of acrylic acid grafted polytetrafluoroethylene fiber as a weak acid cation exchanger for adsorption of Er(III)," *Journal of Hazardous Materials*, vol. 170, no. 2-3, pp. 1125–1132, 2009.

[12] J. F. Wei, Z. P. Wang, J. Zhang, Y. Y. Wu, Z. P. Zhang, and C. H. Xiong, "The preparation and the application of grafted polytetrafluoroethylene fiber as a cation exchanger for adsorption of heavy metals," *Reactive and Functional Polymers*, vol. 65, no. 1-2, pp. 127–134, 2005.

[13] M. Cheng, W. Song, W. Ma et al., "Catalytic activity of iron species in layered clays for photodegradation of organic dyes under visible irradiation," *Applied Catalysis B*, vol. 77, no. 3-4, pp. 355–363, 2008.

[14] N. M. El-Sawy and F. Al Sagheer, "Radiation-induced graft polymerization of acrylic acid onto poly(tetrafluoroethylene-perfluorovinyl ether) copolymer films: complexation with some transition metals and biological activity," *European Polymer Journal*, vol. 37, no. 1, pp. 161–166, 2001.

[15] I. A. Şengil and M. Özacar, "Biosorption of Cu(II) from aqueous solutions by mimosa tannin gel," *Journal of Hazardous Materials*, vol. 157, no. 2-3, pp. 277–285, 2008.

[16] C. Xiong, C. Yao, and X. Wu, "Adsorption of rhenium(VII) on 4-amino-1,2,4-triazole resin," *Hydrometallurgy*, vol. 90, no. 2–4, pp. 221–226, 2008.

[17] G. Henrici-Olive and S. Olive, *Coordination and Catalysis*, Chemie, Weinheim, Germany, 1977.

[18] F. L. Y. Lam, A. C. K. Yip, and X. Hu, "Copper/MCM-41 as a highly stable and pH-insensitive heterogeneous photo-Fenton-like catalytic material for the abatement of organic wastewater," *Industrial and Engineering Chemistry Research*, vol. 46, no. 10, pp. 3328–3333, 2007.

[19] M. Neamtu, A. Yediler, I. Siminiceanu, and A. Kettrup, "Oxidation of commercial reactive azo dye aqueous solutions by the photo-Fenton and Fenton-like processes," *Journal of Photochemistry and Photobiology A*, vol. 161, no. 1, pp. 87–93, 2003.

10

Simulation of Proton Beam Effects in Thin Insulating Films

Ljubinko Timotijevic,[1] Irfan Fetahovic,[2] Djordje Lazarevic,[3] and Milos Vujisic[1]

[1] *Faculty of Electrical Engineering, University of Belgrade, 11060 Belgrade, Serbia*
[2] *State University of Novi Pazar, 36300 Novi Pazar, Serbia*
[3] *Institute of Nuclear Sciences "Vinca", 11000 Belgrade, Serbia*

Correspondence should be addressed to Milos Vujisic; vujsa@ikomline.net

Academic Editor: Predrag Osmokrovic

Effects of exposing several insulators, commonly used for various purposes in integrated circuits, to beams of protons have been investigated. Materials considered include silicon dioxide, silicon nitride, aluminium nitride, alumina, and polycarbonate (Lexan). The passage of proton beams through ultrathin layers of these materials has been modeled by Monte Carlo simulations of particle transport. Parameters that have been varied in simulations include proton energy and insulating layer thickness. Materials are compared according to both ionizing and nonionizing effects produced by the passage of protons.

1. Introduction

Insulating layers are needed in various kinds of microelectronic components. Roles of these insulating layers vary from surface passivation of chips to more specific functions, such as lateral insulation of components in planar technology, capacitor dielectrics, or tunnel oxides in flash memory cells. Depending on their purposes within structures and devices, insulating layers are made out of different materials and with various thicknesses. Since many devices are expected to be operated in radiation environments, it is essential that radiation effects in insulators they comprise are investigated. Actual irradiation at accelerator facilities can be costly and time consuming, which is why radiation hardness of materials and components is often tested through simulations of radiation transport.

The present study investigates effects of proton beams on five insulating materials, commonly encountered in modern day electronic components: silicon dioxide, silicon nitride, aluminium nitride, alumina, and polycarbonate. Radiation effects are predicted and compared on the basis of Monte Carlo simulations of proton transport through ultrathin layers of these insulators.

2. Radiation Effects in Insulators

Insulators are a broad class of materials. They include crystals, amorphous materials, and organics (polymers). The response of insulators to irradiation is determined by their structural properties and electronic configuration [1, 2].

Interaction of incoming particles with matter results in two major effects: ionization energy loss and nonionizing energy loss (NIEL). Interactions of incoming particles which result in electronic excitation or ionization of atoms are referred to as ionization energy loss. In NIEL processes, the energy imparted by the incident particles results in atomic displacements or in collisions where the primary knock-on atom (PKA) remains in its lattice position, in which case the energy is converted to lattice vibrations (phonons). Displaced atoms can also undergo both electronic and displacement energy losses to dissipate their energy inside the medium.

Secondary electrons created by ionization energy losses affect electrical properties of crystalline insulators in a transient manner, except if these electrons get trapped at electrically active point defects in the crystal lattice. Point defects that serve as charge-carrier traps or donors, arise in irradiated insulators as a result of atomic displacements, that is, of NIEL. Polymer insulators exhibit radiation effects that are specific to that kind of materials. They include chain scission and cross-linking, both of which can significantly alter insulator's physical properties.

A heavy charged particle can transfer only a small fraction of its energy in a single electronic collision, and its deflection in the collision is negligible. It therefore has an almost straight trajectory in matter, losing energy continuously in small amounts through collisions with atomic electrons, leaving ionized and excited atoms along its path. Only occasionally does it undergo a substantial deflection, due to elastic scattering from an atomic nucleus [1–4].

For charged particles, such as protons, ionization, and electronic excitation, energy losses are represented by the electronic stopping power (also called collision stopping power, which is a misnomer, since all interactions can be considered collisions) of the material through which they propagate. Stopping power of a medium for a charged particle is the average linear rate of energy loss of the particle in the medium. It is, therefore, equal to the unrestricted linear energy transfer (LET). Several semiempirical stopping power formulas have been devised. The SRIM code, used for simulating proton transport in the present paper, implements the so-called ZBL stopping, which is based on the model given by Ziegler, Biersack, and Littmark.

Displacement damage can occur in crystal insulators when the energy transferred to lattice atoms exceeds the threshold displacement energy (E_d). Irradiation of materials with electrons and light ions introduces predominantly isolated interstitial atoms and vacancies (Frenkel pairs) and small clusters of these point defects, because of the low average recoil atom energies (0.1–1 keV). Energetic heavy ion irradiations, on the other hand, produce energetic displacement cascades that can lead to direct formation of defect clusters within isolated displacement cascades, due to more energetic average recoil atom energies (>10 keV) [5, 6].

The effects of irradiation on the electrical parameters of many materials have been found to display a simple, often linear, relationship with NIEL. NIEL is the rate at which energy is lost to nonionizing events. It is a direct analog of stopping power for ionization events. The units of NIEL are typically MeV/cm, or MeV cm^2/g if mass NIEL is considered. The calculation of NIEL requires information regarding the differential cross section for atomic displacements ($d\sigma/d\Omega$), the average recoil energy of the target atoms (E_r), and a term which partitions the energy into ionizing and nonionizing events, called the Lindhard partition factor (L). NIEL can be written as an integral over solid angle as follows:

$$\text{NIEL}(E) = \frac{N_A}{A}\int_{\theta_{\min}}^{\pi} \frac{d\sigma(\theta,E)}{d\Omega} T(\theta,E)\, L(T(\theta,E))\, d\Omega, \quad (1)$$

where the N_A is Avogadro's number, A is the atomic mass, and $\theta_{\min}$ is the scattering angle for which the recoil energy equals the threshold for atomic displacement. For protons having nonrelativistic energy, the Rutherford differential cross section can be used for elastic scattering at atomic nuclei [7–11].

3. Results of Simulations

Monte Carlo simulations of proton transport through thin layers of insulators were performed in the TRIM module of the SRIM software package [12]. Simulations used monoenergetic unidirectional beams, incident perpendicularly on the film's surface. Proton energies were varied from 10 keV to 10 MeV. At each value of proton energy, the thickness of the insulating film was increased until the whole beam was stopped within it. Film thickness in that case exceeded the maximum range of both protons and any secondary charged particles, which corresponds to a maximum radiation effect in the insulator. Each simulation run followed 10^4 proton histories.

Results in Figures 1–5 present the most illustrative examples of proton beam effects in the studied insulators for the investigated proton energy range. Vacancy depth distribution plots, LET versus depth plots, and NIEL versus depth plots were obtained from SRIM outputs, following the procedure outlined in [13].

The rate of vacancy formation can be converted into NIEL using the modified Kinchin-Pease relationship between the number of atomic displacements N_d and the nonionizing energy E_n as follows:

$$N_d = 0.8\frac{E_n}{2E_d}, \quad (2)$$

where E_d is the threshold energy for atomic displacement. Equation (2) applies for $E_n > 2.5E_d$ [13–15].

Distributions of atomic vacancies, created within the insulating films by the NIEL of both protons and recoils, are shown in total (curves denoted by "Total vacancies" in the graphs), but also for each kind of displaced atom separately (e.g., curves denoted by "Silicon vacancies" or "Oxygen vacancies"). Curves of total vacancy concentration along the depth of the insulating film are, naturally, always the top ones in these graphs.

4. Analysis of the Results

Thin insulating films investigated herein are immune to the passage of protons with energies exceeding 10 MeV. Simulation results (omitted from this paper) have shown that proton beams with 10 MeV energy experience inconsiderable broadening (fanning) for the considered range of insulating film thickness (from 0.4 μm to 20 μm). For these high energy protons, NIEL is also negligible, while LET is at least one order of magnitude lower than for 1 MeV protons.

Of the four investigated insulators, protons went to larger penetration depths in silicon dioxide and lexan than in the other two materials. In SiO_2, there is a somewhat greater concentration of silicon vacancies than oxygen ones, despite the 2 : 1 stoichiometric ratio of oxygen to silicon atoms, which is attributed to the threshold displacement energy of Si being considerably lower ($E_{d\text{Si}}$ = 15 eV, $E_{d\text{O}}$ = 28 eV). Ionizing energy losses (LET) generally dominate NIEL by 1 to 3 orders of magnitude. This difference becomes larger as the energy of the proton beam is increased, which means that it subsides

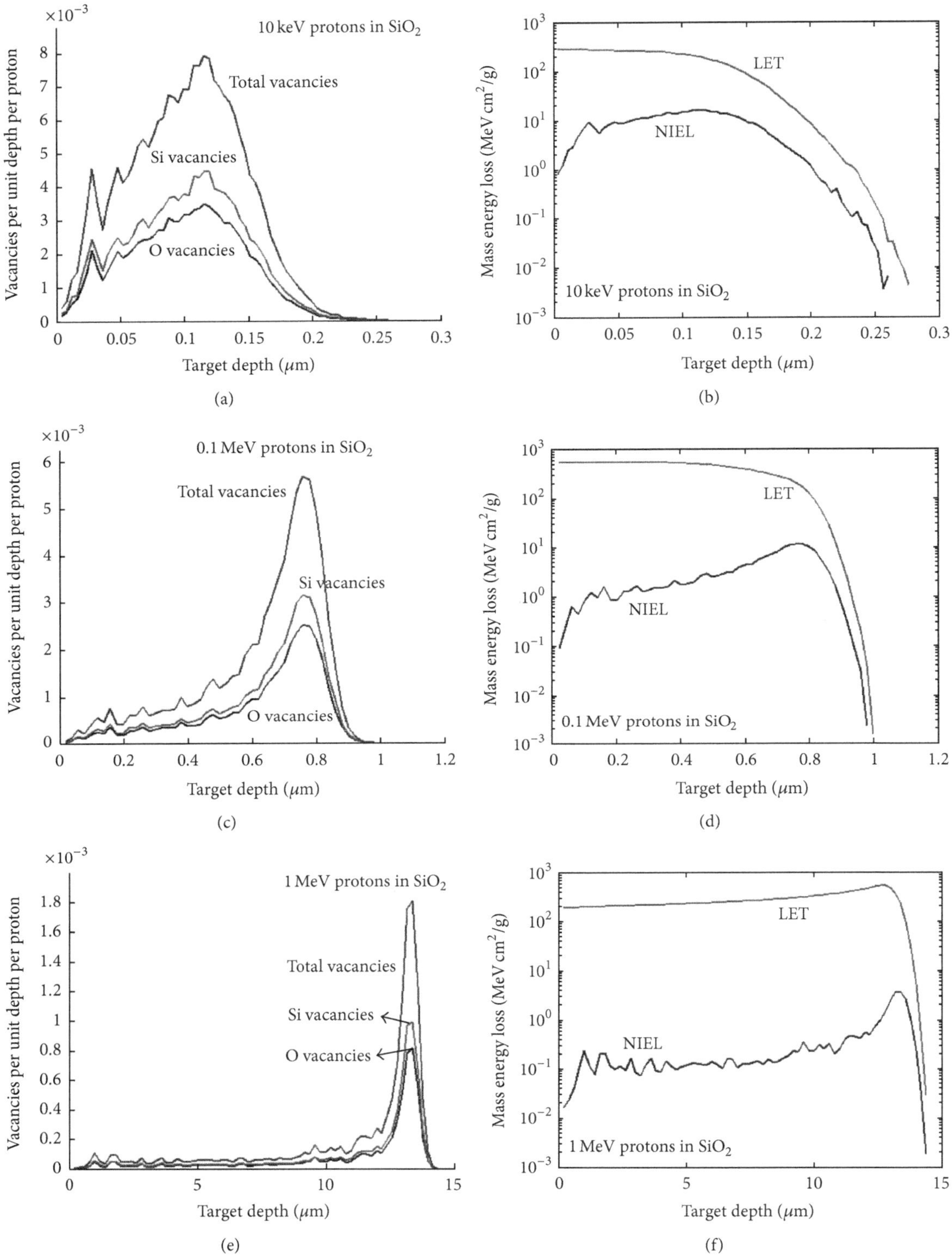

FIGURE 1: Plots of vacancy concentrations and mass energy losses (LET and NIEL divided by material density) along the depth of a silicon dioxide (SiO_2) insulating film. Results were obtained from simulations of 10^4 proton histories, which included subcascades of recoiling atoms.

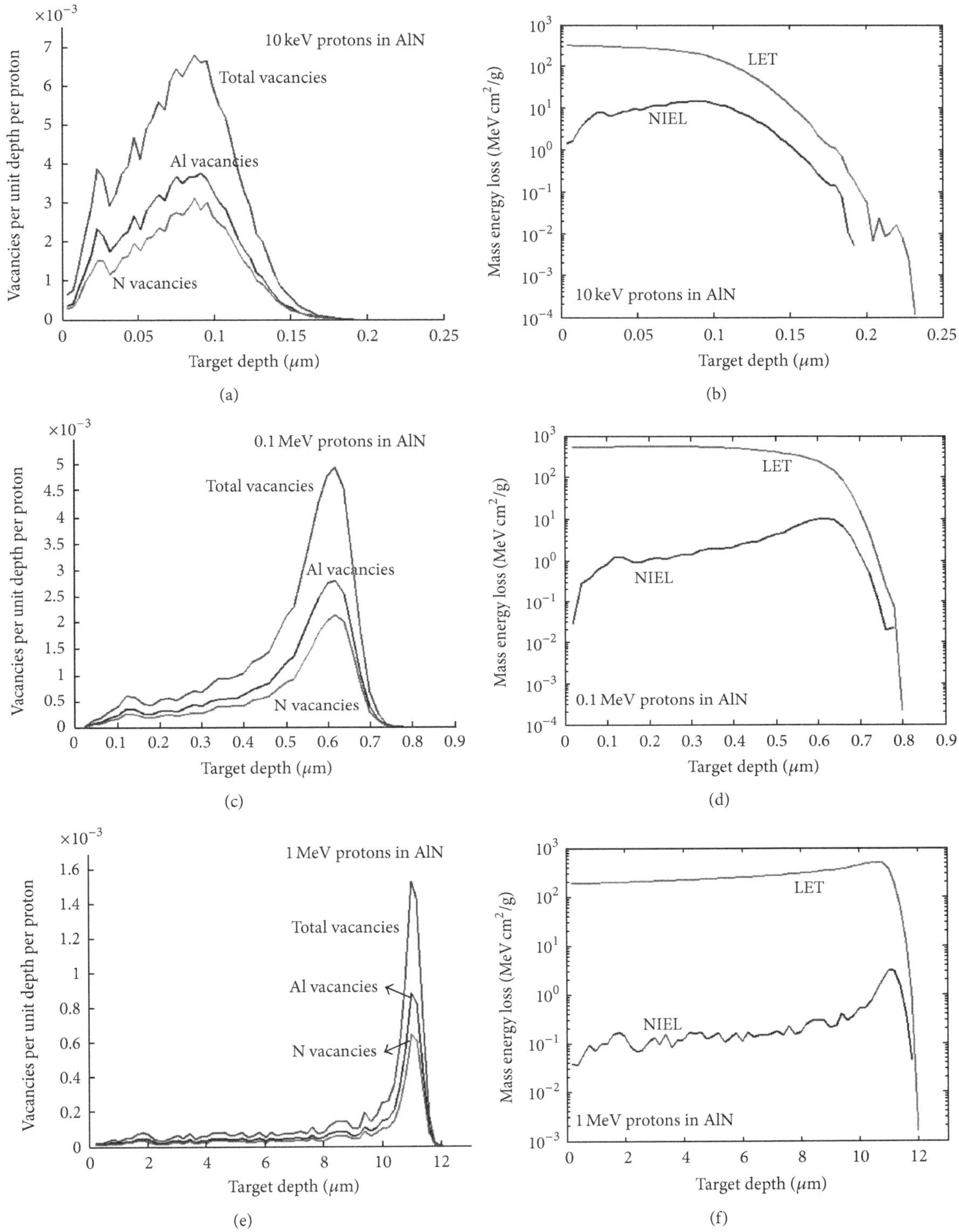

FIGURE 2: Plots of vacancy concentrations and mass energy losses (LET and NIEL divided by material density) along the depth of an aluminium nitride (AlN) insulating film. Results were obtained from simulations of 10^4 proton histories, which included subcascades of recoiling atoms.

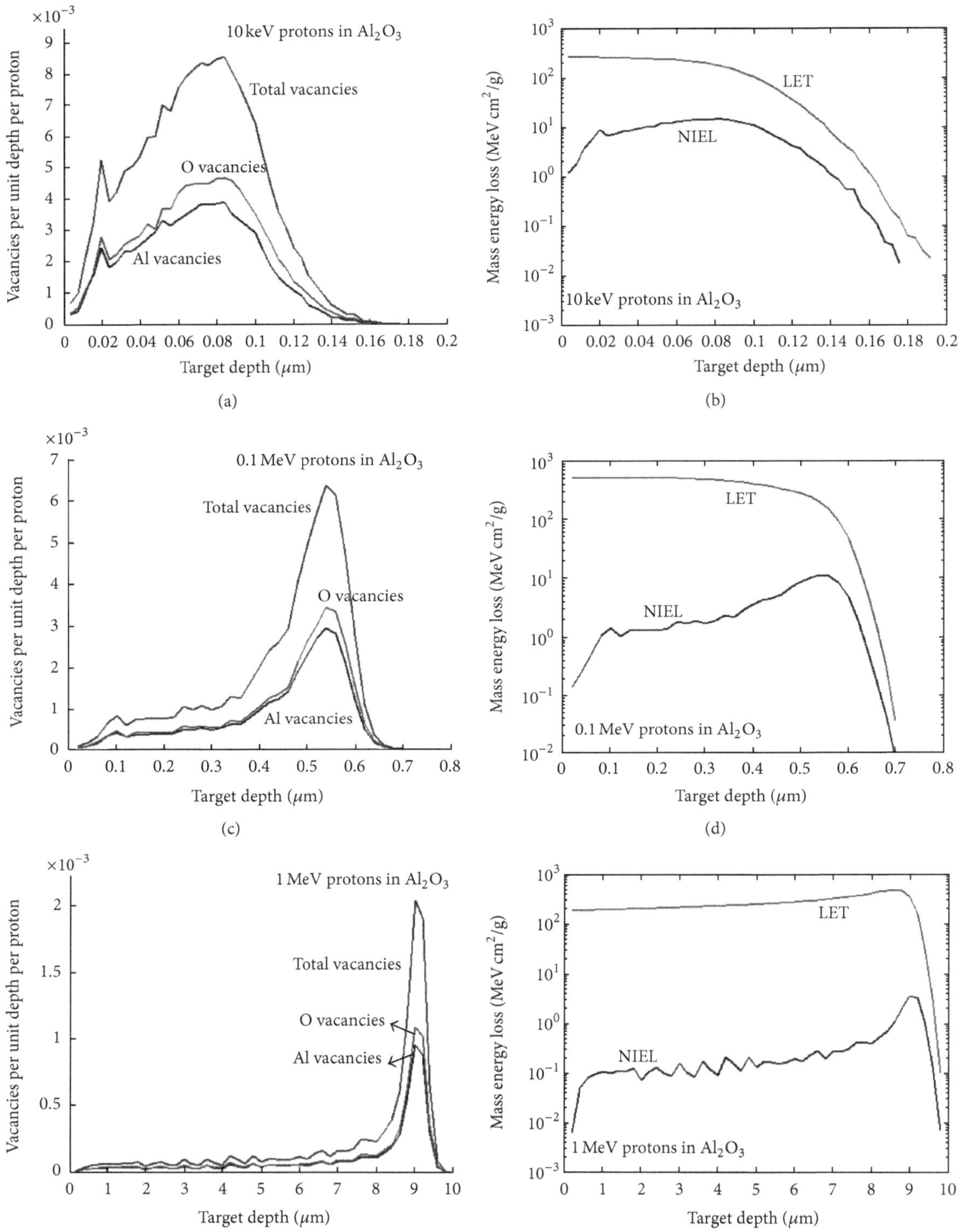

FIGURE 3: Plots of vacancy concentrations and mass energy losses (LET and NIEL divided by material density) along the depth of an alumina (Al_2O_3) insulating film. Results were obtained from simulations of 10^4 proton histories, which included subcascades of recoiling atoms.

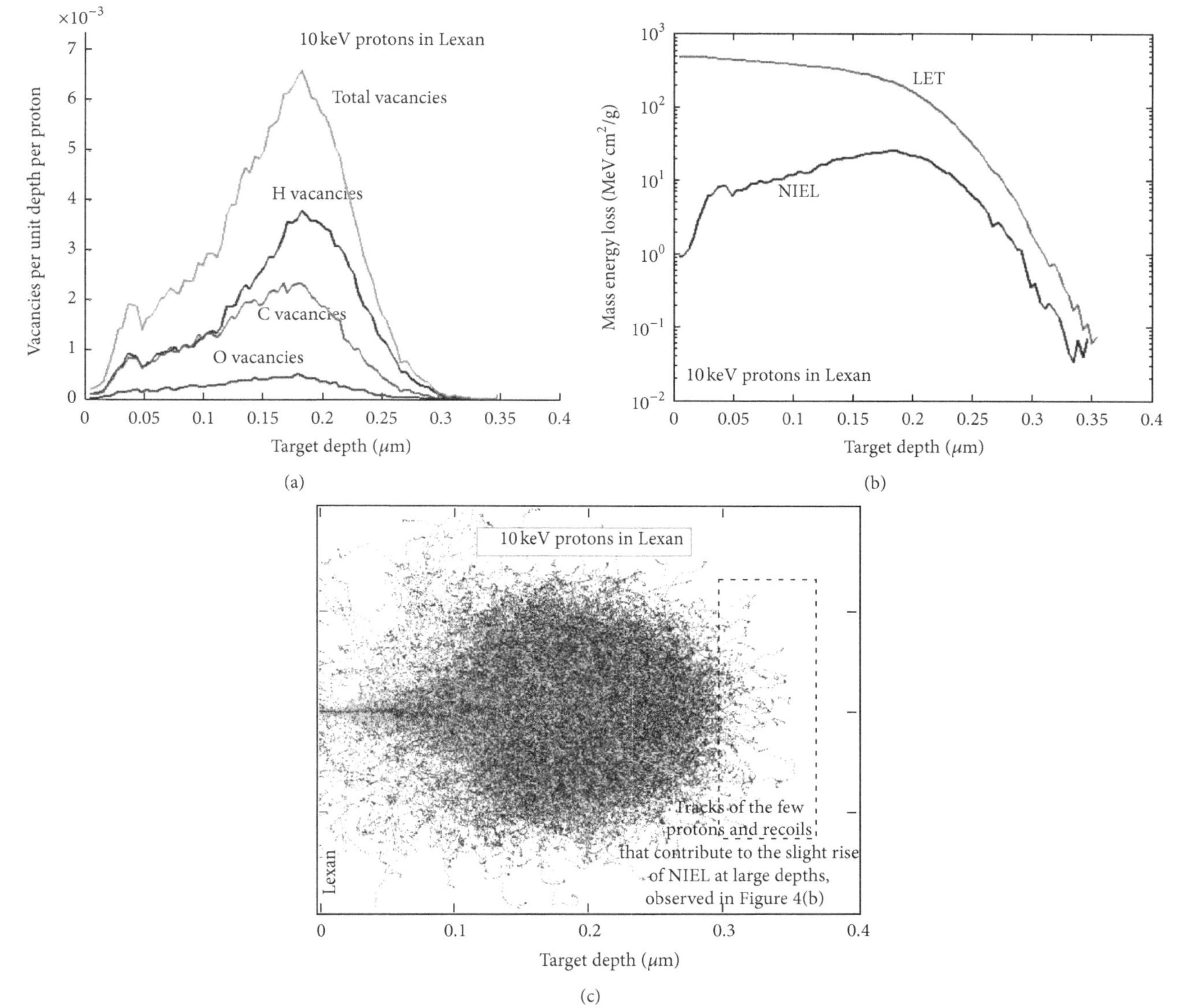

FIGURE 4: Plots of (a) vacancy concentrations and (b) mass energy losses (LET and NIEL divided by material density) along the depth of a 0.4 μm thick lexan (polycarbonate) film. (c) Tracks of protons and recoiling atoms. Results were obtained from simulations of 10^4 proton histories, for a 10 keV incident proton beam.

towards the ends of proton tracks, where nuclear elastic scatterings that give rise to atomic displacements become more probable than electronic collisional events.

The slight increase of NIEL seen at the rightmost part of the lower curve in Figure 4(b) is due to the few protons that reach penetration depths close to 0.35 μm, larger than the mean beam range, and also due to the recoils (mainly hydrogen) that these protons displace. Although the number of such protons is small, they arrive at these large depths without losing much of the original energy in electronic collisions, as most other protons from the beam do. The resulting rise of the NIEL curve is small, but still noticeable in the logarithmic scale. Tracks of protons and recoils that contribute to this rise of NIEL are marked in Figure 4(c).

The energy loss versus depth plot begins to take the classic shape of a Bragg curve only at proton energies >0.5 MeV. As proton energy is increased from ~100 keV, it is first the NIEL curve that starts exhibiting a peak near the maximum penetration depth, as seen in Figures 1(d), 2(d), and 3(d). Around 1 MeV proton energy, both LET and NIEL curves peak toward the end of the penetration depth, giving shape to the Bragg peak, observed much more clearly when the scale of the mass energy loss axis is linear, and not logarithmic. This is illustrated by the plot in Figure 6 for 1 MeV protons in SiO_2.

If one cascade creates more than 8000 recoil atoms, SRIM discards the atoms beyond 8000, which causes certain inaccuracy in vacancy calculations. During our simulations, this happened for protons with energies higher than ~1 MeV

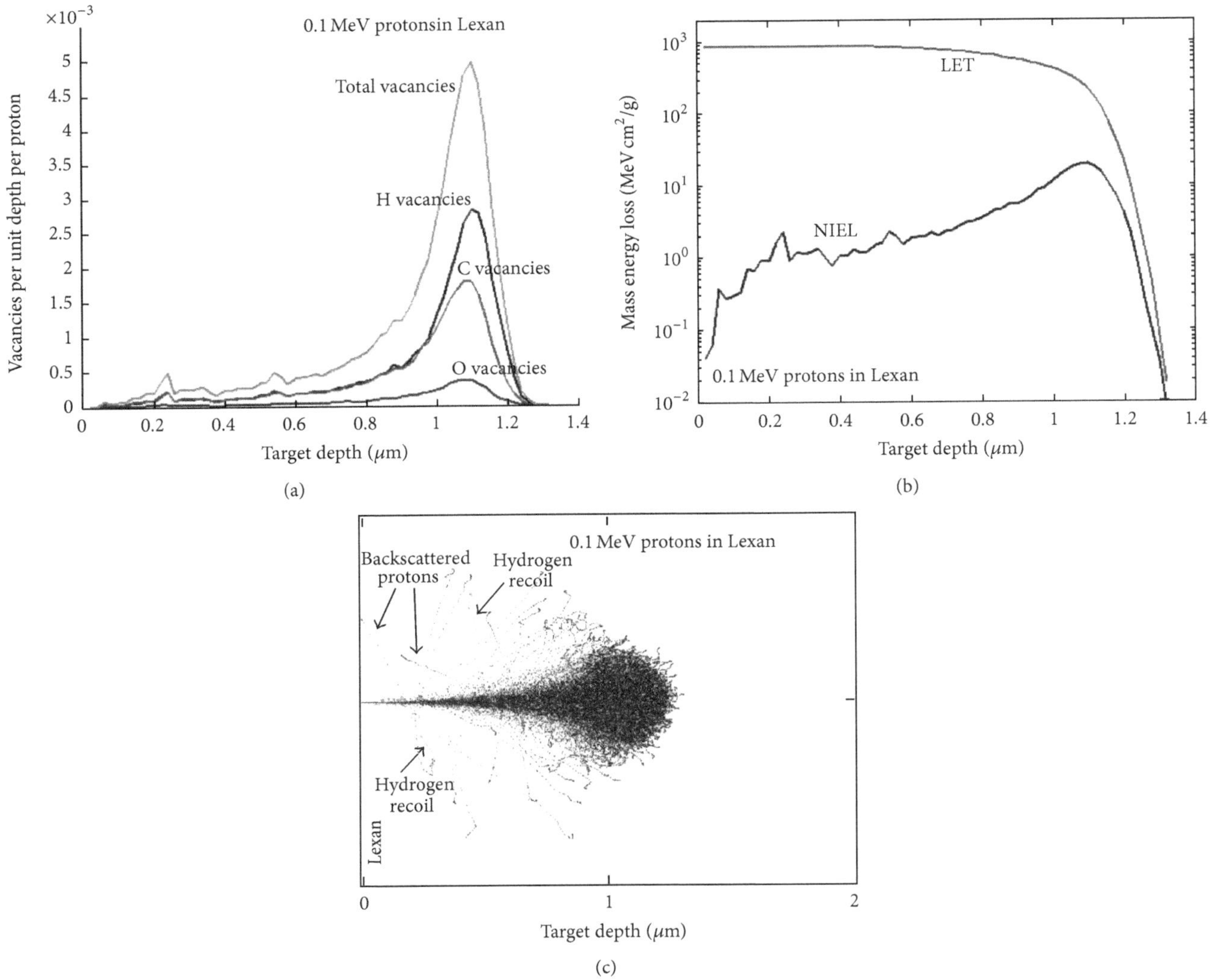

FIGURE 5: Plots of (a) vacancy concentrations and (b) mass energy losses (LET and NIEL divided by material density) along the depth of a 0.2 μm thick lexan (polycarbonate) film. (c) Tracks of protons and recoiling atoms. Results were obtained from simulations of 10^4 proton histories, for a 0.1 MeV incident proton beam.

in lexan films. In a hydrogen-rich medium such as lexan, high energy incident protons give rise to a large number of energetic hydrogen knockons, which results in a multitude of highly branched cascades.

Figure 5(c) clearly demonstrates large-angle backscattering of protons and long tracks of energetic hydrogen recoils in lexan.

Chemical structure unit of polycarbonate is presented in Figure 7. It contains 14 atoms of hydrogen, 16 of carbon, and 3 of oxygen. The stoichiometric ratio of the three elements, along with their threshold displacement energies (E_{dH} = 10 eV, $E_{dC} = E_{dO}$ = 28 eV), accounts for the observed ratios of vacancy concentration curves seen in Figures 4(a) and 5(a).

Secondary electrons, produced by protons through ionization energy losses, further interact with polymer macromolecules, causing their ionization and excitation. Relaxation of excited molecules and locally formed ionization clusters results in a formation of large amounts of free radicals. Highly reactive free radicals cause destruction of polymer chains, by either chain scission (random rupturing of bonds) or cross-linking (formation of large three-dimensional molecular networks). As a result of chain scission, low-molecular-weight fragments, gas evolution, and unsaturated bonds may appear [16].

Many important physical and chemical properties of polymers can be modified by irradiation. Among these are molecular weight, chain length, entanglement, polydispersity, branching, and chain termination. These structural changes also affect the electrical insulating properties of polycarbonate films [17, 18].

Radiation effects in thin insulating layers are relevant due to the increasing miniaturization of electronic components. Miniaturization makes these components sensitive to primary and secondary cosmic rays, which also present one of

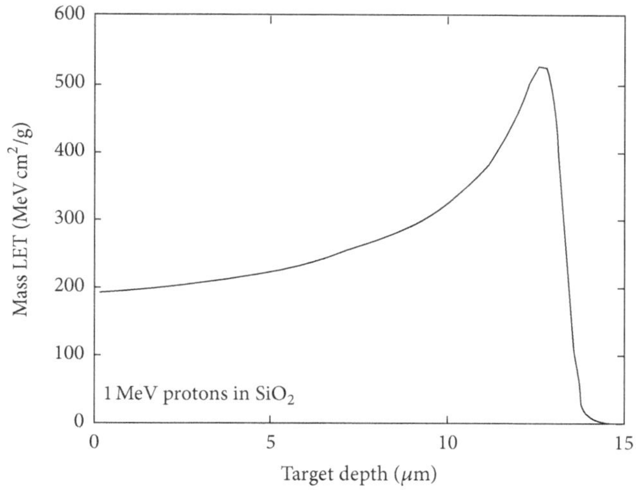

FIGURE 6: Plot of the mass LET for 1 MeV protons along the depth of a 15 μm thick silicon dioxide film. The scale of the energy loss axis is linear, in contrast to Figure 1(f) where it is logarithmic.

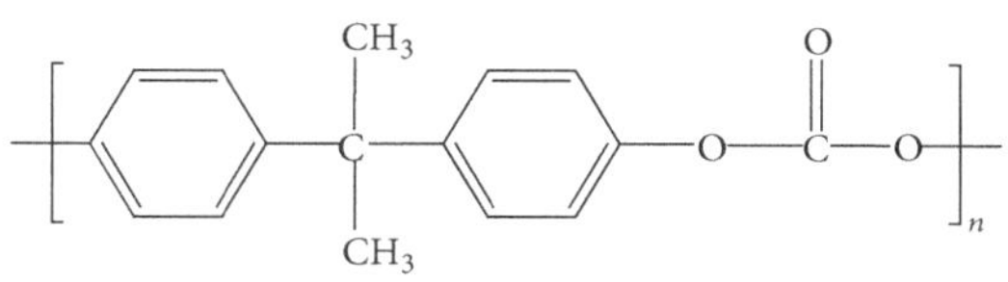

FIGURE 7: Chemical structure unit of polycarbonate, containing 14 atoms of hydrogen, 16 of carbon, and 3 of oxygen.

the most severe obstacles for a widespread use of photovoltaic cells [19–22].

5. Conclusion

Monte Carlo simulations of proton transport through thin films of SiO_2, AlN, Al_2O_3 and polycarbonate have shown that the investigated films are immune to the passage of protons with energies higher than ~10 MeV. Nonionizing energy loss of these high energy protons is low, and they traverse the films without much atomic displacement. In the lower part of the investigated proton energy range (from 10 keV to 1 MeV), however, substantial ionization losses and NIEL are to be expected. Ionization and displacement damage produced by protons could influence the properties of these insulators and compromise their reliability within complex structures and devices. Point defects, some of which are charge-carrier donors, arise in irradiated insulators as a result of atomic displacements. Highly reactive free radicals that can appear in irradiated lexan cause chain scission and/or cross-linking, which then affects the insulating properties of polycarbonate films.

Acknowledgment

The Ministry of Education and Science of the Republic of Serbia supported this work under Contract 171007.

References

[1] A. G. Holmes-Siedle and L. Adams, *Handbook of Radiation Effects*, Oxford University Press, 2nd edition, 2002.

[2] J. R. Schwank, M. R. Shaneyfelt, D. M. Fleetwood et al., "Radiation effects in MOS oxides," *IEEE Transactions on Nuclear Science*, vol. 55, no. 4, pp. 1833–1853, 2008.

[3] S. Stanković, B. Iričanin, D. Nikolić et al., "MSV signal processing system for neutron-gamma discrimination in a mixed field," *Nuclear Technology & Radiation Protection*, vol. 27, no. 2, pp. 165–170, 2012.

[4] I. Knezevic, N. Zdjelarevic, M. Obrenovic, and M. Vujisic, "Absorbed dose assessment in particle-beam irradiated metal-oxide and metal-nonmetal memristors," *Nuclear Technology & Radiation Protection*, vol. 27, no. 3, pp. 290–296, 2012.

[5] P. Sigmund, *Stopping of Heavy Ions*, vol. 4 of *Springer Tracts of Modern Physics*, Springer, Berlin, Germany, 2004.

[6] J. H. Warner, S. R. Messenger, R. J. Walters, and G. P. Summers, "Displacement damage correlation of proton and silicon ion radiation in GaAs," *IEEE Transactions on Nuclear Science*, vol. 52, no. 6, pp. 2678–2682, 2005.

[7] H. García, S. Dueñas, H. Castán et al., "Irradiation effect on dielectric properties of hafnium and gadolinium oxide gate dielectrics," *Journal of Vacuum Science and Technology B*, vol. 27, no. 1, pp. 416–420, 2009.

[8] C. Leroy and P.-G. Rancoita, "Particle interaction and displacement damage in silicon devices operated in radiation environments," *Reports on Progress in Physics*, vol. 70, no. 4, pp. 493–625, 2007.

[9] R. Radosavljević and A. Vasić, "Effects of radiation on solar cells as photovoltaic generators," *Nuclear Technology & Radiation Protection*, vol. 27, no. 1, pp. 28–32, 2012.

[10] Đ. Lazarević, M. Vujisić, K. Stanković, E. Dolićanin, and P. Osmokrović, "Radiation hardness of indium oxide films in the cooper-pair insulator state," *Nuclear Technology & Radiation Protection*, vol. 27, no. 1, pp. 40–43, 2012.

[11] E. Dolićanin, "Gamma ray effects on flash memory cell arrays," *Nuclear Technology & Radiation Protection*, vol. 27, no. 3, pp. 284–289, 2012.

[12] J. F. Ziegler, J. P. Biersack, and M. D. Ziegler, "SRIM—The Stopping and Range of Ions in Matter," http://www.srim.org.

[13] S. R. Messenger, E. A. Burke, G. P. Summers et al., "Nonionizing Energy Loss (NIEL) for heavy ions," *IEEE Transactions on Nuclear Science*, vol. 46, pp. 1595–1602, 1999.

[14] D. Nikolić, A. Vasić, I. Fetahović, K. Stanković, and P. Osmokrović, "Photodiode behavior in radiation environment," *Applied Mathematics, Informatics & Mechanics*, vol. 3, no. 1, pp. 27–34, 2011.

[15] M. Vujisić, N. Marjanović, I. Fetahović, K. Stanković, and P. Osmokrović, "Influence of radiation on titanium dioxide mermistors," *Applied Mathematics, Informatics & Mechanics*, vol. 4, no. 1, pp. 75–82, 2012.

[16] "Polymer durability and radiation effects," in *Proceedings of the Acs Symposium*, M. C. Celina and R. A. Assink, Eds., series 978, 2007.

[17] M. P. Petkov, K. G. Lynn, K. P. Rodbell, W. Volksen, and R. D. Miller, "Radiation effects in low dielectric constant methylsilsesquioxane films," *IEEE Transactions on Nuclear Science*, vol. 49, no. 6, pp. 2724–2728, 2002.

[18] T. Iwahashi, T. Tamagawa, A. Hayato et al., "Heavy ion and proton irradiation of gas electron multipliers with liquid crystal

polymer insulator: evaluation tests for use in space," *IEEE Transactions on Nuclear Science*, vol. 58, no. 5, pp. 2401–2406, 2011.

[19] X. L. Han, H. Ni, K. Zhao, Y.-C. Kong, H. K. Wong, and S. Q. Zhao, "Load-resistance- and voltage-tunable photovoltaic effect in tilting manganite films," *International Journal of Photoenergy*, vol. 2011, Article ID 314085, 2011.

[20] Z. Hu, B. Jiao, J. Zhang, X. Zhang, and Y. Zhao, "Indium-doped zinc oxide thin films as effective anodes of organic photovoltaic devices," *International Journal of Photoenergy*, vol. 2011, Article ID 158065, 5 pages, 2011.

[21] T. T. Chow, G. N. Tiwari, and C. Menezo, "Hybrid Solar: a review on photovoltaic and thermal power integration," *International Journal of Photoenergy*, vol. 2012, Article ID 307287, 17 pages, 2012.

[22] P. Ooshaksaraei, K. Sopian, R. Zulkifli, M. A. Alghoul, and S. H. Zaidi, "Characterization of a bifacial photovoltaic panel integrated with external diffuse and semimirror type reflectors," *International Journal of Photoenergy*, vol. 2013, Article ID 465837, 7 pages, 2013.

11

Multi-Input Converter with MPPT Feature for Wind-PV Power Generation System

Chih-Lung Shen and Shih-Hsueh Yang

Department of Electronic Engineering, National Kaohsiung First University of Science and Technology, Nanzih, Kaohsiung 811, Taiwan

Correspondence should be addressed to Chih-Lung Shen; clshen@ccms.nkfust.edu.tw

Academic Editor: Bhushan Sopori

A multi-input converter (MIC) to process wind-PV power is proposed, designed, analyzed, simulated, and implemented. The MIC cannot only process solar energy but deal with wind power, of which structure is derived from forward-type DC/DC converter to step-down/up voltage for charger systems, DC distribution applications, or grid connection. The MIC comprises an upper modified double-ended forward, a lower modified double-ended forward, a common output inductor, and a DSP-based system controller. The two modified double-ended forwards can operate individually or simultaneously so as to accommodate the variation of the hybrid renewable energy under different atmospheric conditions. While the MIC operates at interleaving mode, better performance can be achieved and volume also is reduced. The proposed MIC is capable of recycling the energy stored in the leakage inductance and obtaining high step-up output voltage. In order to draw maximum power from wind turbine and PV panel, perturb-and-observe method is adopted to achieve maximum power point tracking (MPPT) feature. The MIC is constructed, analyzed, simulated, and tested. Simulations and hardware measurements have demonstrated the feasibility and functionality of the proposed multi-input converter.

1. Introduction

Conventionally, electric power is mainly generated from fossil fuels. However, this kind of energy resources is highly limited and will be exhausted in the near future. With the rapid requirement of electricity and the increase of worse and worse energy crisis, it is of great urgency to replace the fossil fuels with renewable energy. Among the renewable resources, solar energy and wind power attract a great deal of interest owing to their easy acquirement.

In photovoltaic (PV) or wind power generation system, a power converter is needed so as to process renewable energy. In literature [1–5] PV converters are presented while wind power converters are studied in [6–10]. However, these power converters only handle a single kind of renewable energy, that is, which cannot deal with multi-input power. Therefore, some researchers propose multi-input converters for solar-wind hybrid power generation system. As shown in Figure 1, a series double-boost converter is presented to process PV power and wind energy simultaneously [11], in which, as compared with single-boost configuration, power component only imposes one-half power rating. Even though this boost-type converter steps up the voltage and is suitable for a high voltage supply, it cannot be applied to galvanic isolated applications. Double-input buck-boost converter shown in Figure 2 is capable of processing high-/low-voltage sources [12]. Like boost-type converter, this type of configuration is still nonisolated electrically. Instead of combining renewable energy in electricity, the concept of magnetic flux additivity is proposed to design a multi-input isolated converter, as shown in Figure 3, but its structure is complex and control low is complicated [13]. In order to simplify power-stage configuration, the forward-derived configuration is proposed. However, it cannot trap the energy in the leakage inductor and is incapable of applying to high output voltage applications [14].

In this paper, a multi-input converter (MIC) is proposed, which can not only deal with PV power and wind-turbine energy simultaneously/individually but recycles the leakage-inductance energy and steps up the input voltage. As compared with the aforementioned isolated double-input converter, the proposed one has a much simpler structure.

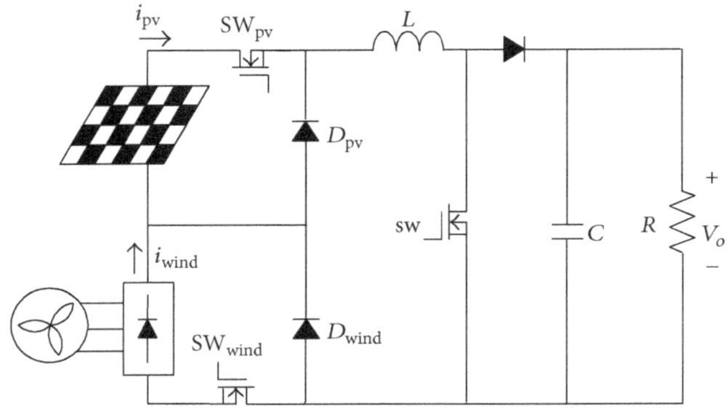

FIGURE 1: An illustration of double-boost converter for wind-PV system.

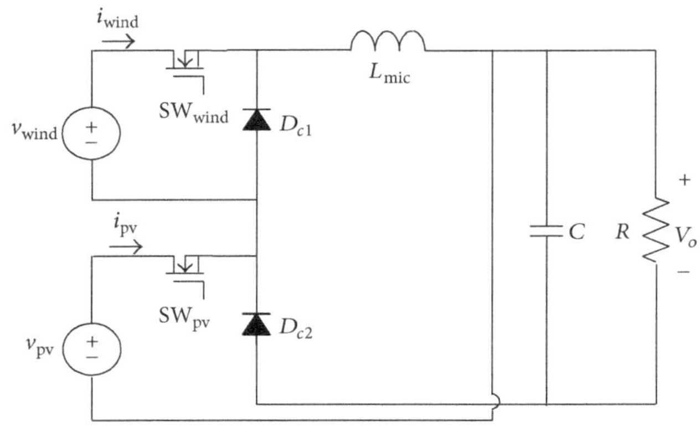

FIGURE 2: An illustration of double-input buck-boost converter.

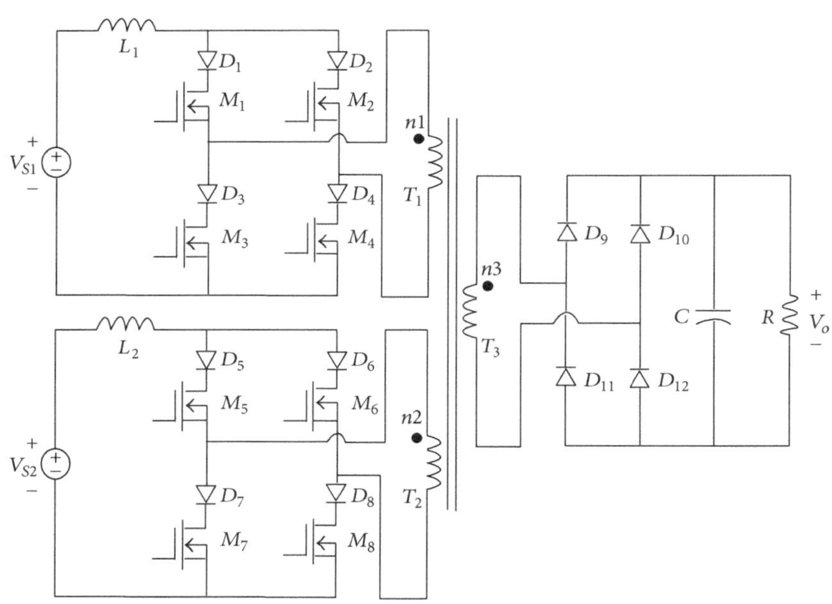

FIGURE 3: An illustration of multi-input converter based on magnetic flux additivity.

Furthermore, the proposed MIC removes the third winding from the conventional forward converter, which releases the energy stored in the magnetizing inductor to a capacitor through the second winding. As a result, its output voltage can be stepped up significantly and the efficiency is improved. Simulated and practical results have validated the proposed PV-wind MIC.

In this paper, system architecture of the proposed MIC is described in Section 2. Section 3 presents the operational principle of the converter, while simulations and practical

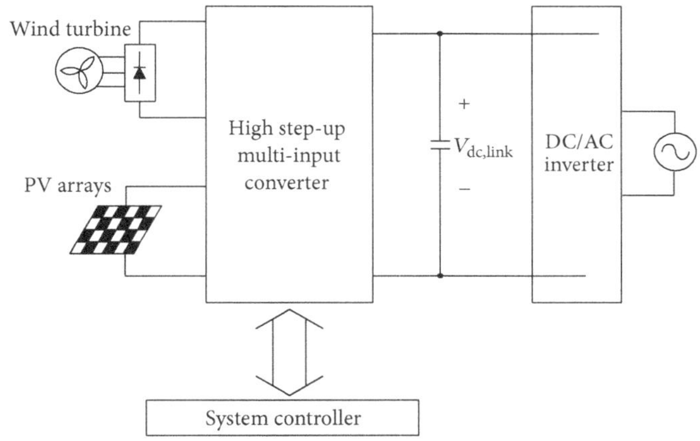

FIGURE 4: A block diagram to represent the configuration of the proposed MIC.

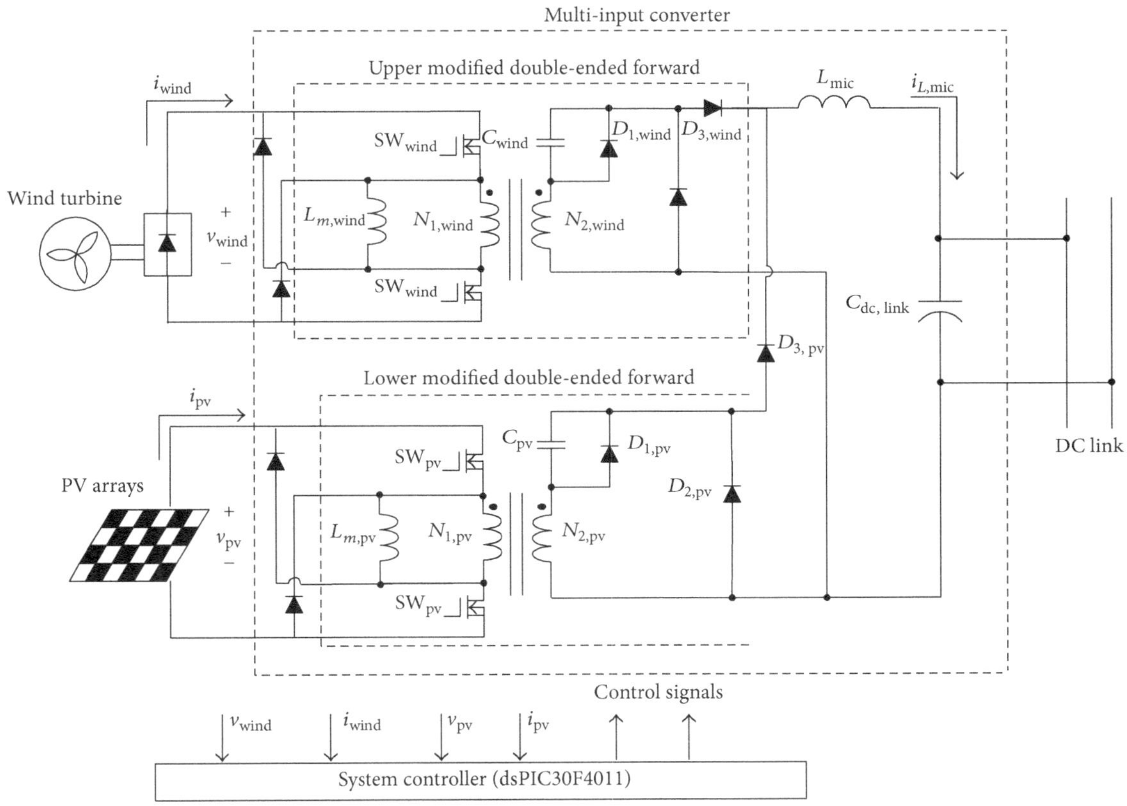

FIGURE 5: The proposed multi-input converter to process PV-wind power.

measurements to verify the feasibility of the MIC are shown in Section 4. Finally, conclusion is given in Section 5.

2. Configuration of the Proposed Converter

Figure 4 illustrates the block diagram of the proposed MIC, which mainly includes PV arrays, a wind turbine, a high step-up multi-input converter, and a system controller. Figure 5 shows the corresponding schematics of the main power stage of the MIC. In Figure 5, the MIC is composed of an upper modified double-ended forward, a lower modified double-ended forward, a common output filter L_{mic}, and a system controller. The upper modified double-ended forward is in charge of dealing with wind-turbine energy while the lower one processes solar power. The both modified double-ended forwards can be operated independently, which expands the degree of control freedom. The active switches in the upper forward or in the lower one are switched synchronously so as to trap the leakage energy and to release the magnetizing inductance energy. The capacitors in the secondary windings, C_{wind} and C_{pv}, will absorb the energy of magnetizing

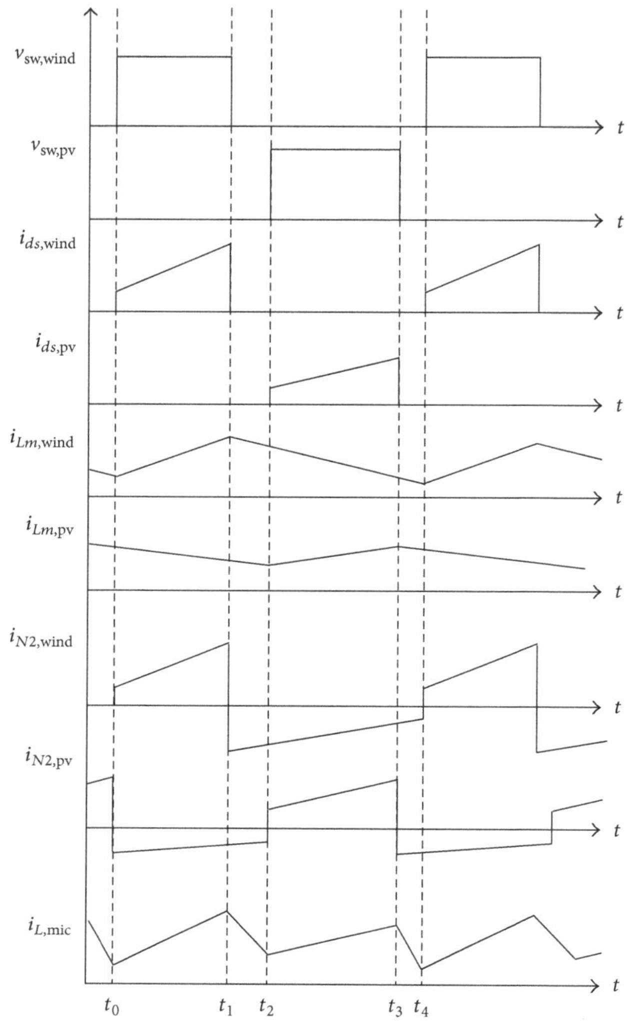

FIGURE 6: Conceptual key waveforms of the proposed MIC.

inductance and then can boost the output voltage. The system controller determines the control signals to perform output power controlling and maximum power point tracking (MPPT). In this paper, perturb-and-observe method is adopted for MPPT.

3. Operation Principle of the Proposed Converter

In Figure 5 the two modified double-ended forwards can operate individually to deal with PV power and wind energy. To achieve better output performance and to lower the output filter volume, the switches SW_{wind} and SW_{pv} are turned on alternatively with a duty ratio less than 0.5 at the same switching frequency. Figure 5 shows conceptual key waveforms. According to the conduction status of the switches SW_{wind} and SW_{pv}, the operation of the MIC over one switching period can be mainly divided into four modes. Corresponding equivalents are presented in Figure 6. Each operation mode is described in the following.

Mode 1 (Figure 7(a), $t_0 \le t < t_1$). During the interval of Mode 1, the status of switches SW_{wind} is on but SW_{pv} off. While SW_{wind} conducts at t_0, this mode begins. The active switch SW_{pv} is in the off-state, and the magnetizing inductor of the lower modified forward discharges energy through the path of $N_{2,pv}$-$D_{2,pv}$-C_{pv}. Meanwhile, wind power is forwarded to output. Therefore, $i_{L,mic}$ is linearly built and can be described as

$$i_{L,mic}(t) = \frac{2 \cdot N_{wind} \cdot k_{wind} \cdot v_{wind}}{L_{mic}} \cdot D_{wind} \cdot \frac{T_s}{2} + i_{L,mic}(t_0), \tag{1}$$

where k_{wind} is the coupling coefficient of the transformer in the upper forward, D_{wind} stands for duty ratio of SW_{wind}, and T_s represents the switching period. The voltage stresses of the switch SW_{pv} and diode $D_{2,wind}$ can be found by

$$\begin{aligned} v_{DS,pv} &= v_{pv}, \\ v_{D2,wind} &= 2 \cdot N_{wind} \cdot k_{wind} \cdot v_{wind}. \end{aligned} \tag{2}$$

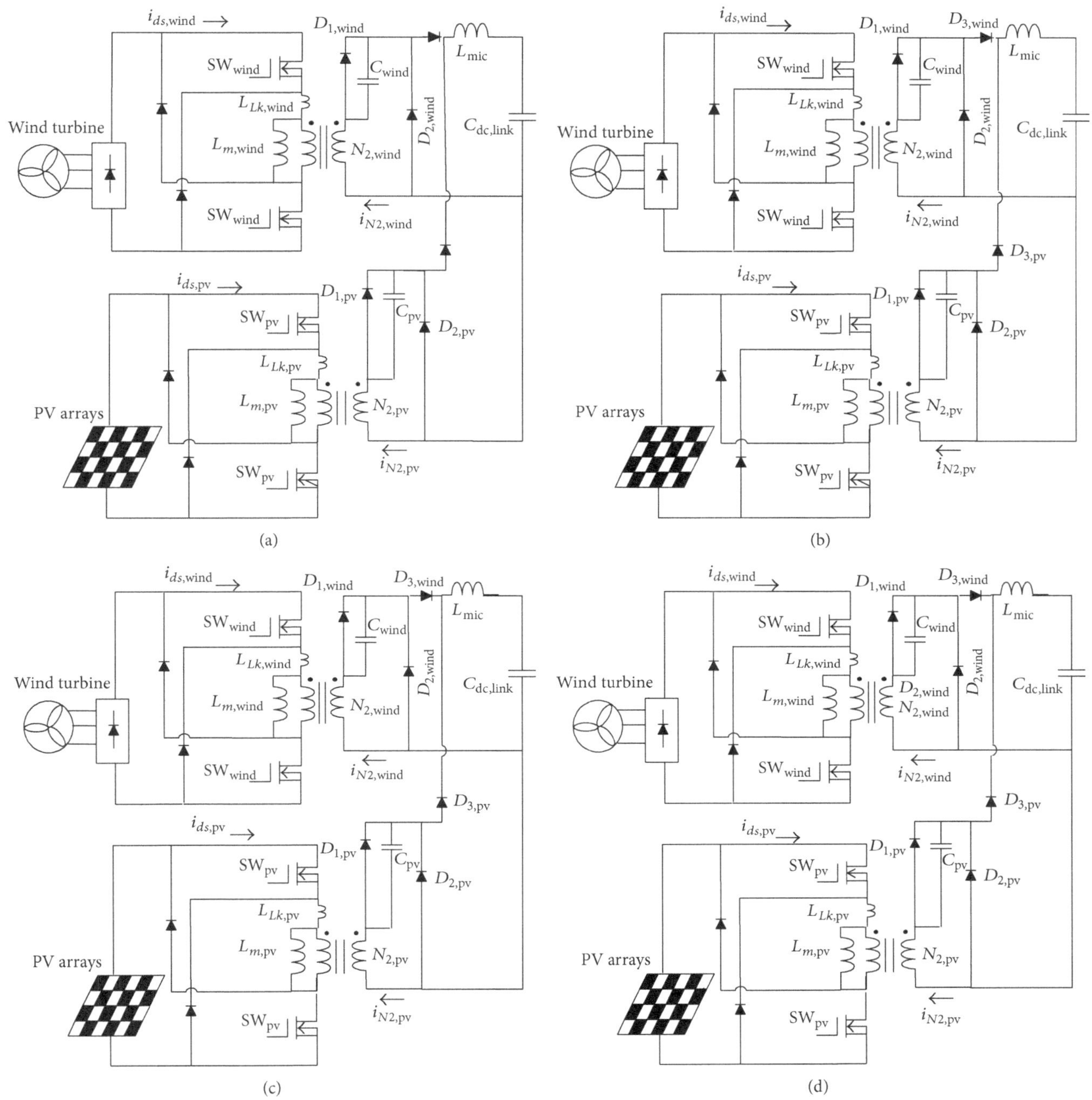

FIGURE 7: Equivalent circuits of the MIC corresponding to the four operation modes over one switching cycle: (a) Mode 1, (b) Mode 2, (c) Mode 3, and (d) Mode 4.

As the current of magnetizing inductor $L_{m,\mathrm{pv}}$ drops to zero, this mode ends.

Mode 2 (Figure 7(b), $t_1 \le t < t_2$). In Mode 2, all the active switches are in off-state. At time t_1, the switch $\mathrm{SW}_{\mathrm{wind}}$ is turned off and $\mathrm{SW}_{\mathrm{pv}}$ in the lower forward still stays in the off-state. The magnetizing inductor in the upper modified forward $L_{m,\mathrm{wind}}$ discharges via the path of $N_{2,\mathrm{wind}}$-$D_{2,\mathrm{wind}}$-C_{wind}, while the energy of leakage inductor $L_{LK,\mathrm{wind}}$ is trapped. The voltages across C_{wind} and C_{pv}, $v_{c,\mathrm{wind}}$ and $v_{c,\mathrm{pv}}$, are obtained by

$$\begin{aligned} v_{c,\mathrm{wind}} &= N_{\mathrm{wind}} \cdot k_{\mathrm{wind}} \cdot v_{\mathrm{wind}}, \\ v_{c,\mathrm{pv}} &= N_{\mathrm{pv}} \cdot k_{\mathrm{pv}} \cdot v_{\mathrm{pv}}, \end{aligned} \tag{3}$$

respectively, in which k_{pv} is the coupling coefficient of the transformer in the lower forward. In addition, the output inductor L_{mic} releases the stored energy to the load by

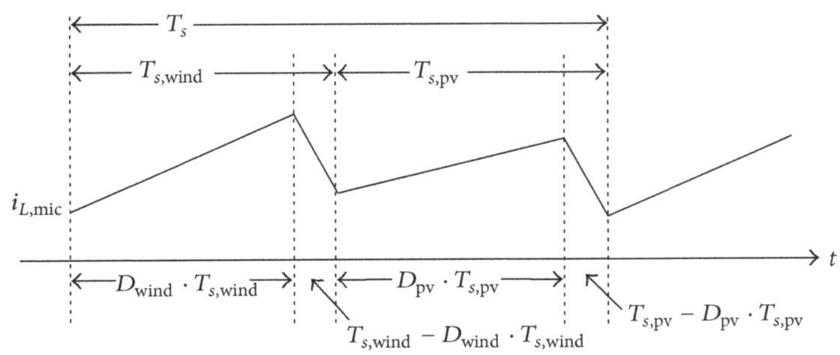

FIGURE 8: The waveform of output inductor current.

the path of L_{mic}-$C_{\text{dc,link}}$-$D_{2,\text{wind}}$, of which the current is linearly decreased and is expressed as

$$i_{L,\text{mic}}(t) = i_{L,\text{mic}}(t_1) - \frac{V_{\text{dc,link}}}{L_{\text{mic}}} \cdot D_{\text{wind}} \cdot \frac{T_s}{2} \cdot (1 - D_{\text{wind}}). \quad (4)$$

Mode 3 (Figure 7(c), $t_2 \leq t < t_3$). During this mode, SW_{pv} is in the on-state but SW_{wind} in off-state. Since the SW_{pv} is turned on at t_2, thus PV energy is dealt with by the lower modified forward. The inductor current $i_{L,\text{mic}}$ increases linearly. The inductor $L_{m,\text{wind}}$ releases the energy to the capacitor C_{wind} through the path of $N_{2,\text{wind}}$-$D_{2,\text{wind}}$-C_{wind}. In this interval, the current of output inductor L_{mic} is linearly built and can be described as

$$i_{L,\text{mic}}(t) = \frac{2 \cdot N_{\text{pv}} \cdot k_{\text{pv}} \cdot v_{\text{pv}}}{L_{\text{mic}}} \cdot D_{\text{pv}} \cdot \frac{T_s}{2} + i_{L,\text{mic}}(t_2), \quad (5)$$

where D_{pv} denotes the duty ratio of SW_{pv}. The voltage stresses of the switch SW_{pv} and diode $D_{2,\text{wind}}$ can be expressed as follows:

$$v_{DS,\text{wind}} = v_{\text{wind}},$$
$$v_{D2,\text{pv}} = 2 \cdot N_{\text{pv}} \cdot k_{\text{pv}} \cdot v_{\text{pv}}. \quad (6)$$

This mode ends at the moment the current flowing through $L_{m,\text{wind}}$ equals zero.

Mode 4 (Figure 7(d), $t_3 \leq t < t_4$). At time t_3, the switch SW_{pv} is turned off and the operation of the MIC enters into Mode 4. That is, in Mode 4 all active switches are off. During this mode, the magnetizing inductor $L_{m,\text{pv}}$ releases energy to capacitor C_{pv} via $N_{2,\text{pv}}$, $D_{2,\text{pv}}$, and C_{pv}. In addition, the energy stored in leakage inductance is recycled. Meanwhile, the output inductor discharges and the current $i_{L,\text{mic}}$ decreases linearly, which can be expressed as

$$i_{L,\text{mic}}(t) = i_{L,\text{mic}}(t_3) - \frac{V_{\text{dc,link}}}{L_{\text{mic}}} \cdot D_{\text{pv}} \cdot \frac{T_s}{2} \cdot (1 - D_{\text{pv}}). \quad (7)$$

A complete switching cycle is terminated at $t = t_4$, at which SW_{wind} is turned on again.

While the proposed MIC operates in continuous conduction mode (CCM), the corresponding waveform of output inductor current is illustrated in Figure 8. The T_s is the switching period and can be expressed as

$$T_{s,\text{wind}} + T_{s,\text{pv}} = T_s. \quad (8)$$

In (8), $T_{s,\text{wind}}$ stands for the intervals that the upper modified forward works, while $T_{s,\text{pv}}$ for the lower modified forward. In the interleaved operation, the following relationship holds:

$$T_{s,\text{pv}} = T_{s,\text{wind}} = \frac{T_s}{2}. \quad (9)$$

Based on volt-second balance criterion, one can obtain the following identity:

$$(2 \cdot N_{\text{wind}} \cdot k_{\text{wind}} \cdot v_{\text{wind}} - V_{\text{dc,link}}) \cdot D_{\text{wind}} \cdot T_{s,\text{wind}} + (2 \cdot N_{\text{pv}} \cdot k_{\text{pv}} \cdot v_{\text{pv}} - V_{\text{dc,link}}) \cdot D_{\text{pv}} \cdot T_{s,\text{pv}} = (1 - D_{\text{wind}}) \cdot T_{s,\text{wind}} \cdot V_{\text{dc,link}} + (1 - D_{\text{pv}}) \cdot T_{s,\text{pv}} \cdot V_{\text{dc,link}}. \quad (10)$$

Rearranging (10) yields

$$V_{\text{dc,link}} = N_{\text{wind}} \cdot k_{\text{wind}} \cdot v_{\text{wind}} \cdot D_{\text{wind}} + N_{\text{pv}} \cdot k_{\text{pv}} \cdot v_{\text{pv}} \cdot D_{\text{pv}}, \quad (11)$$

which reveals that the DC-link voltage can be controlled by the duty ratios of SW_{wind} and SW_{pv}.

In the MIC, perturb-and-observe method is employed to draw maximum power from wind turbine and PV arrays since it is easy to carry out. The perturb-and-observe MPPT is realized by dsPIC30F4011. The related flowchart is presented in Figure 9.

4. Simulations and Practical Measurements

To demonstrate the feasibility of the proposed MIC, a prototype is constructed, simulated, and tested. Important parameters are listed as follows:

(i) PV panel voltage: $v_{\text{pv}} = 40$–56 V,
(ii) wind turbine voltage: $v_{\text{wind}} = 22$–48 V,
(iii) dc-link voltage: 200 V,
(iv) upper modified double-ended forward: 350 W,
(v) lower modified double-ended forward: 350 W,
(vi) switching frequency: 50 kHz for all active switches,
(vii) output inductance: $L_{\text{mic}} = 400\,\mu$H,
(viii) output capacitance: $C_{\text{dc,link}} = 400\,\mu$F,

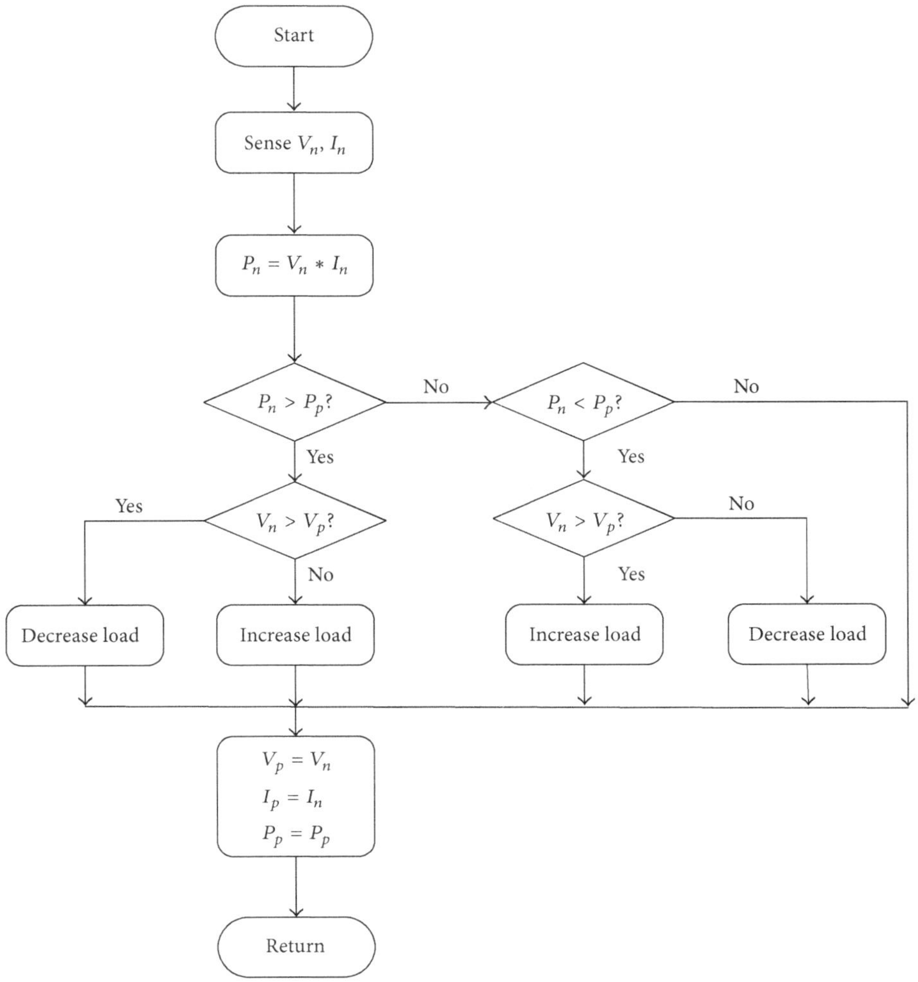

FIGURE 9: The flowchart of perturb-and-observe method to achieve MPPT feature.

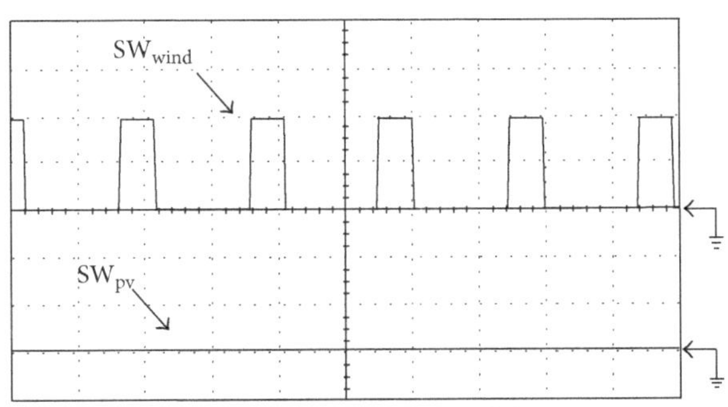

FIGURE 10: Simulated control signals while wind turbine provides 350 W.

(ix) capacitance of upper forward: $C_{wind} = 22\,\mu$F,

(x) capacitance of lower forward: $C_{pv} = 22\,\mu$F.

In the case of only wind turbine providing 350 W, the simulated active switch signals and corresponding output inductor current are shown in Figures 10 and 11, respectively, while Figures 12 and 13 are the practical measurements. With the perturb-and-observe method for maximum power point tracking, the measured result is shown in Figure 14, which has illustrated that the upper modified double-ended forward can draw the maximum power from wind turbine. If only 350 W PV power feeds the MIC, simulations of control signals for active switches and output inductor current are presented in Figures 15 and 16. In addition, Figures 17 and 18 are the

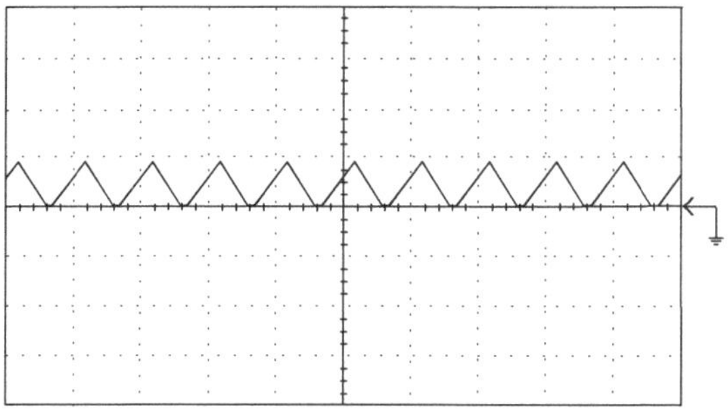

(Current: 5 A/div, time: 20 μs/div)

FIGURE 11: Simulated output inductor current while wind turbine provides 350 W.

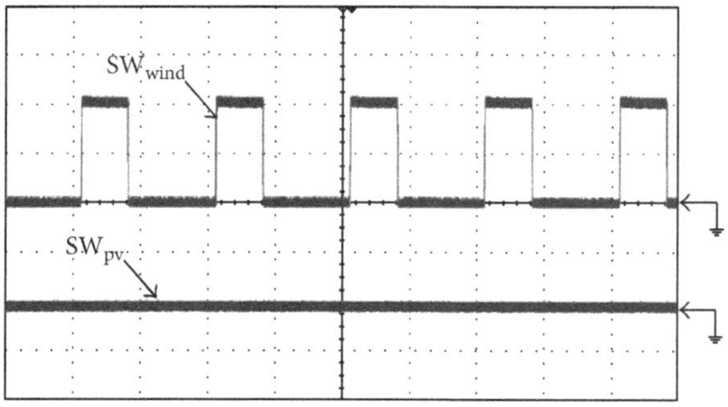

(Voltage: 5 V/div, time: 10 μs/div)

FIGURE 12: Practical measurements of control signals while wind turbine provides 350 W.

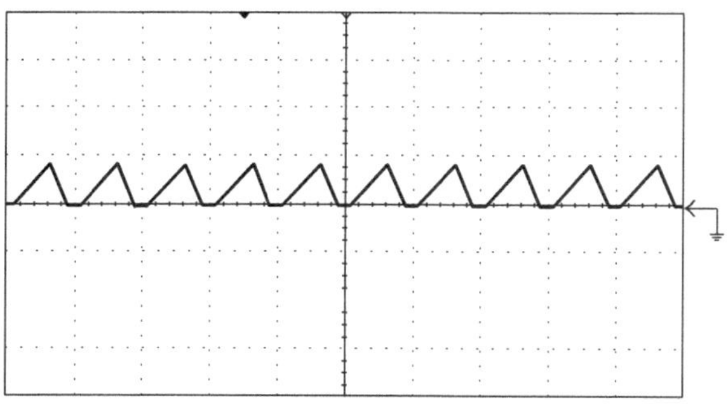

(Current: 5 A/div, time: 20 μs/div)

FIGURE 13: Measured output inductor current while wind turbine provides 350 W.

measured results. For MPPT feature verification, Figure 19 is the relationship between PV power and the terminal voltage of PV panel, from which it can be found that the proposed MIC is able to draw maximum power from PV panel. From Figures 10–19, it has been demonstrated that the proposed MIC is capable of dealing with individual renewable power. As the solar power and wind energy feed the MIC simultaneously, Figures 20 and 21 show the simulations of control signals and output inductor current. Then, Figures 22 and 23 present the hardware measurements. In addition, switch currents are also shown in Figure 24, in which the upper trace and the lower trace are the drain-to-source currents of SW_{wind} and SW_{pv}, respectively. Figure 25 is the hardware measurements of the secondary currents $i_{N2,wind}$ and $i_{N2,pv}$. All the experimental results correspond with the theoretical waveforms in Figure 6. From Figures 20–25, it is verified that the MIC not only can process hybrid wind-PV power but can operate in interleaved mode for current ripple suppression. Additionally, in Figures 21 and 23, the ripple of output inductor current is double the switch frequency, which results in lower volume requirement for output filter inductor. The measured efficiency of the MIC is shown in Figure 26. In the case of wind turbine shutting down from the hybrid power generation system, the output power variation of the MIC is shown in Figure 27. For converse condition, the related output power curve is shown in Figure 28.

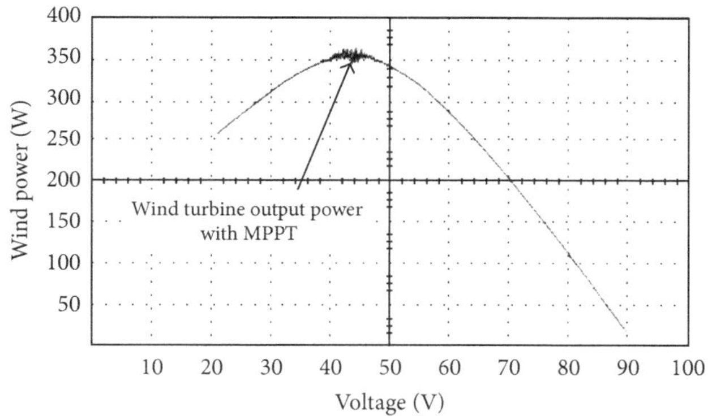

FIGURE 14: Measured result: drawn power from wind turbine with MPPT.

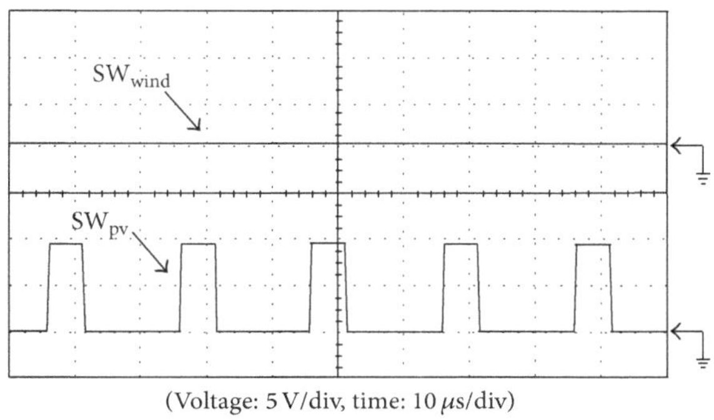

FIGURE 15: Simulated control signals while PV panel provides 350 W.

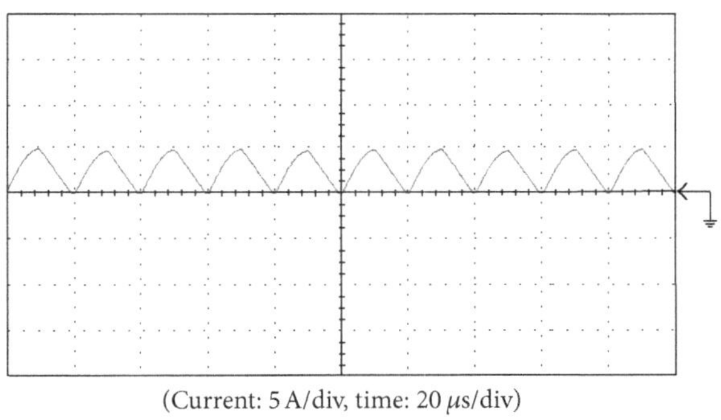

FIGURE 16: Simulated output inductor current while PV panel provides 350 W.

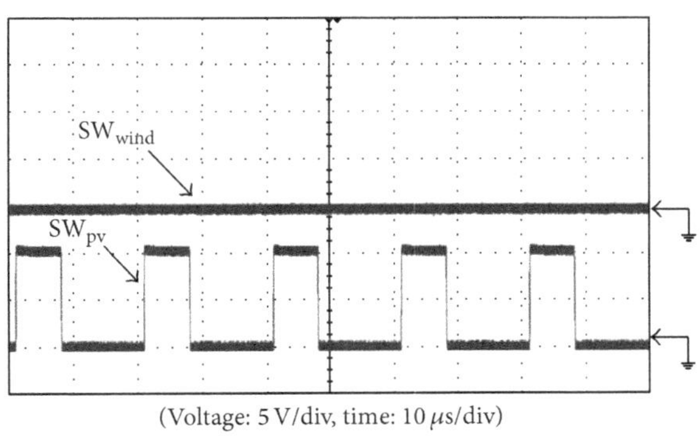

FIGURE 17: Practical measurements of control signals while PV panel provides 350 W.

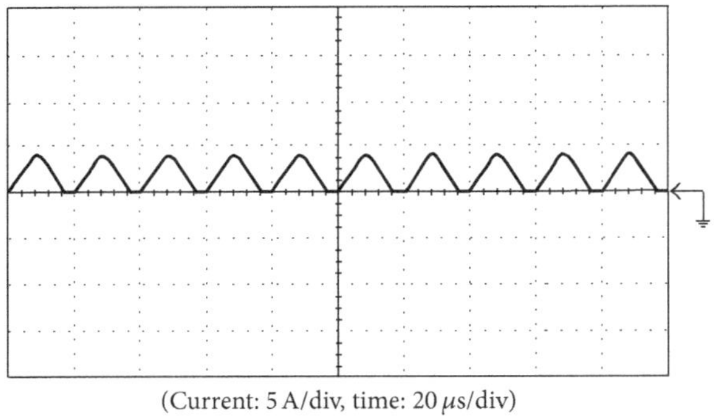

FIGURE 18: Measured output inductor current while PV panel provides 350 W.

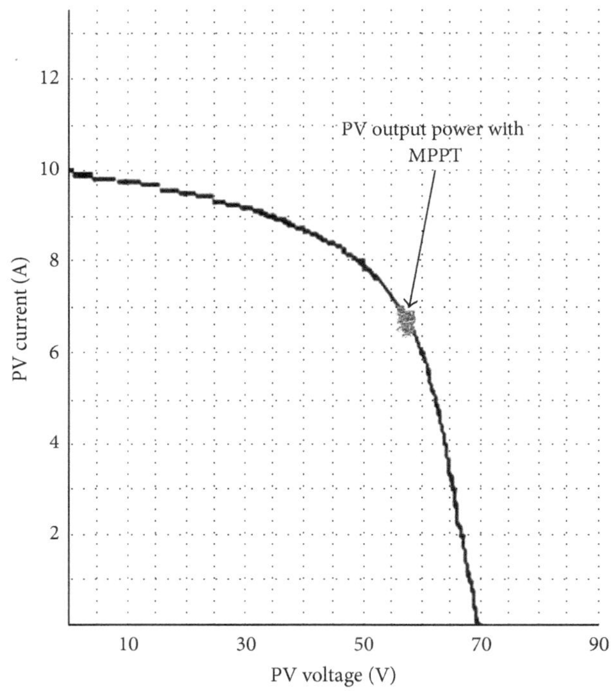

FIGURE 19: Measured result: operation point of PV panel after MPPT.

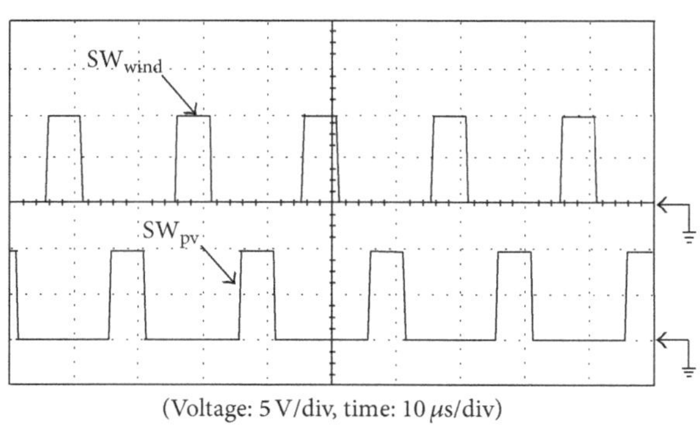

FIGURE 20: Simulated control signals while hybrid wind-PV power is 700 W.

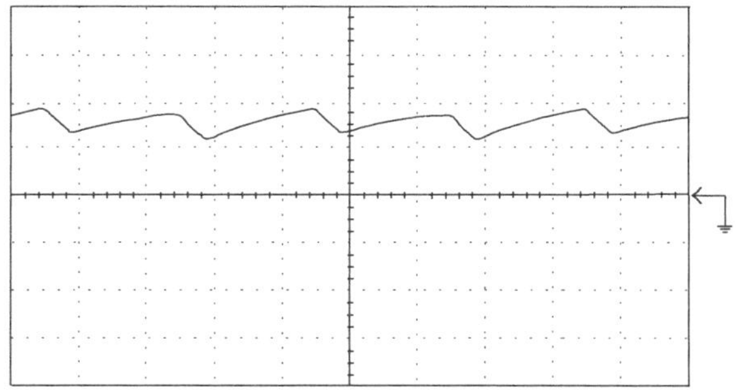

(Current: 2 A/div, time: 5 μs/div)

Figure 21: Simlated output inductor current while hybrid wind-PV power is 700 W.

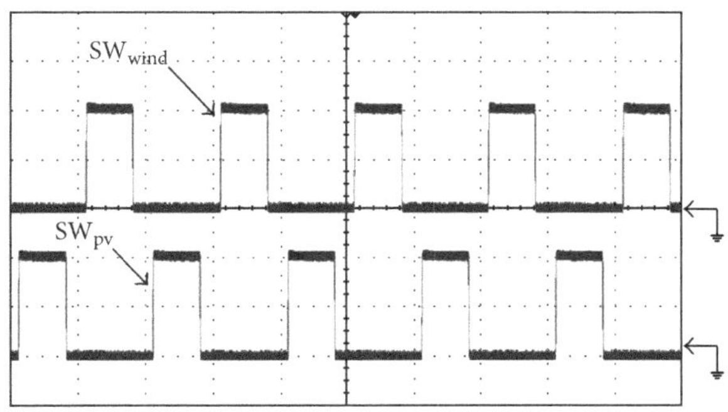

(Voltage: 5 V/div, time: 10 μs/div)

Figure 22: Measured control signals while hybrid wind-PV power is 700 W.

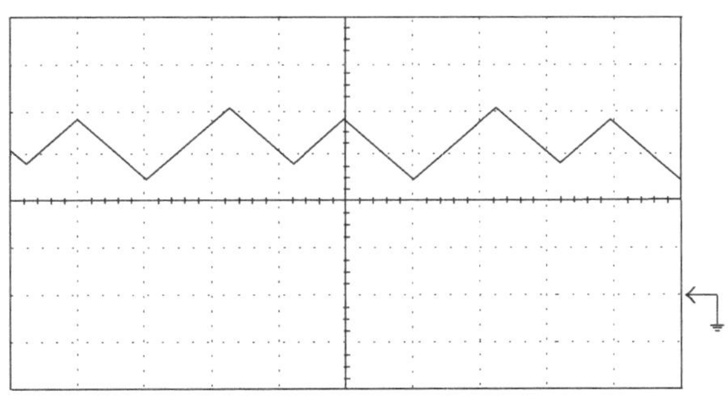

(Current: 1 A/div, time: 5 μs/div)

Figure 23: Measured output inductor current while hybrid wind-PV power is 700 W.

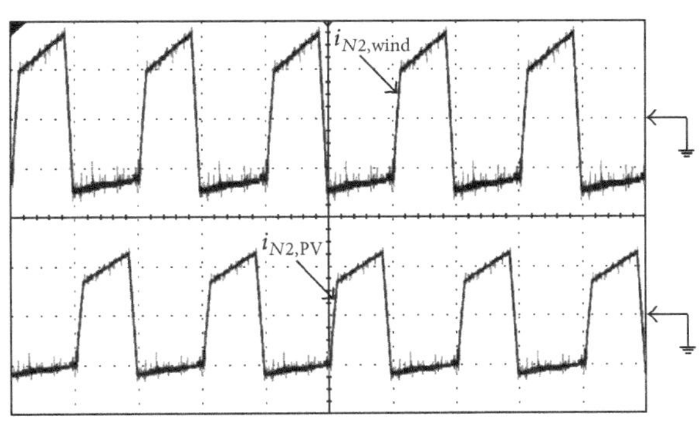

(Current: 2 A/div, time: 10 μs/div)

Figure 24: Measured waveforms of switch currents.

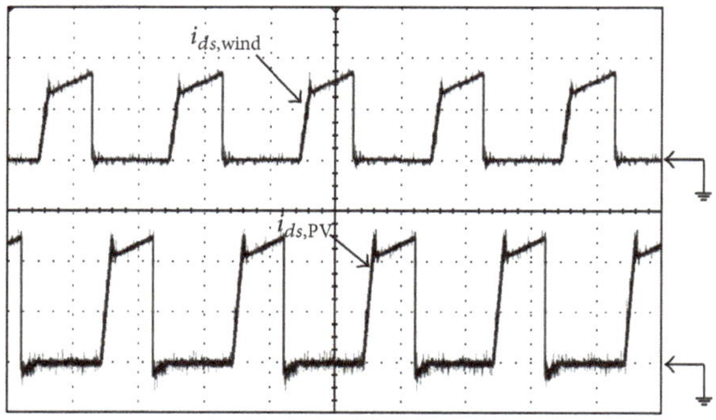

(Current: 10 A/div, time: 10 μs/div)

FIGURE 25: Measured current waveforms of the secondary windings in the MIC.

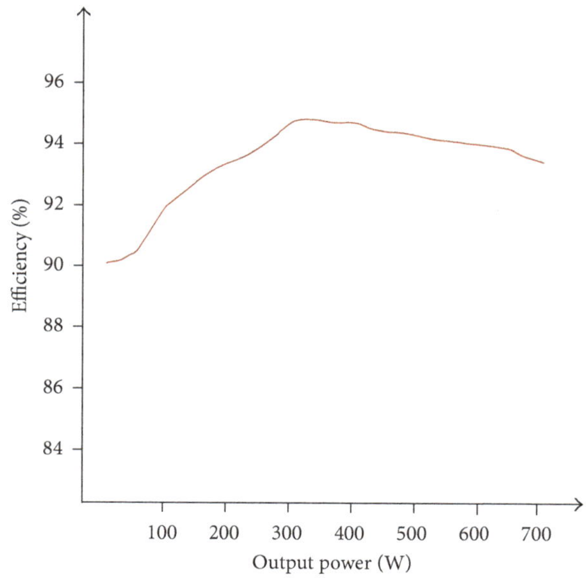

FIGURE 26: The measured efficiency of the proposed MIC.

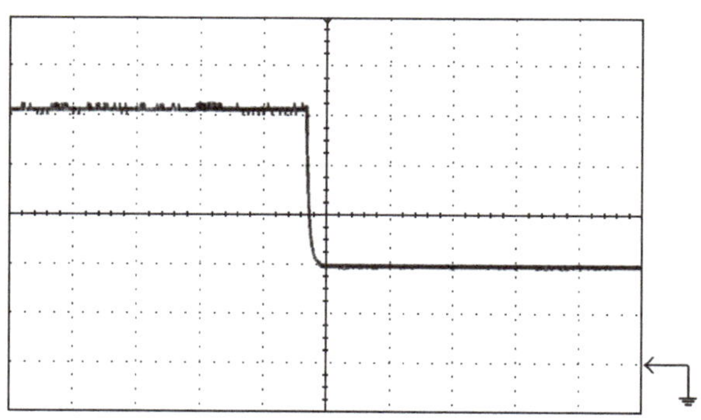

(P_o: 100 W/div, time: 10 ms/div)

FIGURE 27: Output power variation while wind turbine shuts down from the hybrid generation system.

5. Conclusions

This paper proposed a galvanic isolated multi-input converter to deal with wind turbine energy and solar power with MPPT feature. The converter integrates two forward converters and only uses one output inductor. Therefore, the structure of the proposed MIC can lower the volume of the converter. In addition, the MIC can operate in interleaved mode so that the output current ripple is suppressed significantly. The energy stored in leakage inductor can be recycled, which

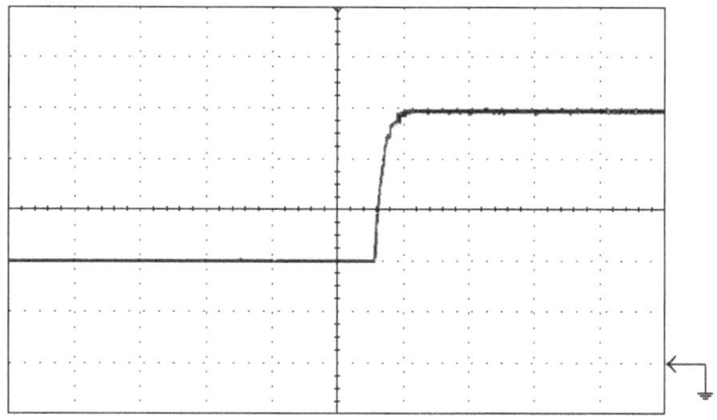

(P_o: 100 W/div, time: 10 ms/div)

FIGURE 28: Output power variation while wind turbine incorporates into the hybrid generation system.

improves efficiency. In this paper, the proposed MIC is analyzed, simulated, and tested. Simulations and hardware measurements have validated the proposed MIC.

References

[1] R.-J. Wai, W.-H. Wang, and C.-Y. Lin, "High-performance stand-alone photovoltaic generation system," *IEEE Transactions on Industrial Electronics*, vol. 55, no. 1, pp. 240–250, 2008.

[2] C.-L. Shen and J.-C. Su, "Grid-connection half-bridge PV inverter system for power flow controlling and active power filtering," *International Journal of Photoenergy*, vol. 2012, Article ID 760791, 8 pages, 2012.

[3] R. Gules, J. D. P. Pacheco, H. L. Hey, and J. Imhoff, "A maximum power point tracking system with parallel connection for PV stand-alone applications," *IEEE Transactions on Industrial Electronics*, vol. 55, no. 7, pp. 2674–2683, 2008.

[4] D. Meneses, O. García, P. Alou, J. A. Oliver, R. Prieto, and J. A. Cobos, "Single-stage grid-connected forward microinverter with constant off-time boundary mode control," in *Proceedings of the 27th Annual IEEE Applied Power Electronics Conference and Exposition (APEC '12)*, pp. 568–574, February 2012.

[5] Z. Liang, R. Guo, J. Li, and A. Q. Huang, "A high-efficiency PV module-integrated DC/DC converter for PV energy harvest in FREEDM systems," *IEEE Transactions on Power Electronics*, vol. 26, no. 3, pp. 897–909, 2011.

[6] X. Yuan, J. Chai, and Y. Li, "A transformer-less high-power converter for large permanent magnet wind generator systems," *IEEE Transactions on Sustainable Energy*, vol. 3, no. 3, pp. 318–329, 2012.

[7] B. Singh and S. Sharma, "Design and implementation of four-leg voltage-source-converter-based VFC for autonomous wind energy conversion system," *IEEE Transactions on Industrial Electronics*, vol. 59, no. 12, pp. 4694–4703, 2012.

[8] R. Li and D. Xu, "Parallel operation of full power converters in permanent-magnet direct-drive wind power generation system," *IEEE Transactions on Industrial Electronics*, vol. 60, no. 4, pp. 1619–1629, 2013.

[9] F. Deng and Z. Chen, "Control of improved full-bridge three-level DC/DC converter for wind turbines in a DC grid," *IEEE Transaction on Power Electronics*, vol. 28, no. 1, pp. 314–324, 2013.

[10] W. Chen, A. Q. Huang, C. S. Li, G. Y. Wang, and W. Gu, "Analysis and comparison of medium voltage high power DC/DC converters for offshore wind energy systems," *IEEE Transaction on Power Electronics*, vol. 28, no. 4, pp. 2014–2023, 2013.

[11] N. Vazquez, A. Hernandez, C. Hernandez, E. Rodríguez, R. Orosco, and J. Arau, "A double input DC/DC Converter for photovoltaic/wind systems," in *Proceedings of the 39th IEEE Annual Power Electronics Specialists Conference (PESC '08)*, pp. 2460–2464, June 2008.

[12] Y.-M. Chen, Y.-C. Liu, S.-C. Hung, and C.-S. Cheng, "Multi-input inverter for grid-connected hybrid PV/wind power system," *IEEE Transactions on Power Electronics*, vol. 22, no. 3, pp. 1070–1077, 2007.

[13] Y.-M. Chen, Y.-C. Liu, and F.-Y. Wu, "Multi-input dc/dc converter based on the multiwinding transformer for renewable energy applications," *IEEE Transactions on Industry Applications*, vol. 38, no. 4, pp. 1096–1104, 2002.

[14] C.-L. Shen, C.-T. Tsai, Y.-E. Wu, and C.-C. Chen, "A modified-forward multi-input power converter for solar energy and wind power generation," in *Proceedings of International Conference on Power Electronics and Drive Systems (PEDS '09)*, pp. 631–636, January 2009.

Effects of Surface Modification of TiO_2 Nanotube Arrays on the Performance of CdS Quantum-Dot-Sensitized Solar Cells

Danhong Li, Nengqian Pan, Jianjun Liao, Xiankun Cao, and Shiwei Lin

Key Laboratory of Ministry of Education for Advanced Materials in Tropical Island Resources, School of Materials and Chemical Engineering, Hainan University, Haikou 570228, China

Correspondence should be addressed to Shiwei Lin; linsw@hainu.edu.cn

Academic Editor: Maria da Graça P. Neves

CdS-sensitized TiO_2 nanotube arrays have been fabricated using the method of successive ionic layer adsorption and reaction and used as a photoanode for quantum-dot-sensitized solar cells. Before being coated with CdS, the surface of TiO_2 nanotube arrays was treated with $TiCl_4$, nitric acid (HNO_3), potassium hydroxide (KOH), and methyltrimethoxysilane (MTMS), respectively, for the purpose of reducing the interface transfer resistance of quantum-dot-sensitized solar cells. The surfaces of the modified samples represented the characteristics of superhydrophilic and hydrophobic which directly affect the power conversion efficiency of the solar cells. The results showed that surface modification resulted in the reduction of the surface tension, which played a significant role in the connectivity of CdS and TiO_2 nanotube arrays. In addition, the solar cells based on CdS/TiO_2 electrode treated by HNO_3 achieved a maximum power conversion efficiency of 0.17%, which was 42% higher than the reference sample without any modification.

1. Introduction

Since the new architecture for the dye-sensitized solar cell (DSSC) was first reported by O'Regan and Grätzel in 1991, it has attracted considerable attention throughout the world from both academic and industrial fields as one of promising candidates for the development of next generation solar cells [1–5]. DSSCs use dyes to sensitize wide band gap semiconductors to convert visible light into electricity, while quantum-dot-sensitized solar cells (QDSSCs) generally use quantum dots (QDs) as sensitized materials instead of conventional organic dyes [6, 7]. Accordingly, semiconductor QDs as light absorbing materials have many fascinating advantages, such as their high extinction coefficients and large intrinsic dipole moment that originated from bulk properties [8, 9]. Besides, the tunable band gap by size controlling and multiple excition generation by impact ionization are derived from the nanoscale dimension as a consequence of the quantum confinement effect [10–12]. However, the quantum-dot-sensitized solar cells have shown comparatively much lower efficiencies than expected and have not been fully systematically explored and evaluated.

To date, one-dimensional TiO_2 nanostructures, particularly TiO_2 nanotube arrays (TNAs), have received great attention in terms of quantum-dot-sensitized solar cells. A large surface area facilitates the creation of a large number of junctions and an increase in the amount of sensitizer material that can be loaded on the TiO_2 surface, thereby leading to favorable light harvesting and collection of photogenerated charge carriers [13]. Additionally, the tubular architecture and vertical orientation of TNAs facilitate the separation of the photoexcited charges and accelerate electron transport [14]. However, at present, the record solar conversion efficiencies of QDSSCs based on TNAs are lower than 5%. QDSSCs based on the CdSe/CdS-sensitized TNAs/Ti photoanode and Cu_2S counterelectrode just had a performance of 3% which was lower than 7% achieved from colloid quantum dot solar cells [14, 15]. Kamat et al. linked CdSe QDs to TiO_2 with bifunctional organic molecules and successfully increased the power conversion efficiency of solar cell due to the acceleration of the electron injection rate from CdSe to TiO_2 [16, 17]. Thus, interface properties greatly influence the interfacial electron transfer process. Surface modification via different approaches, such as surface modification with molecular dipoles and coating quantum-dot-sensitized TNAs with ZnS nanometric barriers, has been demonstrated to be a powerful tool to boost the power conversion efficiencies of the devices [18–20].

In our previous work, we reported the effects of geometric and crystal structures on the photoelectrical properties of TNAs and the enhanced photoelectrochemical property of nitrogen-doped TNAs [21, 22]. In this paper, we investigated the effects of interfacial properties on the QDSSCs performances through surface modification of TiO_2 nanotube arrays. After the modification, the nanotube surfaces showed a superhydrophilic or superhydrophobic behavior, which resulted from various surface tensions. Despite the vast reports on QDSSCs, there is little research on the influence of the characteristics of surface superhydrophilic and superhydrophobic as well as surface tensions. Therefore, we modified the surface of TNAs by $TiCl_4$, nitric acid (HNO_3), and potassium hydroxide (KOH), respectively, to change the surface tension of the TNAs. Subsequently, CdS quantum dots were deposited by the method of successive ionic layer adsorption and reaction (SILAR). After modification, 0.17% power conversion efficiency was achieved by treatment of HNO_3, which was 42% higher than the reference solar cell.

2. Experimental

2.1. Preparation and Modification of TiO_2 Nanotube Arrays. Ti foil (0.5 mm thickness, 99.4% purity) was first cut into 1.5 cm × 3 cm. The foils were ultrasonically cleaned in acetone, ethanol, and deionized water for 30 min, respectively, and dried in ambient air. The Ti foil was anodized in ethylene glycol solution containing 0.3 wt% NH_4F and 2 wt% H_2O by applying a 60 V DC potential for 4 h. The anodization was performed in a two-electrode configuration with the Ti foil as the working electrode and a stainless steel foil as the counterelectrode. After anodization, the TNAs on Ti foil were ultrasonically treated in ethanol for 5 min to remove the electrolyte and debris on top, followed by thermal treatment at 450°C for 3 h. Then the samples were soaked in $TiCl_4$ (0.04 M), HNO_3 (pH = 3), and KOH (pH = 12), respectively, for 30 min at 70°C. And the sample soaked in $TiCl_4$ subsequently re-annealed at 450°C for 30 min.

2.2. Deposition of CdS QDs. The CdS-sensitized TiO_2 photoanodes were achieved by SILAR of CdS QDs on TNAs. The as-prepared photoanodes were firstly immersed into a solution of 0.05 M $Cd(NO_3)_2$ in ethanol for 2 min, then rinsed with ethanol, and then immersed into 0.05 M Na_2S in methanol/water (1 : 1/v : v) for another 2 min, followed by another rinsing with deionized water and methanol. Such an immersion cycle was repeated 8 times until the desired deposition of CdS QDs was achieved. The sensitized photoanodes were passivated with ZnS by alternately dipping into 0.5 M $Zn(NO_3)_2$ ethanol solution and 0.5 M Na_2S methanol/water solution three times for 2 min for each time.

2.3. Fabrication of Counterelectrodes and Solar Cells Assembly. To assemble the solar cell, a Pt-coated counterelectrode was prepared by dropping 5 mM H_2PtCl_6 in isopropanol onto FTO glass and then annealed at 385°C for 30 min. The cells were assembled in a sandwiched structure with a thermal adhesive film (Surlyn, 60 μm, DuPont). A methanol/water (7 : 3 by volume) solution containing 0.5 M Na_2S, 2 M S, and 0.2 M KCl was used as the redox electrolyte. The electrolyte was injected by vacuum backfilling, and then the hole was sealed with a Surlyn film. The effective area of the cell is 0.3 cm^2.

2.4. Characterization. The morphology and microstructure of the photoanodes were characterized by scanning electron microscopy (FESEM, Hitachi S-4800) and transmission electron microscopy (TEM, JOEL JEM2100), and the crystal structure was studied by X-ray diffraction measurements (XRD, Bruker AXS, D8 Advance). The absorption spectrum was obtained by a UV-vis spectrophotometer (Persee, TU-1901). The current and voltage measurements were recorded by Keithley 2400 sourcemeter when the cells were irradiated by a solar simulator (Aulight, CEL-S500) under AM 1.5 illumination of 100 mW/cm^2. All the electrochemical impedance spectroscopy (EIS) measurements of QDSSCs were carried out in dark and under the open circuit potentials. The superimposed signal has the frequency ranging from 0.1 Hz to 100 KHz with AC amplitude of 10 mV (Zahner, Zennium).

3. Results and Discussion

3.1. Contact Angle Analysis. The top-view and side-view images of highly ordered TNAs prepared by anodic oxidation are shown in Figure 1(a). The tube length is measured to be 20 μm while the average inner diameter 150 nm. To observe the surface characteristics of the resulting surfaces, modified TNAs were investigated by equilibrium contact-angle measurements. The TNAs without any surface modification were used as a reference, and the corresponding contact angle is displayed in Figure 1(b). The water contact angle is 6.3 ± 2°, and the surface is practically wetting and represents a superhydrophilic behavior. We doubt that the reason for the phenomenon of the surface derives from the spreading water on the entire surface and into the pores [23]. Figures 1(c)–1(e) exhibit the contact angles of TNAs modified by $TiCl_4$, HNO_3, and KOH, respectively, and the resulting contact angles are 12.2 ± 2°, 1.2 ± 2°, and 9.4 ± 2°, respectively. It can be seen that the contact angles are different and the TNAs treated by HNO_3 represent the best hydrophilicity. Besides the reason that water is spreading into the pores which is similar to the reference sample, the changes of active site and surface property play a key role in the surface wetting behavior of the TNAs modified with HNO_3. Moreover, notice that the contact angles of modified samples are similar to that of the reference, all of which are smaller than 50° and show a superhydrophilic behavior. In order to study the effect of the superhydrophilic/superhydrophobic surface on the QDSSCs performance, a TNAs sample was modified with also methyltrimethoxysilane (MTMS) which is usually used for surface hydrophobic modification. The

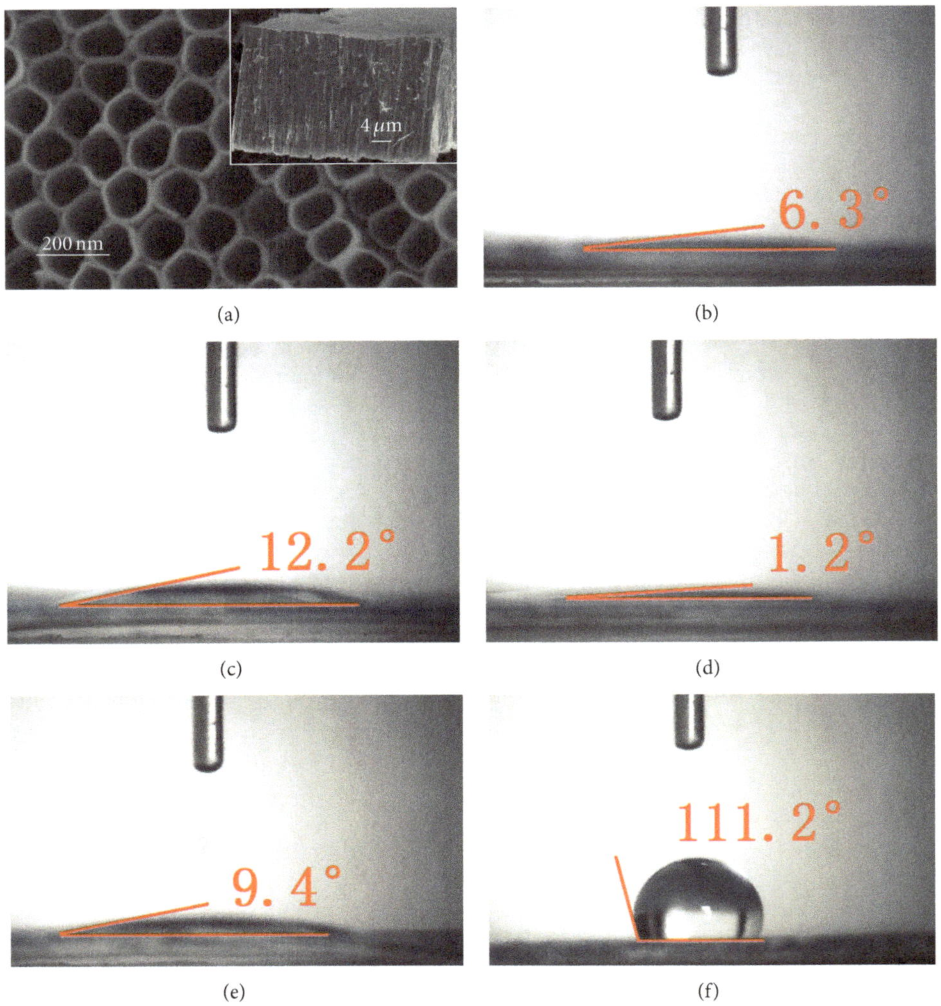

FIGURE 1: (a) Top-view SEM image of the TNAs with the corresponding side-view image in the inset. (b) Contact angle of the reference TNAs sample. Contact angles of the TNAs modified by (c) $TiCl_4$, (d) HNO_3, (e) KOH, and (f) MTMS.

originally superhydrophilic surface becomes hydrophobic with a contact angle of about 111.2±2° as shown in Figure 1(f).

3.2. Morphological Analysis. Figure 2 shows the top-view and side-view SEM images of CdS-sensitized TiO_2 nanotube arrays. In Figure 2(a), CdS quantum dots are well distributed on the top surface of the TNAs without any modification, while some CdS quantum dots are found to aggregate on the openings of TNAs. The side-view SEM image in Figure 2(a) depicts that the vast majority of QDs are found to aggregate and attach to the external tube wall. The results indicate that because of the effect of surface tension, the CdS precursor solutions could not penetrate deeply into the nanotubes and cause large aggregation of CdS on the top surface and external tube wall [24]. Figure 2(b) represents the great majority of CdS quantum dots aggregated in the openings of TiO_2 nanotubes treated by $TiCl_4$. The side-view SEM image shows that CdS QDs are located on the inner tube wall of TNAs and distributed uniformly. As depicted in Figure 2(c), CdS QDs treated by HNO_3 are homogeneous and rarely aggregate in the openings. Moreover, the distribution of CdS QDs in the inner spaces is also uniform and few aggregated as illustrated in the inset of Figure 2(c). In Figure 2(d), a great majority of CdS QDs aggregate in the openings of TNAs treated by KOH, while the side-view image shows a similar distribution of the sample treated by $TiCl_4$. In the case of MTMS-treated TNAs shown in Figure 1(e), large quantity of CdS QDs blocked in the surface openings, and the side-view image in inset shows a uniform distribution in the inner spaces of TNAs. According to the surface characteristics in Figure 1, the results shown in Figure 2 suggest that superhydrophilic surface might have a uniform QDs distribution, while hydrophobic surface might be easy to cause QDs aggregation.

To further clarify whether or not the CdS QDs are introduced into the nanotubes wall under the influence of hydrophilic and hydrophobic modification of TNAs coated with CdS, the samples were investigated by TEM. Figure 3 shows TEM images of the reference sample and CdS/TNAs treated by $TiCl_4$, HNO_3, KOH, and MTMS, respectively. As shown in Figure 3(a), few CdS QDs are deposited in

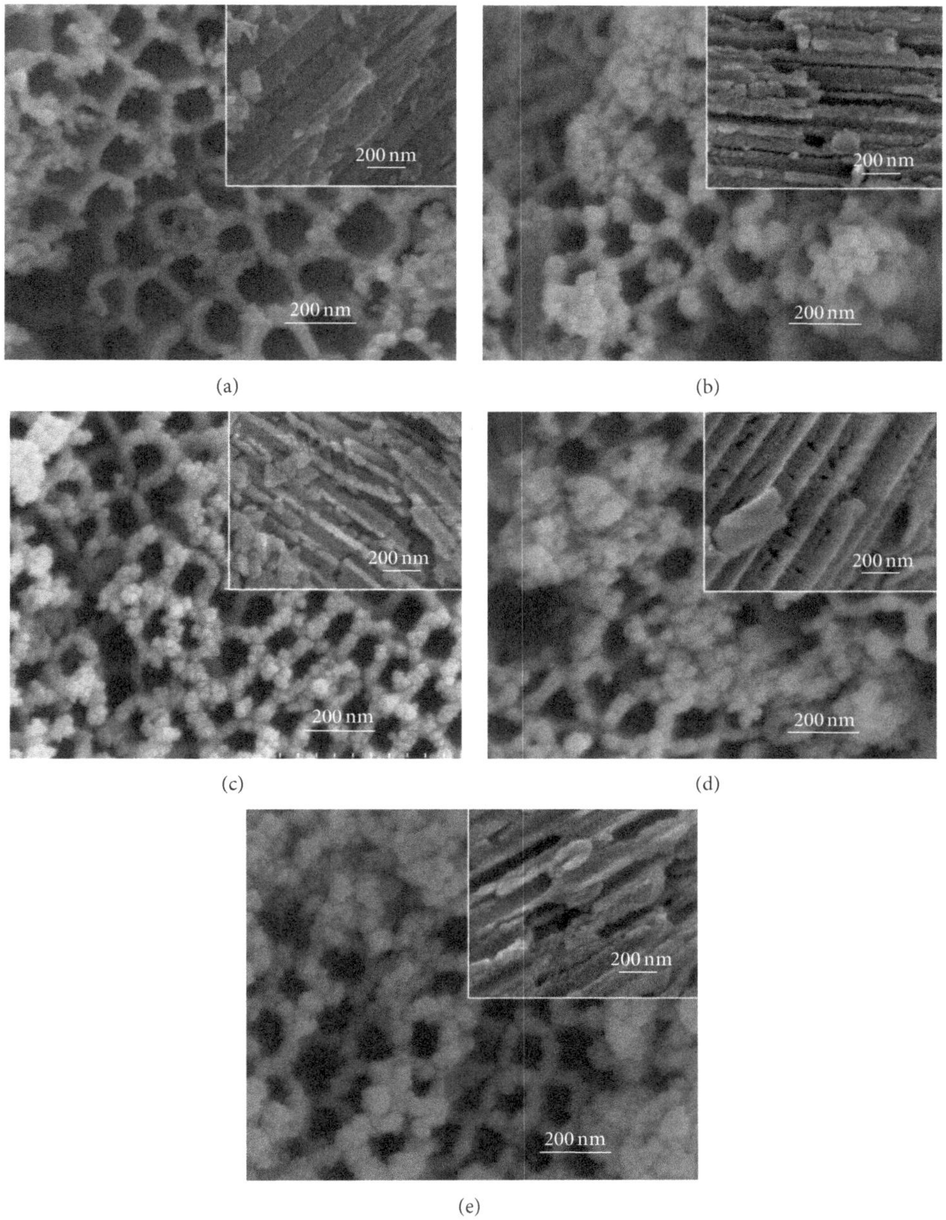

FIGURE 2: (a) Top-view and side-view images of the CdS/TNAs and the TNAs modified by (b) $TiCl_4$, (c) HNO_3, (d) KOH, and (e) MTMS.

the nanotube wall of the reference sample without surface treatment. Compared with the reference sample, many CdS QDs have been attached to the nanotube wall of the sample treated by $TiCl_4$ which is shown in Figure 3(b). And the CdS QDs are found to be aggregated in the tube wall. After being treated by HNO_3, as depicted in Figure 3(c), a great majority of CdS QDs deposited in nanotube wall are better distributed. In addition, the sizes of QDs are more uniform. Similarly, the CdS QDs are also found to be aggregated for the HNO_3-treated sample. As shown in Figure 3(d), CdS QDs apparently aggregate together for the CdS/TiO_2 electrode treated by KOH. Besides, the CdS/TNAs electrode modified by MTMS represents the phenomenon of aggregation and the nonuniform distribution of CdS QDs in Figure 3(e). By combining Figures 2 and 3, we suggest that the CdS QDs are more uniformly distributed on the tube wall of TNAs after the treatment of HNO_3 because of its superhydrophilic characteristic and smaller surface tension. Close observation by high-resolution TEM demonstrates that the size of CdS QDs is about 5–10 nm in diameter, as depicted in Figure 3(f).

3.3. XRD Analysis. The crystal structure of the deposited CdS was examined by X-ray diffraction experiments. For comparison, the X-ray diffraction pattern from the TNAs sample is also given as curve 1 in Figure 4, where diffraction peaks are the anatase phase of TiO_2 (JCPDS card no. 21-1272) and Ti metal phase. Curves 2–6 represent the diffraction peaks of CdS/TNAs treated without and with $TiCl_4$,

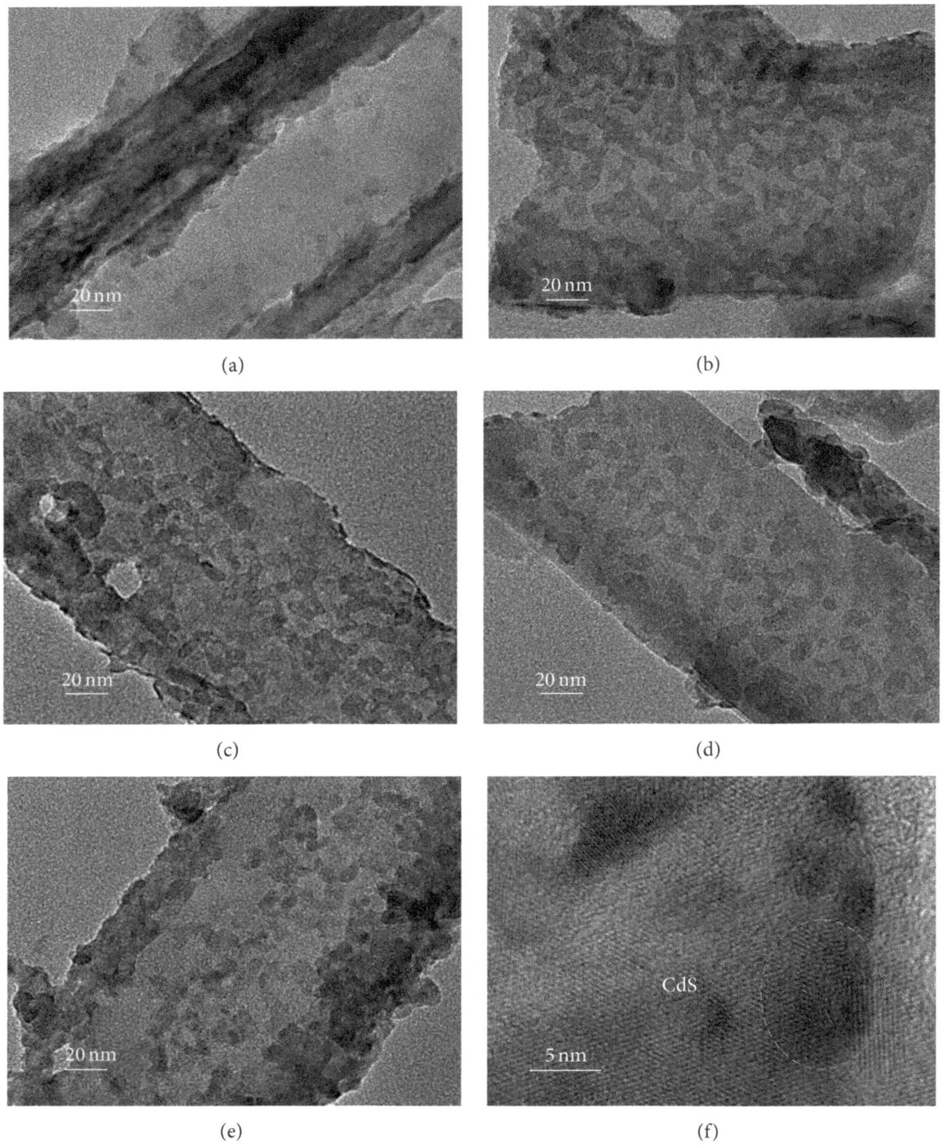

FIGURE 3: TEM images of (a) CdS/TNAs without surface treatment and CdS/TNAs treated by (b) $TiCl_4$, (c) HNO_3, (d) KOH, (e) MTMS, and (f) correspond to high-resolution TEM image of HNO_3-modified CdS/TNAs.

HNO_3, KOH, and MTMS, respectively. Compared to the TiO_2 nanotubes arrays, CdS/TNAs electrodes exhibited new peaks corresponding to (111), (220), (331) planes of CdS, indicating the presence of the cubic phase CdS (JCPDS card no. 80-0019). The XRD patterns of CdS/TNAs electrodes with different treatments are quite similar, which suggests that surface modification could not change the crystal structure of the photoanode.

3.4. UV-Vis Absorption Spectra Analysis. Figure 5(a) shows the UV-vis absorption spectra of the pure TiO_2 nanotube arrays, pure CdS particles, and CdS-sensitized TNAs without any treatment and treated with $TiCl_4$, HNO_3, KOH, and MTMS, respectively. The TNAs sample does not exhibit absorption in the visible light region, although it shows absorption in the UV light region. While compared with the TNAs absorption spectrum, an apparent absorption peak in the visible light region exists for the pure CdS particles, which can be attributed to its narrow band gap properties. Moreover, from Figure 5(b), it can be found that the absorption edge of the CdS appears at about 530 nm by the method of linear extrapolation, which is just making the tangent in the starting absorption point to get the absorption edge. Then the band gap of CdS could be calculated. Since spectra are typically reported in units corresponding to the wavelength of light rather than its energy, the conversion between conventional wavelength and band gap energy units can be carried out by the equation described below [25]:

$$h\nu\,(\mathrm{eV}) = \frac{1239.8\,(\mathrm{eV} \times \mathrm{nm})}{\lambda\,(\mathrm{nm})}. \quad (1)$$

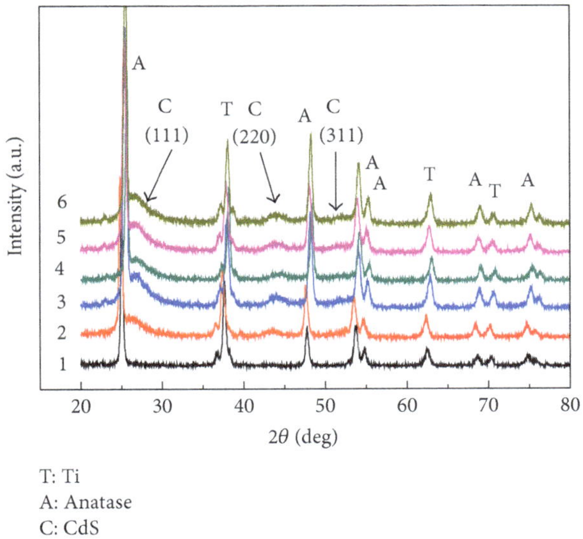

FIGURE 4: XRD patterns of CdS/TNAs without modification (curve 2) and modified with $TiCl_4$ (curve 3), HNO_3 (curve 4), KOH (curve 5), and MTMS (curve 6), respectively. Curve 1 refers to the XRD pattern of the reference TNAs sample.

Herein, the band gap of the pure CdS is 2.34 eV. As depicted in Figure 5(a), curve 1 is the absorption spectrum of reference CdS/TNAs electrode. A considerable absorbance in the visible light region can be observed, and the absorption peak is ascribed to the loading of the CdS nanoparticles. In addition, the absorption spectrum of the CdS/TiO_2 electrode in 400 to 550 nm is very similar to that of the pure CdS. It is implied that CdS QDs have a synergism on TNAs, which basically enhance the absorption intensity in the visible light region in comparison with the pure TNAs. While in the range between 550 and 750 nm, the absorption spectrum of reference electrode is the same as that of TNAs, suggesting that TNAs act synergistically on CdS QDs. Besides, the inhabitation of CdS QDs within the TNAs is beneficial for maintaining more reaction active sites [26]. Curves 2–5 are CdS/TNAs electrodes with the treatment of $TiCl_4$, HNO_3, KOH, and MTMS, respectively. The absorption spectra of these electrodes are similar to that of the reference. And the spectra are close to each other due to the similar CdS particle sizes and CdS contents. From Figures 5(c)–5(e), it can be observed that the band edges of reference electrode and CdS/TiO_2 electrodes treated with HNO_3 and MTMS are 520 nm, 520 nm, and 530 nm, respectively. The corresponding band gaps are about 2.38 eV, 2.38 eV, and 2.34 eV on the basis of (1), which are all slightly higher than that of the bulk CdS (2.25 eV). Furthermore, the sizes of the CdS particles on the TNAs are still within the scale of QDs [27, 28], which have already been testified by the HRTEM image in Figure 3(f). As depicted in Figures 5(c)–5(e), the absorption edges of reference sample and HNO_3-treated electrode are a little lower than that of MTMS-treated sample. Furthermore, the results are in line with that of Figure 2 since the CdS QDs deposited in the reference sample and HNO_3-treated TNAs are more uniformly distributed than other electrodes.

3.5. Photovoltaic Performance and Charge Transfer. The J-V curves of QDSSCs treated without and with $TiCl_4$, HNO_3, KOH, and MTMS are shown in Figure 6. The open circuit potential (V_{oc}), short circuit current density (J_{sc}), fill factor (FF), and conversion efficiency (η) of all cells are listed in Table 1. The J_{sc} and V_{oc} of the reference electrode are 0.943 mA/cm^2 and 0.37 V, respectively, resulting in a very low value of conversion efficiency of 0.12%. And the effect of the $TiCl_4$ treatment TNAs is also evaluated using the same parameters. An increase in the J_{sc} of $TiCl_4$-treated solar cell is observed in Table 1, while the efficiency of the cell represents a small decrease due to the reduction of the V_{oc} and FF. The lower V_{oc} is possibly ascribe to the decrease of the electrons transported from CdS QDs to TiO_2 conduction band, leading to a decrease in the quasi-Fermi level of TiO_2 [29], while lower FF may be attributed to the low driving force for the electron injection [29]. Moreover, these results clearly reflect that the $TiCl_4$ treatment does not have a beneficial effect on QDSSCs. In the case of HNO_3-treated solar cell, a systematic beneficial effect is observed. The J_{sc}, V_{oc}, and η are 1.460 mA/cm^2, 0.36 V, and 0.17%, respectively. The efficiency of 0.17% is 42% higher than that of the reference solar cell. Accordingly, the higher J_{sc} seems to be mostly owing to the better connectivity between TNAs and CdS QDs. This implies that the strong oxidizing property and high acidity of the HNO_3 to some extent help make better contact between TNAs and CdS QDs [30]. Therefore, well-connected structures are able to promote electron transfer and improve J_{sc}. For the solar cell treated with KOH, the photovoltaic performance is similar to the $TiCl_4$-treated solar cell. The J_{sc} is higher while V_{oc} and FF are lower than those of the reference cell. In addition, the MTMS-treated solar cell shows lower V_{oc} and efficiency than that of the reference cell. These results are in good agreement to that of Figures 2 and 3, and the lower efficiencies of the solar cells are due to the blocking QDs in the opening of nanotubes. Moreover, the J_{sc} of HNO_3-treated solar cell increases to a maximum value of 1.460 mA/cm^2, which is possibly because of the increase amount of CdS QDs deposited on TiO_2 and the uniform distribution of the QDs.

In order to find out the reasons why J_{sc} increases in the photoanode after HNO_3 treatment, we further investigated the electron transport and charge recombination processes of the solar cells by applying EIS. Figure 7 shows the Nyquist plots and Bode plots of CdS-sensitized solar cells modified with and without HNO_3, respectively, which are measured under dark conditions and open circuit potentials. EIS is a useful technique to detect kinetic processes in QDSSCs. Two distinct arcs of each plot could be seen from Figure 7(a), representing the electron transport processes at the interface of TNAs/CdS/electrolyte (frequency region 1–1000 Hz) and the feature of polysulfide solution diffusion in the electrolyte (<1 Hz). Generally, the plots of QDSSCs exhibit three arcs, while in our case the interface of the counterelectrode/electrolyte is not so obvious in the frequency region 1–100 kHz which is shown in the inset of Figure 7(a). The electrochemical parameters obtained from equivalent circuit analysis are presented in Table 2. The equivalent circuit of the solar cells is represented in

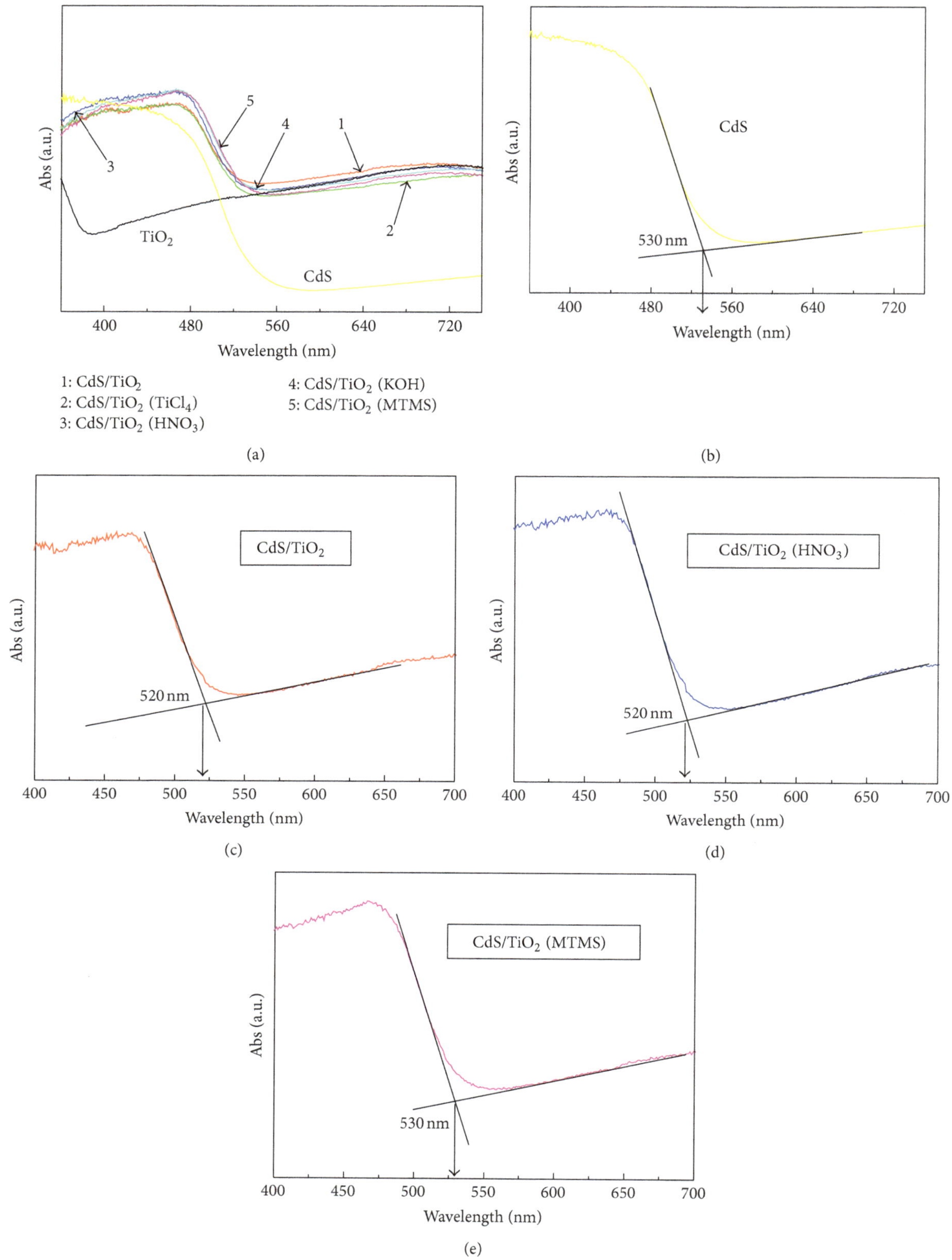

FIGURE 5: (a) UV-vis absorption spectra of TiO_2 nanotube arrays and CdS/TNAs with different modifications. Linear extrapolation is conducted to get absorption edges of (b) CdS particles, (c) CdS/TiO_2, and (d) CdS/TiO_2 treated with HNO_3 and (e) CdS/TiO_2 treated with MTMS.

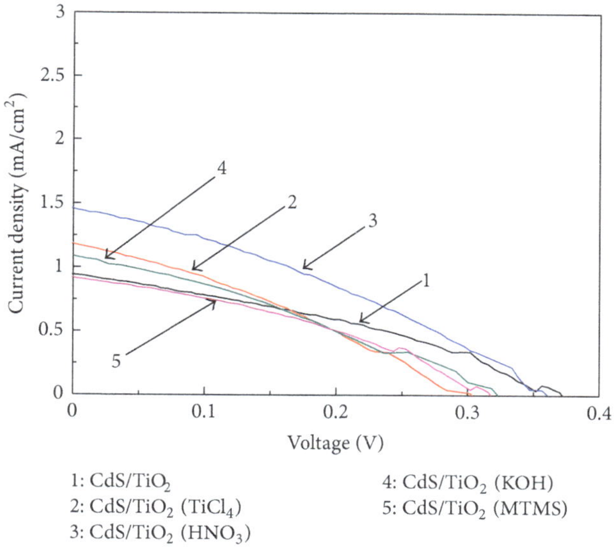

FIGURE 6: J-V curves of CdS-sensitized solar cells with different modifications.

TABLE 1: Photovoltaic parameters of QDSSCs with different modifications.

Sample	J_{sc} (mA/cm^2)	V_{oc} (V)	FF	η (%)
CdS/TiO_2	0.943	0.37	0.35	0.12
CdS/TiO_2 ($TiCl_4$)	1.188	0.30	0.31	0.11
CdS/TiO_2 (HNO_3)	1.460	0.36	0.34	0.17
CdS/TiO_2 (KOH)	1.169	0.32	0.31	0.12
CdS/TiO_2 (MTMS)	0.923	0.32	0.36	0.11

the inset of Figure 7(b). For relatively low resistance, the transfer resistance of counterelectrode/electrolyte interface does not play a key role in the whole resistance of solar cell. Therefore, charge transfer is dominated by the transfer resistance at the interface of TNAs/CdS/electrolyte and the diffusion resistance of electrolyte. From Table 2, it can be seen that the values of R_{CT2} and R_E for reference and HNO_3-treated solar cells are 62.45 and 180.4 Ω, 37.04 and 127.1 Ω, respectively. It is inferred that HNO_3-treated solar cell shows a lower charge transfer resistance which is responsible for the observed higher photocurrent density (J_{sc} = 1.460 mA/cm^2) compared to reference sample. And the results are in good agreement with the photovoltaic performance of solar cells shown in Figure 6. This phenomenon may be due to the better hydrophilicity and lower surface tension of the solar cells treated by HNO_3. Moreover, because of the decrease of excessive CdS hindering the access of electrolyte deep into the inner tube, J_{sc} of the HNO_3 treatment solar cells is larger than other cells. Correspondingly, the characteristic frequency peaks in Bode phase plots are shown in Figure 7(b).

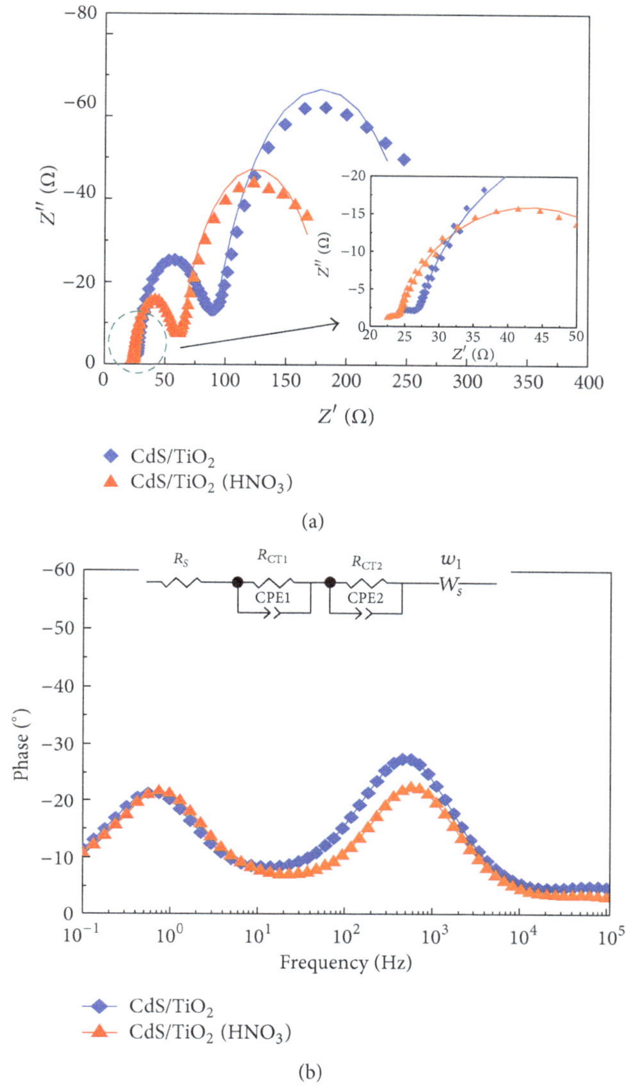

FIGURE 7: Electrochemical impedance spectra of CdS-sensitized solar cells: (a) Nyquist plots and (b) Bode plots.

TABLE 2: Electrochemical parameters of reference and HNO_3-treated solar cells.

Sample	R_s (Ω)	R_{CT1} (Ω)	R_{CT2} (Ω)	R_E (Ω)
CdS/TiO_2	22.98	3.596	62.45	180.4
CdS/TiO_2 (HNO_3)	21.85	2.09	37.04	127.1

4. Conclusions

In summary, CdS-sensitized TiO_2 nanotube arrays have been fabricated by successive ionic layer adsorption and reaction and applied as the photoanode for quantum dots solar cells, in which the TNAs were modified using $TiCl_4$, HNO_3, KOH, and MTMS, respectively, The experimental results indicated that the conversion efficiency of HNO_3-treated electrode was apparent higher than that of the solar cells without any modification, which resulted from the large surface coverage and the uniform distribution of the CdS quantum dots. Furthermore, reducing surface tension leads to a better adsorption and combination and the obvious decrease of interface transfer impedance. As a consequence, the HNO_3-treated solar cells achieve a maximum short current density of 1.460 mA/cm^2 and a power conversion efficiency of 0.17%, which is 42% higher than the reference sample without any surface modification.

Acknowledgments

This work was supported by Program for New Century Excellent Talents in University (NCET-09-0110), National Nature Science Foundation of China (51162007), Hainan Natural Science Foundation (511110), and Hainan Provincial Program for International S&T Cooperation (KJHZ2013-13). The authors acknowledge Mr. Guizhen Wang for the analysis of transmission electron microscopy in the Analytical and Testing Center of Hainan University. They also acknowledge Professor Hong Lin at Tsinghua University for the help of fabrication and characterization of the solar cells.

References

[1] B. O'Regan and M. Grätzel, "A low-cost, high-efficiency solar cell based on dye-sensitized colloidal TiO_2 films," *Nature*, vol. 353, pp. 737–740, 1991.

[2] Y. H. Jang, X. K. Xin, M. Byun, Y. J. Jang, Z. Q. Lin, and D. H. Kim, "An unconventional route to high-efficiency dye-sensitized solar cells via embedding graphitic thin films into TiO_2 nanoparticle photoanode," *Nano Letters*, vol. 12, no. 1, pp. 479–485, 2012.

[3] Q. Zhang, X. Guo, X. Huang et al., "Highly efficient CdS/CdSe-sensitized solar cells controlled by the structural properties of compact porous TiO_2 photoelectrodes," *Physical Chemistry Chemical Physics*, vol. 13, no. 10, pp. 4659–4667, 2011.

[4] G. Zhu, L. Pan, T. Xu, Q. Zhao, B. Lu, and Z. Sun, "Microwave assisted CdSe quantum dot deposition on TiO_2 films for dye-sensitized solar cells," *Nanoscale*, vol. 3, no. 5, pp. 2188–2193, 2011.

[5] A. Braga, S. Giménez, I. Concina, A. Vomiero, and I. Mora-Seró, "Panchromatic sensitized solar cells based on metal sulfide quantum dots grown directly on nanostructured TiO_2 electrodes," *Journal of Physical Chemistry Letters*, vol. 2, no. 5, pp. 454–460, 2011.

[6] S. H. Im, Y. H. Lee, S. I. Seok, S. W. Kim, and S. W. Kim, "Quantum-dot-sensitized solar cells fabricated by the combined process of the direct attachment of colloidal CdSe quantum dots having a ZnS glue layer and spray pyrolysis deposition," *Langmuir*, vol. 26, no. 23, pp. 18576–18580, 2010.

[7] E. M. Barea, M. Shalom, S. Giménez et al., "Design of injection and recombination in quantum dot sensitized solar cells," *Journal of the American Chemical Society*, vol. 132, no. 19, pp. 6834–6839, 2010.

[8] A. Salant, M. Shalom, Z. Tachan, S. Buhbut, A. Zaban, and U. Banin, "Quantum rod-Sensitized solar cell: nanocrystal shape effect on the photovoltaic properties," *Nano Letters*, vol. 12, no. 4, pp. 2095–2100, 2012.

[9] Y. L. Lee and Y. S. Lo, "Highly efficient quantum-dot-sensitized solar cell based on co-sensitization of CdS/CdSe," *Advanced Functional Materials*, vol. 19, no. 4, pp. 604–609, 2009.

[10] Y. C. Park, E. H. Kong, Y. J. Chang, B.-G. Kum, and H. M. Jang, "Tertiary hierarchically structured TiO_2 for CdS quantum-dot-sensitized solar cells," *Electrochimica Acta*, vol. 56, no. 21, pp. 7371–7376, 2011.

[11] P. Sudhagar, J. H. Jung, S. Park, R. Sathyamoorthy, H. Ahn, and Y. S. Kang, "Self-assembled CdS quantum dots-sensitized TiO_2 nanospheroidal solar cells: structural and charge transport analysis," *Electrochimica Acta*, vol. 55, no. 1, pp. 113–117, 2009.

[12] H. Wang, Y. Bai, H. Zhang, Z. Zhang, J. Li, and L. Guo, "CdS quantum dots-sensitized TiO_2 nanorod array on transparent conductive glass photoelectrodes," *Journal of Physical Chemistry C*, vol. 114, no. 39, pp. 16451–16455, 2010.

[13] C. Li, J. Yuan, B. Han, L. Jiang, and W. Shangguan, "TiO_2 nanotubes incorporated with CdS for photocatalytic hydrogen production from splitting water under visible light irradiation," *International Journal of Hydrogen Energy*, vol. 35, no. 13, pp. 7073–7079, 2010.

[14] X. F. Guan, S. Q. Huang, Q. X. Zhang et al., "Front-side illuminated CdS/CdSe quantum dots co-sensitized solar cells based on TiO_2 nanotube arrays," *Nanotechnology*, vol. 22, no. 46, Article ID 465402, 2011.

[15] A. H. Ip, S. M. Thon, S. Hoogland, and E. H. Sargent, "Hybrid passivated colloidal quantum dot solids," *Nanotechnology*, vol. 127, no. 7, pp. 557–582, 2012.

[16] I. Robel, V. Subramanian, M. Kuno, and P. V. Kamat, "Quantum dot solar cells. Harvesting light energy with CdSe nanocrystals molecularly linked to mesoscopic TiO_2 films," *Journal of the American Chemical Society*, vol. 128, no. 7, pp. 2385–2393, 2006.

[17] P. V. Kamat, K. Tvrdy, D. R. Baker, and J. G. Radich, "Beyond photovoltaics: semiconductor nanoarchitectures for liquid-junction solar cells," *Chemical Reviews*, vol. 110, no. 11, pp. 6664–6688, 2010.

[18] H. J. Lee, J. Bang, J. Park, S. Kim, and S. M. Park, "Multilayered semiconductor (CdS/CdSe/ZnS)-sensitized TiO_2 mesoporous solar cells: all prepared by successive ionic layer adsorption and reaction processes," *Chemistry of Materials*, vol. 22, no. 19, pp. 5636–5643, 2010.

[19] M. Seol, E. Ramasamy, J. Lee, and K. J. Yong, "Highly efficient and durable quantum dot sensitized ZnO nanowire solar cell using noble metal-free counter electrode," *The Journal of Physical Chemistry C*, vol. 115, no. 44, pp. 22018–22024, 2011.

[20] Y. L. Lee, C. F. Chi, and S. Y. Liau, "CdS/CdSe Co-Sensitized TiO_2 photoelectrode for efficient hydrogen generation in a photoelectrochemical cell," *Chemistry of Materials*, vol. 22, no. 3, pp. 922–927, 2010.

[21] D. H. Li, S. W. Lin, S. P. Li, X. Huang, X. K. Cao, and J. B. Li, "Effects of geometric and crystal structures on the photoelectrical properties of highly ordered TiO_2 nanotube arrays," *The Journal of Materials Research*, vol. 27, no. 7, pp. 1029–1036, 2012.

[22] S. P. Li, S. W. Lin, J. J. Liao, N. Q. Pan, D. H. Li, and J. B. Li, "Nitrogen-doped TiO_2 nanotube arrays with enhanced photo-electrochemical property," *International Journal of Photoenergy*, vol. 2012, Article ID 794207, 7 pages, 2012.

[23] E. Balaur, J. M. MacAk, L. Taveira, and P. Schmuki, "Tailoring the wettability of TiO_2 nanotube layers," *Electrochemistry Communications*, vol. 7, no. 10, pp. 1066–1070, 2005.

[24] L. Yang, S. Luo, R. Liu et al., "Fabrication of cdse nanoparticles sensitized long TiO_2 nanotube arrays for photocatalytic degradation of anthracene-9-carbonxylic acid under green monochromatic light," *Journal of Physical Chemistry C*, vol. 114, no. 11, pp. 4783–4789, 2010.

[25] J. D. Dow and D. Redfield, "Electroabsorption in semiconductors: the excitonic absorption edge," *Physical Review B*, vol. 1, no. 8, pp. 3358–3371, 1970.

[26] M. Samadpour, P. P. Boix, S. Giménez et al., "Fluorine treatment of TiO_2 for enhancing quantum dot sensitized solar cell performance," *Journal of Physical Chemistry C*, vol. 115, no. 29, pp. 14400–14407, 2011.

[27] Y. L. Lee, B. M. Huang, and H. T. Chien, "Highly efficient CdSe-sensitized TiO_2 photoelectrode for quantum-dot-sensitized solar cell applications," *Chemistry of Materials*, vol. 20, no. 22, pp. 6903–6905, 2008.

[28] S. C. Hayden, N. K. Allam, and M. A. El-Sayed, "TiO_2 nanotube/CdS hybrid electrodes: extraordinary enhancement in the inactivation of escherichia coli," *Journal of the American Chemical Society*, vol. 132, no. 41, pp. 14406–14408, 2010.

[29] J. Chen, D. W. Zhao, J. L. Song et al., "Directly assembled CdSe quantum dots on TiO_2 in aqueous solution by adjusting pH value for quantum dot sensitized solar cells," *Electrochemistry Communications*, vol. 11, no. 12, pp. 2265–2267, 2009.

[30] P. P. Boix, G. Larramona, A. Jacob, B. Delatouche, and J. Bisquert, "Hole transport and recombination in all-solid Sb2S3-sensitized TiO_2 solar cells using CuSCN as hole transporter," *The Journal of Physical Chemistry C*, vol. 116, no. 1, pp. 1579–1587, 2012.

13

Annealing Effect of ZnO Seed Layer on Enhancing Photocatalytic Activity of ZnO/TiO_2 Nanostructure

Woo-Young Kim,[1] Soon-Wook Kim,[1] Dae-Hwang Yoo,[1,2] Eui Jung Kim,[3] and Sung Hong Hahn[1,2]

[1] *Department of Physics, University of Ulsan, Ulsan 680-749, Republic of Korea*
[2] *Basic Science Research Institute (Energy Harvest-Storage Research Center), University of Ulsan, Ulsan 680-749, Republic of Korea*
[3] *Department of Chemical Engineering, University of Ulsan, Ulsan 680-749, Republic of Korea*

Correspondence should be addressed to Sung Hong Hahn; shhahn@ulsan.ac.kr

Academic Editor: David Lee Phillips

Zinc oxide (ZnO)/titanium dioxide (TiO_2) nanorods have been synthesized via a hydrothermal method for ZnO nanorods and an electron-beam deposition for TiO_2 nanorods. This work examined the effect of annealing ZnO seed layer on the photocatalytic activity of the ZnO/TiO_2 nanorods which was determined from photodecomposition of methylene blue under UV irradiation. The photocatalytic activity of the ZnO/TiO_2 nanorods was improved with increasing annealing temperature of the seed layer from 300°C to 500°C. Annealing the seed layer at 500°C showed the best photocatalytic activity resulting from high UV absorption ability, a large surface area with flower structure and copious oxygen defects which promote separation of electron-hole pairs reducing electron recombination. The prepared nanorods were characterized by field emission-scanning electron microscopy (FE-SEM), X-ray diffraction (XRD), photoluminescence (PL), and UV-visible spectroscopy.

1. Introduction

The synthesis of semiconductor nanomaterial is of great worldwide interest. The nanomaterials have enhanced optical and electronic properties, finding their application in various fields such as optoelectronic nanodevices and optical excitations [1–3]. Titanium dioxide (TiO_2) is the most promising material for a photocatalyst, because it has good photoactivity, long-term stability, and low cost. Anatase phase, one of TiO_2 crystal structures, shows higher photoactivity than rutile [4, 5]. Zinc oxide (ZnO), which has a direct band gap of 3.37 eV and large exciton binding energy of 60 meV, has a high mechanical and a thermal stability [6–8]. In addition, it can be utilized as a source which makes blue/UV optoelectronic devices due to its photochemical properties [9]. It is known that ZnO/TiO_2 nanomaterials promote charge separation decreasing electron-hole recombination [10]. Furthermore, they have high chemical and thermal stability and high absorption in the UV region [11]. As a result, the ZnO/TiO_2 nanorods show an enhanced photocatalytic activity. The fabrication of one-dimensional (1D) nanostructures has attracted considerable attention due to their great optical and electrical properties [12]. Various 1D of ZnO nanostructures such as nanorods, nanowires, and nanorings have been synthesized by physical and chemical methods including molecular beam epitaxy (MBE) [13], plasma-enhanced chemical vapor deposition (PECVD) [14], thermal evaporation [15], and template based method [16]. The wet chemical processes are a cost-effective low-temperature technique [17–20]. The hydrothermal technique is an effective method to control the 1D ZnO nanostructure [21], which is surface independent [22]. The physical properties of the ZnO/TiO_2 nanorods relate to their nanostructure which is dependent on deposition conditions and annealing. Especially, the annealing of the ZnO seed layer improved adhesion on substrate, which contributed to well-aligned ZnO nanorods [23]. The annealing process is important to obtain desirable structural and optical properties. The main contributions of seed layer annealing may come from the improvement of the crystallinity and absorption which could give an significantly effect on the photocatalytic acitivity of ZnO/TiO_2 nanorods [24].

In this study, we investigate the effect of seed layer annealing of the ZnO/TiO_2 nanorods on the photocatalytic activity. The TiO_2 nanorods were prepared by electron-beam evaporation. ZnO seed solution was then spin-coated on the TiO_2 nanorods, which was followed by annealing at 300°C, 400°C, and 500°C. Subsequently, the ZnO nanorods have been grown by hydrothermal method.

2. Experiments

2.1. Preparation of TiO_2 Nanorods. TiO_2 nanorods were obliquely deposited on quartz substrate using electron-beam evaporation. TiO_2 tablets with high purity (99.99%) were used as a raw material. After the deposition chamber was evacuated to a base pressure of 6×10^{-6} torr, the TiO_2 films were deposited at a working oxygen gas pressure of 5×10^{-5} torr. The working current was 200 mA and the electron gun voltage was 7.0 kV. The electron beam was irradiated at an incident angle of 75° to obtain the TiO_2 nanorods. The prepared nanorods were annealed at 600°C for 1 hour to form anatase structure.

2.2. Growth of ZnO Nanorods. ZnO nanorods were grown on the prepared TiO_2 nanorods via a hydrothermal method. ZnO seed solution was prepared by dissolving 5 mM of zinc acetate dihydrate ($C_4H_6O_4Zn{\cdot}2H_2O$) in 100 mL of methoxy ethanol and then stirring for 24 hours. The ZnO seed solution was spin-coated on the TiO_2 nanorods/quartz substrate twice. The ZnO seed coated sample was annealed at 300°C, 400°C, and 500°C, which are referred to as ZT-300, ZT-400, and ZT-500, respectively. The ZnO growth solution was prepared by dissolving 16 mM of zinc nitrate hexahydrate ($Zn\ (NO_3)_2{\cdot}6H_2O$) and 25 mM of hexamethylenetetramine (HMT) into 200 mL of distilled water. The ZnO seed coated sample was immersed into the growth solution at 90°C for 4 hours. The sample was then removed from the solution, washed with distilled water, and annealed at 350°C for 1 hour. For comparison, ZnO nanorods were grown without TiO_2, and their seed layers were annealed at 300°C, 400°C, and 500°C, which are labeled as Z-300, Z-400, and Z-500, respectively.

2.3. Characterization and Photocatalytic Activity. The surface morphology of the ZnO/TiO_2 and TiO_2 nanorods was examined by field emission-scanning electron microscopy (FE-SEM, Supra 40, Carl Zeiss, Swiss). The X-ray diffraction (XRD) patterns were measured with a Rint 2000 vertical goniometer using Cu Kα radiation at 40 kV and 100 mA. The photoluminescence (PL) was measured by a SPEX 750 M PL spectrometer with 325 nm He-Cd laser with 4 mW output power at room temperature. UV-vis absorbance spectra were examined to investigate absorption property of the ZnO/TiO_2 nanorods. The photocatalytic activities of the ZnO/TiO_2 and ZnO nanorods were examined using a UV-visible spectrophotometer (HP 8453) by measuring the photodegradation rate of methylene blue (MB) with an initial concentration of 1×10^{-5} mol/L. The thin films were immersed in the MB solution in a tubular quartz reactor under four surrounding UV lamps of 20 W (wavelength = 352 nm). After dark adsorption equilibrium for 30 min, photodegradation of the MB was measured by UV-vis spectrophotometer (HP 8453) at λ_{max} = 664 nm.

3. Results and Discussion

Figure 1 shows the SEM images of TiO_2 nanorods (as the buffer layer), ZT-300, ZT-400, and ZT-500. The obliquely deposited TiO_2 nanorod arrays have a high surface area as shown in Figure 1(a). Figure 1 observed that the size and structure of the ZnO nanorods depend on the annealing temperature of the seed layer. From the SEM results, the sizes of the TiO_2 nanorods and ZT-300, ZT-400, and ZT-500 were evaluated to diameter with (a) 21.1 nm, (b) 306.8 nm, (c) 210.2 nm, and (d) 96.6 nm, and length with (e) 60.0 nm, (f) 2.98 μm, (g) 2.48 μm, and (h) 1.91 μm. In the case of ZT-500, a flower-like structure composed of thin nanorods was formed, which relates to the polycrystalline structure of the large-size ZnO seed as shown in Figure 1(c). The big grains are polycrystalline ZnO seeds consisting of plenty of small crystallites [24]. The ZnO nanorods have been grown on the TiO_2 nanorod arrays with a porous surface. As a result, the agglomerate of ZnO nanorods with various directions was formed [19].

Figure 2 displays the XRD patterns of the Z-300, ZT-300, ZT-400, and ZT-500. In case of the ZnO (JCPDS card no. 79-2205) nanorods, the (100), (002), (101), and (102) peaks appeared. The high intensity (002) peak indicates that the prepared ZnO nanorods have hexagonal wurtzite structure and are grown along the *c*-axis direction to the substrate surface [25, 26]. With the increase of the seed annealing temperature, the intensity of the (002) peak decreased due to a decrease in the reinforced atomic diffusion between the ZnO and TiO_2. When the seed layer of the ZnO/TiO_2 nanorods was annealed at high temperature, the atoms can acquire energy enough to move, leading to strengthened interdiffusion motion between the ZnO and TiO_2 [27]. The apparent (101) peaks of the TiO_2 (JCPDS card no. 84-1286) nanorods indicate anatase TiO_2. It is well known that the anatase TiO_2 exhibits a better photocatalytic activity than rutile TiO_2 [4]. Due to the presence of a ZnO/TiO_2 heterojunction, the recombination of electron-hole pairs will be restrained. This improves the availability of the electrons to transfer to the TiO_2 surface of the ZnO/TiO_2 nanorods and consequently causes a redox reaction [28].

In Figure 3, the PL spectra of the ZT-300, ZT-400, and ZT-500 show that the UV emission peak of the ZnO/TiO_2 nanorods appeared at 382 nm due to the recombination of free exciton [29, 30]. The peak of the ZT-500 was higher than those of other samples, implying that the crystallinity of the ZnO/TiO_2 nanorods was improved with increasing annealing temperature of the seed layers. A visible peak of the ZnO/TiO_2 nanorods appeared at 600 nm corresponding to

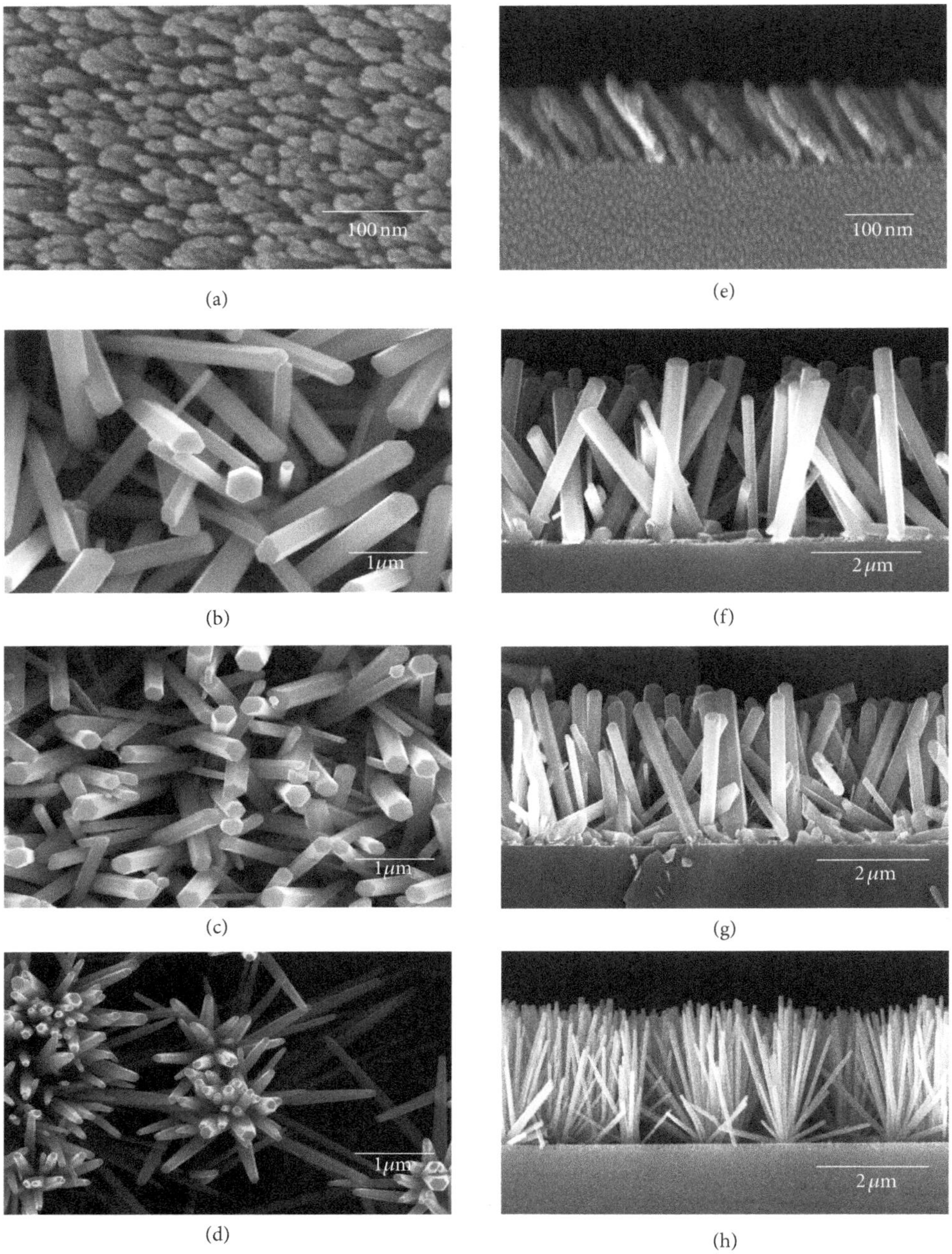

Figure 1: SEM images of the top view of (a) TiO_2 nanorods, (b) ZT-300, (c) ZT-400, and (d) ZT-500 and cross-sectional view of (e) TiO_2 nanorods, (f) ZT-300, (g) ZT-400, and (h) ZT-500.

the orange emission. It is related to the levels within band gap caused by defects, and in this case it is associated with oxygen vacancy defects [30]. The orange emission is due to the deep defects and it means that oxygen vacancy defects were deeply inserted in the band gap [31, 32]. The emission intensity of the ZnO/TiO_2 nanorods was enhanced with increasing annealing temperature of the seed layers. The inset of Figure 3 shows that the intensities of the visible peaks of the ZT-500 and Z-500 at 600 nm are greatly different. This implies that TiO_2 nanorods with ZnO seed layers annealed at 500°C work as supportive materials in the intense increase of orange emission in many ways [29]. The first thing is that they made the radius of ZnO nanorods reduce which is supported from SEM analysis in Figure 1. It leads to the increase of surface area active to the irradiating light and contribute to the increase of PL peak [33]. The second thing is to cause the increase of defects. Atomic diffusion between the ZnO and TiO_2 nanorods becomes vigorous at high temperature, resulting in the increase of defects [27]. A difference in the photocatalytic activity is related to the concentration of oxygen defects

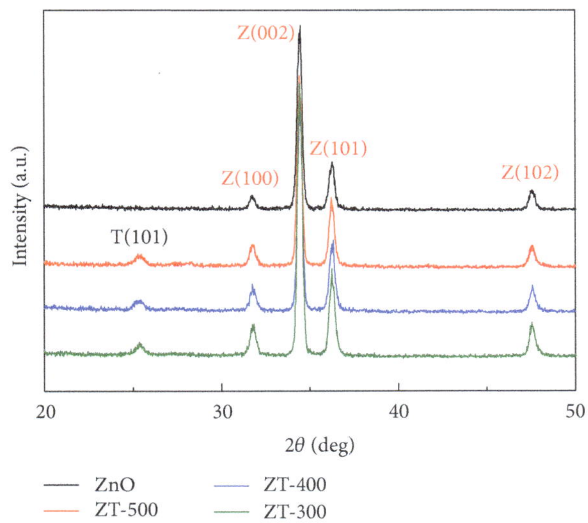

FIGURE 2: XRD patterns of ZT-300, ZT-400, ZT-500, and ZnO.

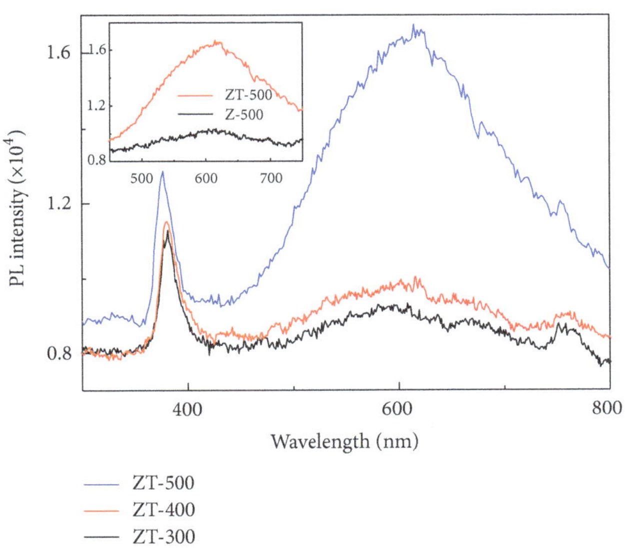

FIGURE 3: Room temperature PL spectra of ZT-300, ZT-400, and ZT-500. The inset shows visible emission of Z-500, and ZT-500.

on the surface layer as well as surface adsorption ability. The oxygen vacancy defects work as an electron acceptor and trap photogenerated electrons decreasing the surface recombination of electron-hole pairs. As described above, the oxygen vacancy and the interstitial oxygen defects promote separation of electron-hole pairs to minimize the electron recombination, leading to higher photocatalytic activity [30, 34, 35].

Figure 4 shows the UV-visible absorbance spectra of the ZT-300, ZT-400, and ZT-500. The absorption edge was observed to be located at 371 nm for the crystalline ZnO. The absorbance increased significantly with increasing the annealing temperature of the seed layers. It results from a large surface area of ZT-500 with flower structure composed of thin nanorods as shown in Figure 1. It is that

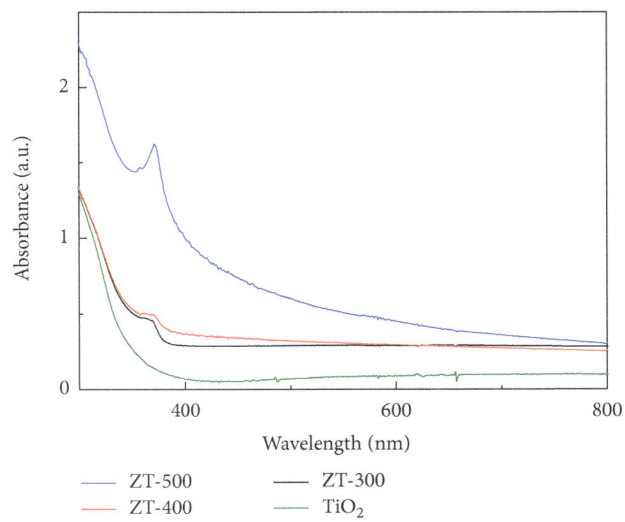

FIGURE 4: UV-vis absorbance spectra of ZT-300, ZT-400, ZT-500, and TiO_2.

the amorphous phase is reduced with increasing the seed layer annealing temperature since more energy is supplied for crystalline growth, thus resulting in an improvement in crystallinity of the ZT-500. The relationship of the absorption coefficient and the incident photon energy of ZnO and TiO_2 nanorods is given by the equation $\alpha(h\upsilon) \propto (h\upsilon - E_0)^n$, where α is the absorption coefficient, $h\upsilon$ is the energy of the incident photon, n is 0.5 for the semiconductor with direct band gap, and E_0 is value of energy gap [36]. According to the above equation, the band gap of the ZnO nanorods was determined as 3.22 eV and that of the TiO_2 nanorods as 3.19 eV.

The photocatalytic activities of the ZnO/TiO_2 and ZnO nanorods were evaluated by the degradation of the MB under the UV irradiation. Figure 5(a) shows the UV-visible absorption spectra of the MB for the ZT-500 with various UV irradiation times. The MB showed a major absorption band at 665 nm which became transparent after 120 min irradiation, resulting from facile destruction of the chromophoric structure of the MB [37].

The photocatalytic activity of the samples can be represented by a plot of C/C_o versus UV irradiation time [38]. Here C is the concentration at irradiation time t and C_o is the initial concentration. Figure 5(b) shows the photocatalytic activity of the ZnO/TiO_2 and ZnO nanorods by the photodegradation of the MB under the UV irradiation. The MB was decomposed by 76.2%, 83.3%, 87.0%, and 95.5% after 4 h of the UV irradiation for Z-500, ZT-300, ZT-400, and ZT-500, respectively. The samples with a heterojunction of the ZnO/TiO_2 nanorods showed an enhanced photocatalytic activity than the ZnO nanorods. Improved photocatalytic activity comes from a high surface area of the TiO_2 nanostructure [27], and coupling effect of ZnO and TiO_2, which increases interfacial charge transfer and decreases the recombination of the electrons and holes by promoting their separation [39–41]. The ZT-500 showed

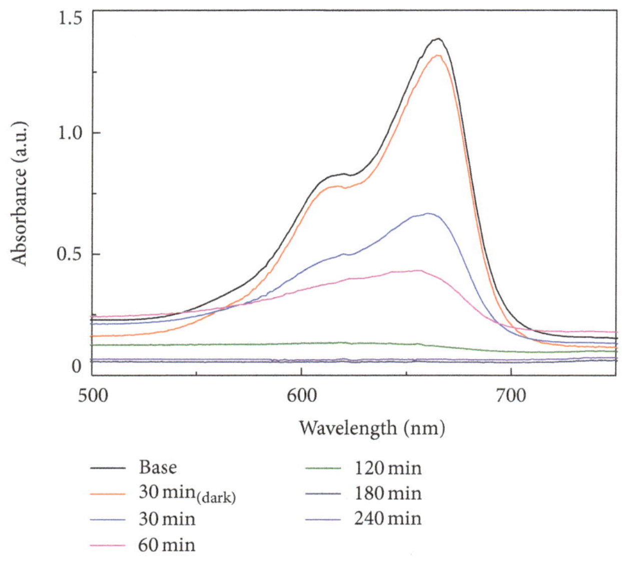

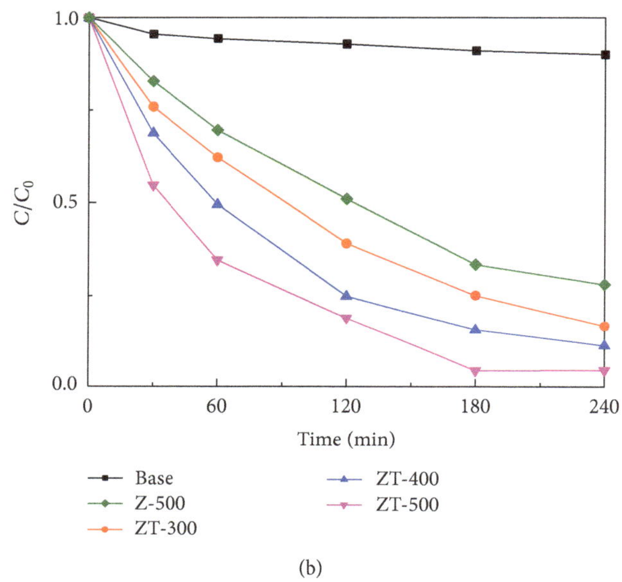

FIGURE 5: (a) UV-vis absorbance spectra of MB for the ZT-500. (b) Photodegradation of the MB for ZT-300, ZT-400, ZT-500, and Z-500.

the best photocatalytic activity, indicating that defects on the ZnO nanorods can act as a dominant factor for enhancing photocatalytic activity. In addition, the ZT-500 has a high surface area leading to high absorption ability compared with those of other samples and a high ability of the generating photoinduced electron-hole pairs [37].

Figure 6 shows the diagram of the band gap structure of the ZnO/TiO_2 nanorods. The photogenerated electrons transfer from the conduction band of ZnO to the conduction band of TiO_2. This leads to a decrease in the electron-hole pairs recombination by promoting their separation, and the lifetime of the charge carriers is increased [39, 40]. The charge separation increases the efficiency of the interfacial charge

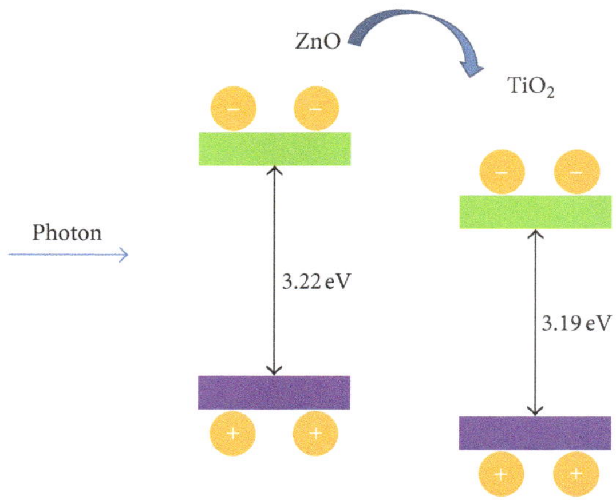

FIGURE 6: A schematic diagram of the electron and hole transfers for the ZnO/TiO_2 nanorods.

transfer and improves the redox process [40]. Therefore, the ZnO/TiO_2 nanorods have a strongly positive effect on the photocatalytic activity.

4. Conclusion

TiO_2 nanorod structure was prepared by electron-beam evaporation method through oblique beam incidence. ZnO seed layers were coated on TiO_2 and annealed at 300–500°C, which was followed by ZnO nanorod growth to form heterojunction structure of ZnO/TiO_2. The annealed ZnO seed layer was found to have a significant influence on the morphology of the ZnO/TiO_2 nanorod arrays. With increasing the annealing temperature of the ZnO seed layers, the size of the ZnO nanoparticles was gradually increased, leading to different morphology of ZT-500 composed of the thin nanorods which have a large surface area and a high absorption ability. The heterostructure of ZnO/TiO_2 can promote the separation of the electron-hole pairs and decrease their recombination enhancing photocatalytic reaction. When the annealing temperature of the seed layers reaches a relatively high temperature of 500°C, the atomic diffusion between the ZnO and TiO_2 nanorods is significantly enhanced, resulting in more interface defects which reduce the electron-hole recombination. As a result, ZnO/TiO_2 nanorods with high annealing temperature of the seed layers exhibited excellent photocatalytic activities for the decomposition of the MB.

Acknowledgments

This research was supported by the Human Resource Training Project for Regional Innovation (2012H1B8A2026179) and Basic Science Research Program (2012R1A1A4A010-15466) through the National Research Foundation of Korea (NRF) funded by the Ministry of Education Science and Technology (MEST).

References

[1] Y. Sun and M. N. R. Ashfold, "Photoluminescence from diameter-selected ZnO nanorod arrays," *Nanotechnology*, vol. 18, no. 24, Article ID 245701, 2007.

[2] M. Yang, G. Pang, L. Jiang, and S. Feng, "Hydrothermal synthesis of one-dimensional zinc oxides with different precursors," *Nanotechnology*, vol. 17, no. 1, pp. 206–212, 2006.

[3] J. Hu, T. W. Odom, and C. M. Lieber, "Chemistry and physics in one dimension: synthesis and properties of nanowires and nanotubes," *Accounts of Chemical Research*, vol. 32, no. 5, pp. 435–445, 1999.

[4] T. Kawahara, Y. Konishi, H. Tada, N. Tohge, J. Nishii, and S. Ito, "A patterned TiO_2(Anatase)/TiO_2(Rutile) bilayer-type photocatalyst: effect of the anatase/rutile junction on the photocatalytic activity," *Angewandte Chemie*, vol. 114, no. 15, pp. 2935–2937, 2002.

[5] K. S. Lee and I. S. Park, "Anatase-phase titanium oxide by low temperature oxidation of metallic Ti thin film," *Scripta Materialia*, vol. 48, no. 6, pp. 659–663, 2003.

[6] B. H. Kong and H. K. Cho, "Formation of vertically aligned ZnO nanorods on ZnO templates with the preferred orientation through thermal evaporation," *Journal of Crystal Growth*, vol. 289, no. 1, pp. 370–375, 2006.

[7] M. H. Huang, S. Mao, H. Feick et al., "Room-temperature ultraviolet nanowire nanolasers," *Science*, vol. 292, no. 5523, pp. 1897–1899, 2001.

[8] Y. Chen, D. M. Bagnall, H. J. Koh et al., "Plasma assisted molecular beam epitaxy of ZnO on c-plane sapphire: growth and characterization," *Journal of Applied Physics*, vol. 84, no. 7, pp. 3912–3918, 1998.

[9] R. S. Mane, W. J. Lee, H. M. Pathan, and S. H. Han, "Nanocrystalline TiO_2/ZnO thin films: fabrication and application to dye-sensitized solar cells," *Journal of Physical Chemistry B*, vol. 109, no. 51, pp. 24254–24259, 2005.

[10] Z. Zhang, Y. Yuan, Y. Fang, L. Liang, H. Ding, and L. Jin, "Preparation of photocatalytic nano-ZnO/TiO_2 film and application for determination of chemical oxygen demand," *Talanta*, vol. 73, no. 3, pp. 523–528, 2007.

[11] L. Xu, L. Shi, and X. Li, "Effect of TiO_2 buffer layer on the structural and optical properties of ZnO thin films deposited by E-beam evaporation and sol-gel method," *Applied Surface Science*, vol. 255, no. 5, pp. 3230–3234, 2008.

[12] Y. Xia, P. Yang, Y. Sun et al., "One-dimensional nanostructures: synthesis, characterization, and applications," *Advanced Materials*, vol. 15, no. 5, pp. 353–389, 2003.

[13] R. Calarco, M. Marso, T. Richter et al., "Size-dependent photoconductivity in MBE-grown GaN—nanowires," *Nano Letters*, vol. 5, no. 5, pp. 981–984, 2005.

[14] X. Liu, X. Wu, H. Cao, and R. P. H. Chang, "Growth mechanism and properties of ZnO nanorods synthesized by plasma-enhanced chemical vapor deposition," *Journal of Applied Physics*, vol. 95, no. 6, pp. 3141–3147, 2004.

[15] Z. W. Pan, Z. R. Dai, L. Xu, S. T. Lee, and Z. L. Wang, "Temperature-controlled growth of silicon-based nanostructures by thermal evaporation of SiO powders," *Journal of Physical Chemistry B*, vol. 105, no. 13, pp. 2507–2514, 2001.

[16] B. Cheng and E. T. Samulski, "Hydrothermal synthesis of one-dimensional ZnO nanostructures with different aspect ratios," *Chemical Communications*, vol. 10, no. 8, pp. 986–987, 2004.

[17] B. Liu and H. C. Zeng, "Hydrothermal synthesis of ZnO nanorods in the diameter regime of 50 nm," *Journal of the American Chemical Society*, vol. 125, no. 15, pp. 4430–4431, 2003.

[18] X. Y. Zhang, J. Y. Dai, H. C. Ong, N. Wang, H. L. W. Chan, and C. L. Choy, "Hydrothermal synthesis of oriented ZnO nanobelts and their temperature dependent photoluminescence," *Chemical Physics Letters*, vol. 393, no. 1–3, pp. 17–21, 2004.

[19] S. Baruah and J. Dutta, "Effect of seeded substrates on hydrothermally grown ZnO nanorods," *Journal of Sol-Gel Science and Technology*, vol. 50, no. 3, pp. 456–464, 2009.

[20] A. Sugunan, H. C. Warad, M. Boman, and J. Dutta, "Zinc oxide nanowires in chemical bath on seeded substrates: role of hexamine," *Journal of Sol-Gel Science and Technology*, vol. 39, no. 1, pp. 49–56, 2006.

[21] M. K. Hossain, S. C. Ghosh, Y. Boontongkong, C. Thanachayanont, and J. Dutta, "Growth of Zinc Oxide nanowires and nanobelts for gas sensing applications," *Journal of Metastable and Nanocrystalline Materials*, vol. 23, pp. 27–30, 2005.

[22] L. E. Greene, M. Law, J. Goldberger et al., "Low-temperature wafer-scale production of ZnO nanowire arrays," *Angewandte Chemie*, vol. 42, no. 26, pp. 3031–3034, 2003.

[23] J. Chung, J. Lee, and S. Lim, "Annealing effects of ZnO nanorods on dye-sensitized solar cell efficiency," *Physica B*, vol. 405, no. 11, pp. 2593–2598, 2010.

[24] M. Wang, E. J. Kim, J. S. Chung et al., "Influence of annealing temperature on the structural and optical properties of sol-gel prepared ZnO thin films," *Physica Status Solidi A*, vol. 203, no. 10, pp. 2418–2425, 2006.

[25] C. W. Zou, X. D. Yan, J. Han et al., "Preparation and enhanced photoluminescence property of ordered ZnO/TiO_2 bottlebrush nanostructures," *Chemical Physics Letters*, vol. 476, no. 1–3, pp. 84–88, 2009.

[26] N. Wang, C. Sun, Y. Zhao, S. Zhou, P. Chen, and L. Jiang, "Fabrication of three-dimensional ZnO/TiO_2 heteroarchitectures via a solution process," *Journal of Materials Chemistry*, vol. 18, no. 33, pp. 3909–3911, 2008.

[27] L. Xu, H. Shen, X. Li, and R. Zhu, "Influence of annealing temperature on the photoluminescence property of ZnO thin film covered by TiO_2 nanoparticles," *Journal of Luminescence*, vol. 130, no. 11, pp. 2123–2127, 2010.

[28] H. Y. Yang, S. F. Yu, S. P. Lau, X. Zhang, D. D. Sun, and G. Jun, "Direct growth of ZnO nanocrystals onto the surface of porous TiO_2 nanotube arrays for highly efficient and recyclable photocatalysts," *Small*, vol. 5, no. 20, pp. 2260–2264, 2009.

[29] J. H. Hyung, M. S. Akhtar, D. J. Kim, and S. K. Lee, "ZnO-nanowire-covered TiO_2 thin-film electrodes for improving the photovoltaic properties of dye-sensitized solar cells," *Journal of the Korean Physical Society*, vol. 55, no. 1, pp. 89–93, 2009.

[30] Y. Zheng, C. Chen, Y. Zhan et al., "Luminescence and photocatalytic activity of ZnO nanocrystals: correlation between structure and property," *Inorganic Chemistry*, vol. 46, no. 16, pp. 6675–6682, 2007.

[31] S. Parida, S. K. Rout, L. S. Cavalcante et al., "Structural refinement, optical and microwave dielectric properties of $BaZrO_3$," *Ceramics International*, vol. 38, no. 3, pp. 2129–2138, 2012.

[32] A. T. de Figueiredo, V. M. Longo, R. O. da Silva et al., "Structural XANES characterization of $Ca_{0.99}Sm_{0.01}TiO_3$ perovskite and correlation with photoluminescence emission," *Chemical Physics Letters*, vol. 544, pp. 43–48, 2012.

[33] A. E. Souza, G. T. A. Santos, B. C. Barra et al., "Photoluminescence of $SrTiO_3$: influence of particle size and morphology," *Crystal Growth & Design*, vol. 12, no. 11, pp. 5671–5679, 2012.

[34] J. Wang, P. Liu, X. Fu, Z. Li, W. Han, and X. Wang, "Relationship between oxygen defects and the photocatalytic property of zno nanocrystals in nafion membranes," *Langmuir*, vol. 25, no. 2, pp. 1218–1223, 2009.

[35] J. Lv, W. Gong, K. Huang et al., "Effect of annealing temperature on photocatalytic activity of ZnO thin films prepared by sol-gel method," *Superlattices and Microstructures*, vol. 50, no. 2, pp. 98–106, 2011.

[36] Z. Zhang, Y. Yuan, L. Liang, Y. Cheng, G. Shi, and L. Jin, "Preparation and photoelectrocatalytic activity of ZnO nanorods embedded in highly ordered TiO_2 nanotube arrays electrode for azo dye degradation," *Journal of Hazardous Materials*, vol. 158, no. 2-3, pp. 517–522, 2008.

[37] D. Chen, H. Zhang, S. Hu, and J. Li, "Preparation and enhanced photoelectrochemical performance of coupled bicomponent ZnO-TiO_2 nanocomposites," *Journal of Physical Chemistry C*, vol. 112, no. 1, pp. 117–122, 2008.

[38] L. Palmisano, M. Schiavello, A. Sclafani, C. Martin, I. Martin, and V. Rives, "Surface properties of iron-titania photocatalysts employed for 4-nitrophenol photodegradation in aqueous TiO_2 dispersion," *Catalysis Letters*, vol. 24, no. 3-4, pp. 303–315, 1994.

[39] M. A. Kanjwal, N. A. M. Barakat, F. A. Sheikh, S. J. Park, and H. Y. Kim, "Photocatalytic activity of ZnO-TiO_2 hierarchical nanostructure prepared by combined electrospinning and hydrothermal techniques," *Macromolecular Research*, vol. 18, no. 3, pp. 233–240, 2010.

[40] X. Yan, C. Zou, X. Gao, and W. Gao, "ZnO/TiO_2 core-brush nanostructure: processing, microstructure and enhanced photocatalytic activity," *Journal of Materials Chemistry*, vol. 22, pp. 5629–5640, 2012.

[41] M. Zhang, T. An, X. Liu, X. Hu, G. Sheng, and J. Fu, "Preparation of a high-activity ZnO/TiO_2 photocatalyst via homogeneous hydrolysis method with low temperature crystallization," *Materials Letters*, vol. 64, no. 17, pp. 1883–1886, 2010.

Fabrication of a $Cu(InGa)Se_2$ Thin Film Photovoltaic Absorber by Rapid Thermal Annealing of CuGa/In Precursors Coated with a Se Layer

Chun-Yao Hsu,[1] Peng-Cheng Huang,[1] Yu-Yao Chen,[2] and Dong-Cherng Wen[3]

[1] *Department of Mechanical Engineering, Lunghwa University of Science and Technology, Taoyuan 33306, Taiwan*
[2] *Department of Industrial Engineering, Shanghai Dianji University, Shanghai 201306, China*
[3] *Department of Mechanical Engineering, China University of Science and Technology, Taipei 11581, Taiwan*

Correspondence should be addressed to Dong-Cherng Wen; dcwen@cc.cust.edu.tw

Academic Editor: Ho Chang

$Cu(InGa)Se_2$ (CIGS) thin film absorbers are prepared using sputtering and selenization processes. The CuGa/In precursors are selenized during rapid thermal annealing (RTA), by the deposition of a Se layer on them. This work investigates the effect of the Cu content in precursors on the structural and electrical properties of the absorber. Using X-ray diffraction, field emission scanning electron microscopy, Raman spectroscopy, and Hall effect measurement, it is found that the CIGS thin films produced exhibit facetted grains and a single chalcopyrite phase with a preferred orientation along the (1 1 2) plane. A Cu-poor precursor with a Cu/(In + Ga) ratio of 0.75 demonstrates a higher resistance, due to an increase in the grain boundary scattering and a reduced carrier lifetime. A Cu-rich precursor with a Cu/(In + Ga) ratio of 1.15 exhibits an inappropriate second phase ($Cu_{2-x}Se$) in the absorber. However, the precursor with a Cu/(In + Ga) ratio of 0.95 exhibits larger grains and lower resistance, which is suitable for its application to solar cells. The deposition of this precursor on Mo-coated soda lime glass substrate and further RTA causes the formation of a $MoSe_2$ layer at the interface of the Mo and CIGS.

1. Introduction

The use of polycrystalline $Cu(InGa)Se_2$ (CIGS) thin films as the absorber material for thin film solar cells allows easier commercial production because of the suitability of its bandgap and its high absorption coefficient for solar radiation [1]. An important feature of CIGS is that high-efficiency solar cells can be processed with a large range of Cu contents. The Cu content of the final film has a significant influence on the modular and cellular properties. At low Cu contents ($Cu/(In+Ga) < 0.7$) the conductivity of the CIGS is very low, and therefore shunt paths, either within a single solar cell or at the boundaries between cells in a module, are suppressed. Cu-rich CIGS ($Cu/(In+Ga) > 1$) is inappropriate for use as an absorber material, since the formation of a highly conductive $Cu_{2-x}Se$ phase produces shunt paths.

Although several manufacturing processes have been developed [2–4], the two-stage method is the cheapest and the easiest to perform on an industrial scale [5, 6]. For a two-stage method, the compositional uniformity and surface morphology of metallic precursors affect the quality of the absorber layer. To ensure the compositional uniformity of precursor films, many procedures using stacked metal or alloy layers, such as In/CuGa, CuGa/In, CuGa/In/CuGa, or In/CuGa/In films, have been proposed [7, 8]. In traditional two-stage growth processes, Cu–Ga–In metallic precursors are selenized in an elemental Se vapor or a H_2Se/Ar gas mixture to form CIGS. However, Ga accumulation near the Mo side of the substrate/Mo/CIGS structure is often observed during the selenization of Cu–Ga–In precursors, yielding phase-separated $CuInSe_2$ (CIS) and $CuGaSe_2$ (CGS), which subsequently result in lower open-circuit voltages [9]. Dejene [10] reported that the concentration of Ga in the quaternary alloys is altered by variation in the relative thickness of GaSe with respect to InSe and Cu, during the formation of the precursors. Koo et al. [11] successfully employed the rapid thermal annealing (RTA) of glass/Mo/$(In,Ga)_2Se_3$/CuSe stacked precursors that results in improved Ga homogeneity and greater control of $MoSe_2$ formation.

Compared to the other selenization methods, the progressive RTA treatment minimizes the defect density in the CIGS thin film by reducing the thermal budget and reaction time; thus, the conversion efficiency of solar cells is remarkably improved [12]. Therefore, good control of precursor structure, process parameters, and reaction routes can improve the quality of photovoltaic absorber [13, 14]. This study investigates the effect of stacked CuGa/In precursors with different Cu contents on the growth of CIGS thin films. The precursors are selenized during RTA, by deposition of a 2 μm thick Se layer on them.

2. Experimental Details

CIGS films were prepared using a two-stage process. The first stage involved the deposition of multilayer precursors on either thin soda lime glass (SLG) or Mo-coated SLG substrates, using DC-magnetron sputtering of CuGa alloy with 25 wt.% Ga and elemental In targets and the thermal evaporation of Se. Based on the preliminary experiments and our previous study [15], the bottom layer of a 300-nm-thick CuGa film was deposited at a power of 100 W and at room temperature. The In layer was prepared at room temperature, using a power of 40 W. Thickness of the In layer was adjusted by adjusting the deposition time, to produce precursors with Cu/(In +Ga) atomic ratios of 0.75, 0.95, and 1.15. The top 2 μm of the Se layer was deposited at 250°C, after preheating to 150°C for 10 min at a rate of 15°C/s. The substrate was rotated at 20 rpm during deposition, to improve the film uniformity.

The second stage involved the reaction of the Se-coated metallic precursors to the CIGS semiconductor in a RTA system that consisted of a quartz tube reactor with an inner diameter of 62 mm, a quartz sample tray, and an infrared (IR) heater. The quartz sample tray held up to six 2.5 cm × 2.5 cm samples. The IR heater produced a rapid increase in temperature, requiring only 1 min to reach 1000°C from room temperature. The benefits of high heating rates are that they prevent dewetting of the elemental Se layer at the initial stage of the heating process, and they allow short thermal cycles [14, 16]. As no external Se-vapor source was used during the reaction process, the deposition of an excess of Se onto the precursor stack compensates for the natural loss of a portion of Se during heating. To prevent the peel off of film from the substrate during selenization, the reactor was heated at a rate of 1°C/s. The heating profile for the selenization process is shown in Figure 1, which has three annealing steps. In the first stage, the temperature of the substrate was maintained at 150°C for 5 min, in order to homogenize the precursors. The temperature was then increased to 350°C, for 5 min, to form the chalcopyrite phase. Finally, the as-prepared films were selenized at 550°C for 5 min, to allow recrystallization and grain growth. Schematic of the fabrication process and phase evolution during the rapid thermal annealing of CuGa/In precursors coated with a Se layer is shown in Figure 2.

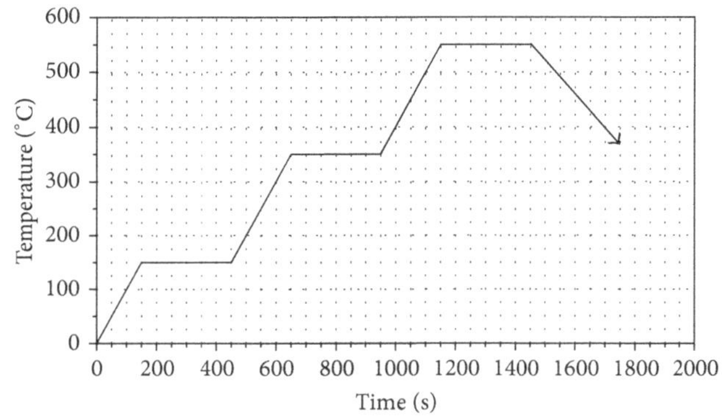

FIGURE 1: The heating profile for rapid thermal annealing.

The surface morphologies of the films were analyzed using field emission scanning electron microscopy (SEM, JEOL, JSM-6500F). Electrical resistivity was measured by the four-point probe method (Mitsubishi Chemical MCP-T600). The carrier concentration and Hall mobility were measured using the van der Pauw method, and Hall effect measurement (Nanometrics/Accent/NAN-HL5500PC) was performed at room temperature. The phases and crystal structure were determined by X-ray diffraction (Rigaku 2000 X-ray diffractometer), using Cu-Kα radiation and an angle incidence of 1°. The elemental concentrations and detailed compositional uniformity of CIGS films were determined by X-ray fluorescence (XRF) intensity measurements of the $K_{\alpha1,2}$ lines. The scanned $K_{\alpha1,2}$ lines were recorded using a sequential XRF wavelength dispersive spectrometer (SRS3000, Bruker-AXS, Rh-anode, 60 kV). In determining the detailed compositional uniformity the absorbers were repeatedly etched in bromine methanol, followed by XRF $K_{\alpha1,2}$ line intensity measurements. Chemical etching was conducted at room temperature, and the samples were rinsed in water and blow-dried with nitrogen, before XRF measurements were performed [16]. Raman scattering measurements were performed using a Horiba's LabRAM HR high-resolution spectrometer with a multichannel detection system, in the backscattering configuration.

3. Results and Discussion

Firstly, the precursors were deposited on a bare SLG substrate, to determine the proper Cu content for the fabrication of solar cells. Table 1 lists the elemental compositions of the as-deposited precursors and the CIGS films. The Cu/(In +Ga) ratios of three precursors were matched to the designed values. However, the selenization process caused this ratio to become slightly increased for samples A and C. Only sample B reached the expected ratio of 0.95 after selenization. In addition, the Ga/(In +Ga) ratio increases as the Cu/(In +Ga) ratio increases.

Figure 3 shows the XRD patterns for the precursors and CIGS films. As reported by Park et. al. [7], a pure In peak and intermetallic Cu_2In and Cu_3Ga as equilibrium phases at room temperature were observed on the precursor

TABLE 1: Elemental composition of the as-deposited precursors and CIGS films.

Sample	Precursor layer					CIGS layer					
	Compositions (at %)			Ratios		Compositions (at %)				Ratios	
	Cu	In	Ga	Cu/(In + Ga)	Ga/(In + Ga)	Cu	In	Ga	Se	Cu/(In + Ga)	Ga/(In + Ga)
A	42.94	44.96	12.10	0.753	0.212	22.78	21.96	6.88	48.38	0.790	0.239
B	48.49	37.04	13.83	0.953	0.272	24.26	18.46	7.07	50.3	0.950	0.277
C	53.52	31.55	14.93	1.151	0.321	26.89	14.38	7.71	47.62	1.217	0.349

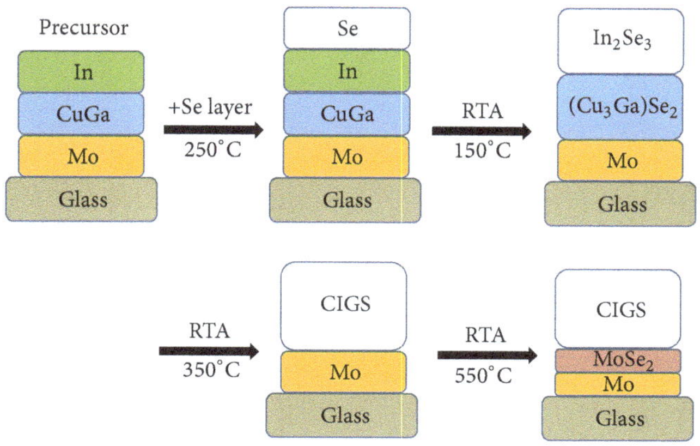

FIGURE 2: Schematic of the fabrication process and phase evolution during the rapid thermal annealing of CuGa/In precursors coated with a Se layer.

samples (Figure 3(a)). As the Cu/(In +Ga) ratio increases, the intensity of the Cu_3Ga phase increases, whereas the Cu_2In phase begins to disappear as the Cu/(In +Ga) ratio increases to 1.15. Figure 3(b) shows that all of the films exhibit the basic chalcopyrite crystal structure and very similar XRD patterns, the diffraction peak (1 1 2) is the strongest, only single-phase chalcopyrite CIGS is detected, and the intensity of (1 1 2) peak increases as the Cu/(In +Ga) ratio increases. The position of the diffraction peak (1 1 2) is shown in Figure 4 and demonstrates a clear shift to higher diffraction angles, because of a decrease in the lattice parameters due to the incorporation of Ga, which is expected when comparatively smaller Ga atoms replace the larger In atoms in the chalcopyrite lattice. It is also important to mention that this observation on a structural level is in line with from the results of other groups and demonstrates the variation in the electronic properties of CIGS films with changes in the Ga/(In +Ga) ratios [10, 17].

Figure 5 shows the surface SEM images of the CIGS thin films. In all three cases, dense and uniform films were produced. The film morphologies were characterized by the presence of facetted chalcopyrite grains, which are typical for device quality material and confirm the successful use of RTA in CuGa/In precursors coated with a Se layer to form CIGS films. It is apparent that the grain sizes become smaller as the Cu/(In +Ga) ratio decreases. This causes an increase in the full width at half maximum (FWHM) as the Cu content decreases, as shown in Figure 4. The FWHM is wider for Cu-poor absorber, because of the presence of defects and the reduced crystallinity of the films. This observation is in good agreement with the results of other related studies [18, 19].

Despite the chalcopyrite peak being clearly visible in the XRD spectra (Figure 3(b)), the Cu–Se compound peaks overlap with the chalcopyrite peak making a proper identification difficult. To obtain more precise information about the films' composition, Raman analyses of the films were performed. Figure 6 shows the Raman spectra of the CIGS absorber films with various Cu contents, measured at room temperature. A CIGS peak was observed at $177\,cm^{-1}$ only for samples A and B, with Cu/(In +Ga) < 1, indicating the presence of single-phase chalcopyrite in the film. For sample C, with Cu/(In +Ga) = 1.15, an additional peak appears at $260\,cm^{-1}$, which is attributable to the presence of Cu–Se compounds such as CuSe or Cu_2Se and is labeled as $Cu_{2-x}Se$ [20]. Kessler et al. [21] noted that the efficiency of solar cells decreases when a second phase ($Cu_{2-x}Se$) is present in the absorber.

Table 2 shows the sheet resistance, carrier concentration, and Hall mobility of the CIGS films. These selenized films demonstrate p-type conductivity. The resistance is a combined result of both the Hall mobility and the carrier concentration. As the Cu/(In +Ga) ratio increases from 0.75 to 0.95, the grain size increases. A larger grain size reduces the grain boundary scattering and increases the carrier lifetime, which results in an increase in conductivity, due to an increase in Hall mobility and carrier concentration. As a result, the sheet resistance of the CIGS films decreases. When the Cu/(In +Ga) ratio increases to 1.15, the film exhibits the lowest sheet resistance. However, a decrease in the resistance results in the formation of a highly conductive $Cu_{2-x}Se$ phase, rather than an increase in grain size, rendering it inappropriate for use as an absorber material.

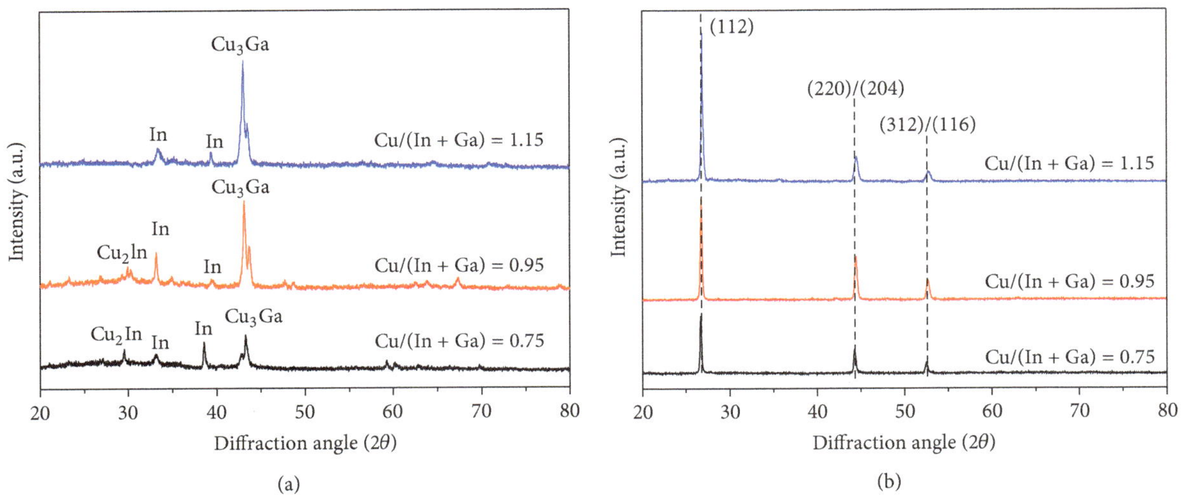

FIGURE 3: XRD patterns of (a) CIG precursors and (b) CIGS films. JCPDS: Cu_2In (00-026-0552), Cu_3Ga (00-044-1117), In (03-065-9292), CIGS (00-035-1102).

TABLE 2: The resistivity, carrier concentration, and Hall mobility of CIGS films.

Sample	Cu/(In + Ga)	Type	Sheet resistance (Ω/□)	Carrier concentration (cm^{-3})	Mobility ($cm^2 V^{-1} s^{-1}$)
A	0.790	P	1.7346E + 6	2.9969E + 14	8.1816E + 2
B	0.950	P	4.5617E + 5	2.3531E + 14	4.8521E + 2
C	1.217	P	9.4192E + 3	1.3566E + 14	5.9954E + 3

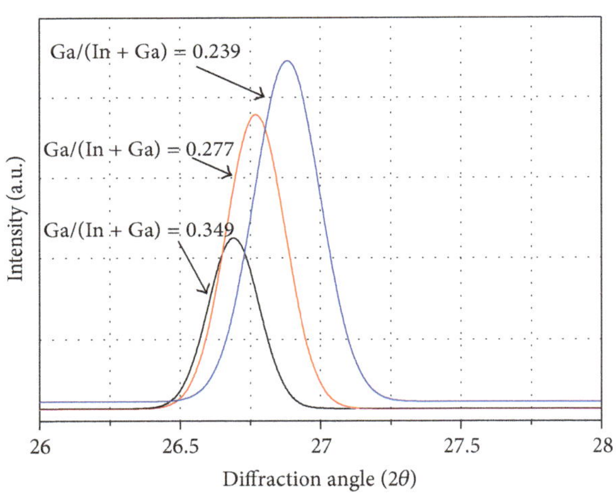

FIGURE 4: XRD patterns for the (1 1 2) reflection peak of the CIGS films.

Based on the above results, the growth conditions for sample B, with a Cu/(In +Ga) ratio of 0.95, were used for deposition onto the Mo-coated SLG substrate to produce solar cells. As with a previous study by the authors [15], the Mo back contact was grown using a RF power of 125 W, a working pressure of 3 mTorr, a substrate temperature of 200°C, and a deposition time of 50 min.

Figure 7 shows the depth profile of sample B, prepared on a Mo-coated SLG substrate by RTA of CuGa/In precursors covered with a Se layer. It is clearly evident that this process produces CIGS films with a high degree of detailed compositional uniformity, with no evidence of phase segregation. The corresponding XRD pattern of this CIGS film is shown in Figure 8. After RTA, the precursor is completely transformed into a chalcopyrite CIGS layer, and the Mo layer is partially converted to $MoSe_2$. The main diffraction peak of the preferred orientation is (1 1 2) at $2\theta = 26.9°$. No other obvious secondary phases are evident.

Figure 9 shows the cross-sectional SEM and surface morphology for the CIGS film from the precursor with a Cu/(In +Ga) ratio of 0.95. The columnar Mo back contact with narrow grains extending the full thickness of the film provides an effective pathway for Na diffusion from the SLG substrate to the absorber during selenization [22]. A thin $MoSe_2$ layer is formed at the interface of the Mo and the CIGS. This significantly improves the structural quality of the CIGS films and the electrical contact at the Mo/CIGS interface [23]. A dense CIGS layer with a thickness of about 1.2–1.3 μm, large grains (~1.2 μm), and no cracking or peeling phenomena is observed. The obtained results confirm that a qualified CIGS absorber layer was successfully fabricated when a precursor with Cu/(In +Ga) ratio of 0.95 was prepared onto the Mo-coated SLG substrate and selenized by RTA. These results allow a better understanding of how precursors influence the structural and electrical properties

FIGURE 5: The surface morphology of CIGS absorber films with Cu/(In +Ga) ratios of (a) 0.75, (b) 0.95, and (c) 1.15.

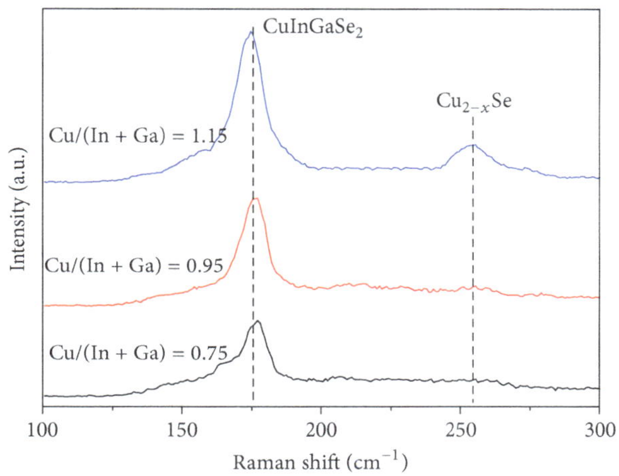

FIGURE 6: Raman spectra of CIGS absorber films with various Cu/(In +Ga) ratios.

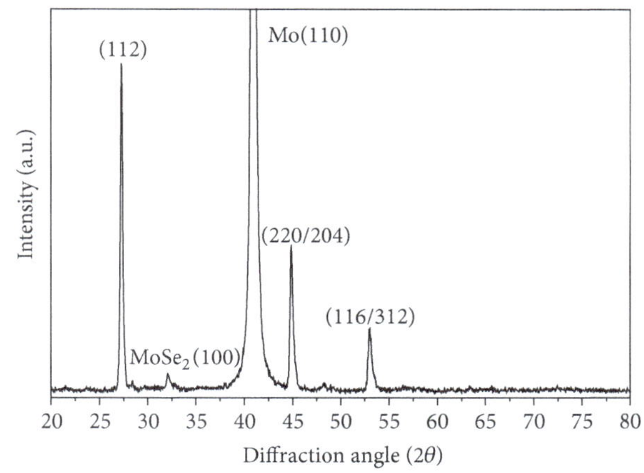

FIGURE 8: XRD patterns of the CIGS film from the precursor with a Cu/(In +Ga) ratio of 0.95 deposited on a Mo-coated SLG substrate.

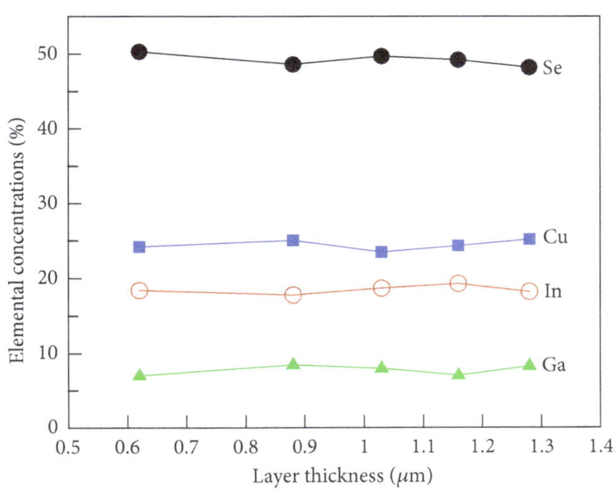

FIGURE 7: The XRF detailed compositional uniformity of Cu, In, Ga, and Se in the selenized CIGS film from the precursor with a Cu/(In +Ga) ratio of 0.95.

of an absorber that is selenized by RTA which in turn will enable a better control of the final performance of the solar cells.

4. Conclusions

CIGS thin film absorbers were fabricated by RTA of CuGa/In precursors with various Cu contents. The precursors were prepared by the sequential sputtering of CuGa and In targets and then coated with a Se layer. The results demonstrate that the structural and electrical properties of the absorber are heavily dependent on the Cu/(In +Ga) atomic ratio. Cu-poor precursor with a Cu/(In +Ga) ratio of 0.75 exhibits an initial decrease in CIGS grain size, which results in a higher resistance, due to increased grain boundary scattering and a reduction in carrier lifetime. Cu-rich precursor with a Cu/(In +Ga) ratio of 1.15 exhibits a second phase ($Cu_{2-x}Se$) in the absorber. It is inappropriate for use as an absorber material, even though the film has the largest grain size and the lowest resistance. A CIG precursor with a Cu/(In +Ga)

(a) (b)

Figure 9: (a) The cross-sectional SEM image and (b) surface morphology of the CIGS film from the precursor with a Cu/(In +Ga) ratio of 0.95, deposited on a Mo-coated glass substrate.

ratio of 0.95 exhibits a single chalcopyrite structure with larger grains and lower resistance, which is suitable for application to solar cells. The preparation of this precursor on Mo-coated SLG substrate and RTA results in a preferential (1 1 2) orientation of the single chalcopyrite structure CIGS film with a $MoSe_2$ layer at the interface of the Mo and the CIGS.

Conflict of Interests

The authors certify that there is no conflict of interests with any financial organization regarding the material discussed in the paper.

Acknowledgments

The authors gratefully acknowledge the support of the Solar Applied Materials Technology Corp. and the National Science Council of the Republic of China, through Grant No. NSC 99-2221-E-262-015-MY2, and the Chung-Shan Institute of Science & Technology (Armaments Bureau). material for thin film solar cells allows easier commercial production because

References

[1] M. A. Contreras, K. Ramanathan, J. Abushama et al., "Diode characteristics in state-of-the-art ZnO/CdS/ $Cu(In_{1-x}Ga_x)Se_2$ solar cells," *Progress in Photovoltaics*, vol. 13, no. 3, pp. 209–216, 2005.

[2] I. Repins, M. A. Contreras, B. Egaas et al., "19.9%-efficient ZnO/CdS/$CuInGaSe_2$ solar cell with 81.2% fill factor," *Progress in Photovoltaics*, vol. 16, no. 3, pp. 235–239, 2008.

[3] Y. C. Lin, J. H. Ke, W. T. Yen, S. C. Liang, C. H. Wu, and C. T. Chiang, "Preparation and characterization of $Cu(In,Ga)(Se,S)_2$ films without selenization by co-sputtering from $Cu(In,Ga)Se_2$ quaternary and In_2S_2 targets," *Applied Surface Science*, vol. 257, no. 9, pp. 4278–4284, 2011.

[4] E. Tzvetkovaa, N. Stratievaa, M. Gancheva, I. Tomovb, K. Ivanovaa, and K. Kocheva, "Preparation and structure of annealed CuInSe electrodeposited films," *Thin Solid Films*, vol. 311, no. 1-2, pp. 101–106, 1997.

[5] R. Caballero, C. Guillén, M. T. Gutiérrez, and C. A. Kaufmann, "$CuIn_{1-x}Ga_xSe_2$-based thin-film solar cells by the selenization of sequentially evaporated metallic layers," *Progress in Photovoltaics*, vol. 14, no. 2, pp. 145–153, 2006.

[6] V. Probst, W. Stetter, W. Riedl et al., "Rapid CIS-process for high efficiency PV-modules: development towards large area processing," *Thin Solid Films*, vol. 387, no. 1-2, pp. 262–267, 2001.

[7] H. Park, S. C. Kim, S. H. Lee et al., "Effect of precursor structure on $Cu(InGa)Se_2$ formation by reactive annealing," *Thin Solid Films*, vol. 519, no. 21, pp. 7245–7249, 2011.

[8] H. K. Song, J. K. Jeong, H. J. Kim, S. K. Kim, and K. H. Yoon, "Fabrication of $CuIn_{1-x}Ga_xSe_2$ thin film solar cells by sputtering and selenization process," *Thin Solid Films*, vol. 435, no. 1-2, pp. 186–192, 2003.

[9] G. M. Hanket, W. N. Shafarman, B. E. McCandless, and R. W. Birkmire, "Incongruent reaction of Cu-(InGa) intermetallic precursors in H2 Se and H2 S," *Journal of Applied Physics*, vol. 102, no. 7, Article ID 074922, 2007.

[10] F. B. Dejene, "The structural and material properties of $CuInSe_2$ and $Cu(In,Ga)Se_2$ prepared by selenization of stacks of metal and compound precursors by Se vapor for solar cell applications," *Solar Energy Materials and Solar Cells*, vol. 93, no. 5, pp. 577–582, 2009.

[11] J. Koo, S. C. Kim, H. Park, and W. K. Kim, "$Cu(InGa)Se_2$ thin film photovoltaic absorber formation by rapid thermal annealing of binary stacked precursors," *Thin Solid Films*, vol. 520, no. 5, pp. 1484–1488, 2011.

[12] D. S. Chen, J. Yang, F. Xu et al., "Effect of rapid thermal annealing on the compositional ratio and interface of $Cu(In,Ga)Se_2$ solar cells by XPS," *Applied Surface Science*, vol. 264, no. 1, pp. 459–463, 2013.

[13] W. Li, Y. Sun, W. Liu, and L. Zhou, "Fabrication of $Cu(In,Ga)Se_2$ thin films solar cell by selenization process with Se vapor," *Solar Energy*, vol. 80, no. 2, pp. 191–195, 2006.

[14] X. Wang, S. S. Li, W. K. Kim et al., "Investigation of rapid thermal annealing on $Cu(In,Ga)Se_2$ films and solar cells," *Solar Energy Materials and Solar Cells*, vol. 90, no. 17, pp. 2855–2866, 2006.

[15] S. S. Wang, C. Y. Hsu, F. J. Shiou, P. C. Huang, and D. C. Wen, "Properties of the Mo back contact for the formation of a thin

film photovoltaic absorber," *Journal of Electronic Materials*, vol. 42, no. 1, pp. 71–77, 2013.

[16] J. Horzel, C. Allebe, J. Szlufcik, and S. Sivoththaman, "Development of RTP for industrial solar cell processing," *Solar Energy Materials and Solar Cells*, vol. 72, no. 1–4, pp. 263–269, 2002.

[17] J. Wang, Y. F. Zhang, F. Dong, and J. Zhu, "Ga-grading profiles formed by incorporation of gallium into Cu($In_{1-x}Ga_x$)Se_2 absorber thin films," *Applied Surface Science*, vol. 258, no. 22, pp. 8636–8640, 2012.

[18] Z. H. Li, E. S. Cho, and S. J. Kwon, "Properties of the Cu(In,Ga)Se_2 absorbers deposited by electron-beam evaporation method for solar cells," *Current Applied Physics*, vol. 11, no. 1, pp. 28–33, 2011.

[19] W. Witte, R. Kniese, and M. Powalla, "Raman investigations of Cu(In,Ga)Se_2 thin films with various copper contents," *Thin Solid Films*, vol. 517, no. 2, pp. 867–869, 2008.

[20] I. H. Choi, "Raman spectroscopy of $CuIn_{1-x}Ga_xSe_2$ for in-situ monitoring of the composition ratio," *Thin Solid Films*, vol. 519, no. 13, pp. 4390–4393, 2011.

[21] J. Kessler, C. Chityuttakan, J. Lu, J. Schöldström, and L. Stolt, "Cu(In,Ga)Se_2 thin films grown with a Cu-poor/rich/poor sequence: growth model and structural considerations," *Progress in Photovoltaics*, vol. 11, no. 5, pp. 319–331, 2003.

[22] M. Bodegård, K. Granath, L. Stolt, and A. Rockett, "Behaviour of Na implanted into Mo thin films during annealing," *Solar Energy Materials and Solar Cells*, vol. 58, no. 2, pp. 199–208, 1999.

[23] T. Wada, N. Kohara, S. Nishiwaki, and T. Negami, "Characterization of the Cu(In,Ga)Se_2/Mo interface in CIGS solar cells," *Thin Solid Films*, vol. 387, no. 1-2, pp. 118–122, 2001.

15

Photocatalytic Properties of Nitrogen-Doped $Bi_{12}TiO_{20}$ Synthesized by Urea Addition Sol-Gel Method

Jiyong Wei, Baibiao Huang, Peng Wang, Zeyan Wang, Xiaoyan Qin, Xiaoyang Zhang, Xiangyang Jing, Haixia Liu, and Jiaoxian Yu

State Key Laboratory of Crystal Materials, Institute of Crystal Materials, Shandong University, Jinan 250100, China

Correspondence should be addressed to Baibiao Huang, bbhuang@sdu.edu.cn

Academic Editor: Jiaguo Yu

Undoped and nitrogen-doped $Bi_{12}TiO_{20}$ materials were synthesized by urea addition sol-gel method. By adding urea, undoped, and N-doped gel-type precursors were synthesized by low-temperature dehydrolyzation. Nitrogen-doped and undoped nanocrystalline $Bi_{12}TiO_{20}$ were prepared by annealing at 600∘C for 30 minutes. From UV-Vis absorption and diffuse reflection spectrum, the absorbing band shifted from 420 to 500 nm by nitrogen doping. The bonds of Ti–N and N–O were identified by XPS spectra from the prepared materials, and the enhancement of visible light absorption was attributed to nitrogen's substitution of oxygen. Photocatalytic properties of prepared materials were characterized by the decomposition of Rhodamine B illuminated by whole spectra of 300 W Xe light. The photocatalyst $Bi_{12}TiO_{20-y}N_y$ ($y = 0.03$) with N/(N+O) mole ratio about 3% shows better performance than that of heavily doped $Bi_{12}TiO_{20-z}N_z$ ($z = 0.06$), undoped $Bi_{12}TiO_{20}$, and light-doped $Bi_{12}TiO_{20-x}N_x$ ($x = 0.01$) photocatalysts due to its better crystalline morphology.

1. Introduction

Environmental and energetic problems are great challenges for human beings now. The photosensitized electrolytic oxidation [1] and the electrochemical photolysis of water [2] on TiO_2 electrode performed by Fujishima and Honda offered some solutions. In recent years, TiO_2 and other photocatalysts have been extensively studied for environmental pollutant treatment [3] such as water disinfection [4].

However, wide bandgap semiconductor photocatalysts, such as TiO_2, can only absorb UV light, which only take 4% of the whole solar spectra owing to the wide bandgap. In order to expand the absorption spectra region of TiO_2, doping with nonmetallic atoms, such as nitrogen, fluorine and sulfur has been developed [5–18]. The conventional doping processes reported were commonly by the annealing of prepared TiO_2 materials in NH_3 or other dopant atmospheres [5–8]. Rengifo-Herrera et al. have reported N, S codoped commercial TiO_2 powders [17] which were doped by the decomposition of thiourea as a nitrogen and sulfur source. Effects of calcination temperatures [19], morphologies [20], and composite [21] on photocatalytic activity were also studied.

Recently, photocatalysts containing ion units of $(Bi_2O_2)^{2+}$, such as Bi_2WO_6 [22, 23] and bismuth titanate were extensively studied for the photocatalytic splitting of water and the photodegradation of organic pollutants. Bismuth titanate was a wide bandgap semiconductor with several crystal phases: ferroelectric perovskite phase ($Bi_4Ti_3O_{12}$), dielectric pyrochlorite phase ($Bi_2Ti_2O_7$), refractive sillenite phase ($Bi_{12}TiO_{20}$), and so forth. Kudo and Hijii have studied bismuth titanate as potential photocatalyst [22]. Yao et al. [24–27] and other groups [28–36] explored nano crystalline materials of $Bi_{12}TiO_{20}$ as visible light photocatalysts, and the as-prepared materials showed high photocatalytic activity for decomposing organic dyes under ultraviolet light irradiation. The further expansion of $Bi_{12}TiO_{20}$ material's absorption to visible region is very important for the visible light responsive photocatalytic activity, because $Bi_{12}TiO_{20}$ can only absorb light with wavelength below 420 nm (about 2.9 eV) [21–24]. Little work has been performed for nonmetallic (such as nitrogen and carbon) doping of bismuth titanate crystals, though the doping of $Bi_{12}TiO_{20}$ nano- or single crystals with metal element [25, 26] and nonmetallic element such as P for the influence of refractive optical properties has been studied [37].

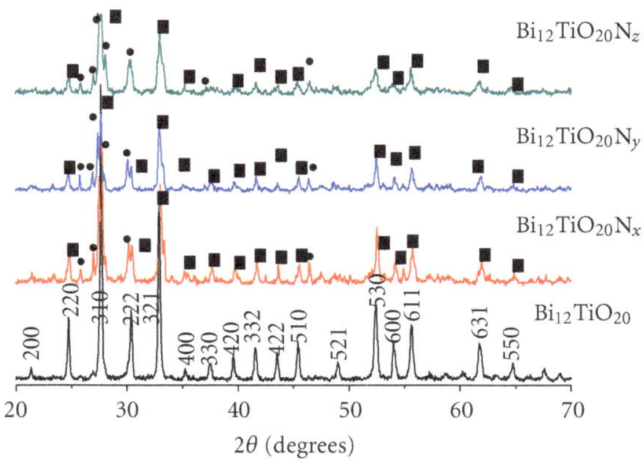

FIGURE 1: XRD spectrum of bismuth titanate: $Bi_{12}TiO_{20}$, $Bi_{12}TiO_{20-x}N_x$ ($x = 0.01$), $Bi_{12}TiO_{20-y}N_y$ ($y = 0.03$), and $Bi_{12}TiO_{20-z}N_z$ ($z = 0.06$).

In this paper, nitrogen-doped bismuth titanate powders and films were prepared by sol-gel method and dip-coating method, respectively, the nitrogen dopant was introduced by the adding of urea in precursor solutions as additives, which is widely used as fertilizer. Urea can be dissolved in water and other polar solvents, and can release free $NH_2^{\bullet}$ radicals as nitrogen sources during thermal decomposition. After preparation of urea included sol-gel precursors, high-temperature annealing was applied to undoped and doped $Bi_{12}TiO_{20}$ precursors for crystallization and also for the in-place nitrogen doping by urea decomposing. The results of whole-spectra photocatalytic activities of nitrogen-doped and undoped $Bi_{12}TiO_{20}$ were presented by the photodegradation of rhodamine b (RB) solutions.

2. Experimental Details

Undoped and nitrogen-doped $Bi_{12}TiO_{20}$ nanocrystalline powders and films were prepared by sol-gel method and dip coating. Chemical agents of analytical grade were used in the experiments, such as acetic acid, $Bi(NO_3)_3{\cdot}5H2O$ and $Ti(OC_4H_9)_4$. Reagents were dissolved in ethylene glycol monomethyl ether solution while urea was used as nitrogen additive. Crystalline materials of undoped and nitrogen-doped $Bi_{12}TiO_{20}$ were prepared by annealing of the sol-gel precursors at temperature of 600°C for about 30 minutes.

2.1. Preparation of $Bi_{12}TiO_{20}$ and Nitrogen-Doped $Bi_{12}TiO_{20}$ Crystalline Materials. The mole ratio of reagents $Bi(NO_3)_3{\cdot}5H_2O$ and $Ti(OC_4H_9)_4$ was 12 : 1 in the precursor solution for preparation of $Bi_{12}TiO_{20}$ and $Bi_{12}TiO_{20-x}N_x$ ($x = 0.01$), $Bi_{12}TiO_{20-y}N_y$ ($y = 0.03$), and $Bi_{12}TiO_{20-z}N_z$ ($z = 0.06$) sol-gel precursors, with additional urea added for nitrogen-doped materials (molecule ratio of $NH_2^{\bullet} : Bi^{3+}$ about 0 : 1, 1 : 1, 2 : 1, and 3 : 1 in sol-gel). The reagents were blended in acetic acid and ethylene glycol monomethyl ether solutions, and were dried by infrared light to get the gel. $Bi_{12}TiO_{20}$ and $Bi_{12}TiO_{20-x}N_x$ ($x = 0.01$), $Bi_{12}TiO_{20-y}N_y$ ($y = 0.03$), and $Bi_{12}TiO_{20-z}N_z$ ($z = 0.06$) sillenite phase crystalline powders were obtained by annealing the gel precursors at 600°C for about 30 minutes.

2.2. Preparation of $Bi_{12}TiO_{20}$ Crystalline Films for UV-Vis Absorption Characterization. Crystalline films of $Bi_{12}TiO_{20}$, $Bi_{12}TiO_{20-x}N_x$ ($x = 0.01$), $Bi_{12}TiO_{20-y}N_y$($y = 0.03$), and $BBi_{12}TiO_{20-z}N_z$ ($z = 0.06$) were prepared by dipping glass chips into the prepared sol-gel precursor solutions, and then were dried by infrared light, and annealed at 600°C for 30 minutes in nitrogen atmosphere.

2.3. Characterization Methods. The crystal phases of prepared powders were identified by X-ray diffraction (XRD, Cu K_α, D/max-ra X-ray). Morphology of the doped and undoped $Bi_{12}TiO_{20}$ nanocrystals was characterized by scanning electron microscopy (SEM, JEOL JSM6700F). The compositions and the electron bonding states of nitrogen-doped $Bi_{12}TiO_{20}$ were characterized by X-ray photoelectron spectroscopy (XPS, ESCALAB 250 of Thermal Fisher Scientific). To study the red shift of absorption wavelength of nitrogen-doped materials, UV-Vis absorption spectra of $Bi_{12}TiO_{20}$ crystalline films were performed by UV-Vis spectrophotometer (U-3500, 187 nm–3500 nm). The UV-Vis diffuse reflection spectra of doped and undoped $Bi_{12}TiO_{20}$ powders were characterized by UV/Vis spectroscopy (UV-2550, Shimadzu). Photocatalytic activity of undoped and nitrogen-doped $Bi_{12}TiO_{20}$ was characterized by the photodegradation of rhodamine b (RhB) solutions under the irradiation of 300 W Xe arc lamp (focused through a shutter window). The efficiency of the degradation processes was evaluated by monitoring the dye decolorization at the maximum absorption around 557 nm as a function of irradiation time in the separated RhB solution with a UV-vis spectrophotometer (UV-7502PC, Xinmao, Shanghai).

3. Results and Discussion

3.1. X-Ray Characterization. The crystal phases of synthesized doped and undoped nano crystalline materials were identified by X-ray diffraction. The XRD spectra of annealed $Bi_{12}TiO_{20}$, $Bi_{12}TiO_{20-x}N_x$ ($x = 0.01$), $Bi_{12}TiO_{20-y}N_y$ ($y = 0.03$), and $Bi_{12}TiO_{20-z}N_z$ ($z = 0.06$) were shown in Figure 1. No phase transformation of $Bi_{12}TiO_{20}$ structure (JCPDS no.78–1158) was observed by adding of urea as nitrogen

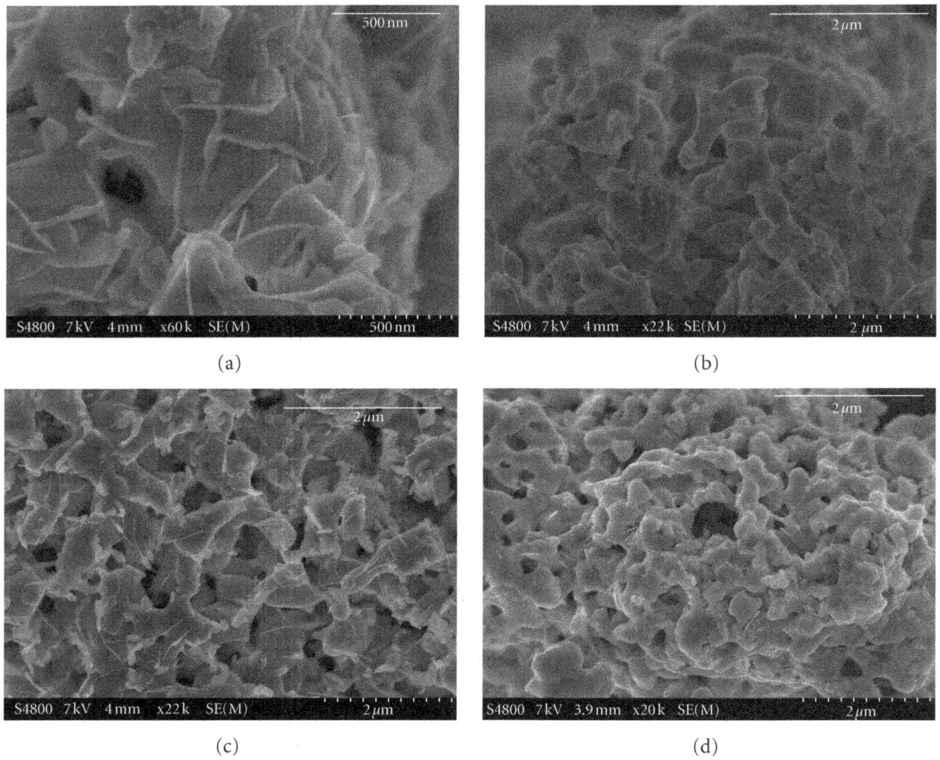

FIGURE 2: Morphologies of annealed undoped and nitrogen-doped $Bi_{12}TiO_{20}$ micro- and nano crystals observed by SEM: (a) $Bi_{12}TiO_{20}$; (b) $Bi_{12}TiO_{20-x}N_x$ ($x = 0.01$); (c) $Bi_{12}TiO_{20-y}N_y$ ($y = 0.03$); (d) $Bi_{12}TiO_{20-z}N_z$ ($z = 0.06$).

additive. The slight shift and broadening of XRD peaks showed the effect of nitrogen doping in $Bi_{12}TiO_{20}$ structure materials. From Figure 1, the sharp diffraction peak and strong intensity of $Bi_{12}TiO_{20}$ indicates good crystallinity; while after nitrogen doping, the diffraction peak intensity became weaker and the peaks broadened slightly. The (310) and (222) peak split into two peaks and shifted to lower angles with the increase of nitrogen dopant, which shows us there must be an increase for the interlamellar spacing of (310) and (222) crystal facets as nitrogen is doped in $Bi_{12}TiO_{20}$.

3.2. Morphology of Nanocrystalline Bismuth Titanate Materials. Morphologies of undoped and nitrogen-doped $Bi_{12}TiO_{20}$ materials prepared by sol-gel synthesis and annealing were characterized by SEM as show in Figure 2. The undoped $Bi_{12}TiO_{20}$ microgel like bulks dotted with nanoflakes were observed in Figure 2(a), with few holes due to the decomposition and elution of sol-gel precursor. With addition of urea in precursors, the decomposition eluted more gas and the microbulk became fractal bulks piled up as shown in Figures 2(b), 2(c), and 2(d); while with too much urea added, there are only fractal bulks and few nanoflakes, which may decrease the surface and refrain the photocatalytic performance of heavily nitrogen-doped $Bi_{12}TiO_{20-z}N_z$ ($z = 0.06$). The morphology of proper nitrogen-doped material is $Bi_{12}TiO_{20-y}N_y$ ($y = 0.03$), as shown in Figure 2(c), which is composed of separated nano bulks dotted with nanosheet-like crystals. The morphology of $Bi_{12}TiO_{20-y}N_y$ ($y = 0.03$) has more nano crystalline facets and larger surface areas than other samples, which would show better photocatalytic performance than the other samples.

3.3. Element Content and Doping of Nanocrystalline Bismuth Titanate Materials. The chemical bonding states of as-prepared materials were characterized by XPS as shown in Figure 3. As shown in Figure 3(a), the XPS peaks of Bi4f, C 1s, N 1s, and Ti 2p were detected. The split of nitrogen 1s peak and carbon 1s peak shown in Figures 3(b) and 3(c) indicate the variety of bonding states. The split of N1s peaks in Figure 3(b) was assigned to N 1s peaks of N–O bond (401.6 eV) [38] and N–Ti bond (396.4 eV) [39, 40]. The N/(N+O) molar ratio calculated from the XPS peak is about 3%. The broadening and split of C 1s peak shown in Figure 3(c) were assigned to CO/Bi_2O_3 (286.6 eV) and CO/TiO_2 (288.5 eV) physical absorb circumstances [41]. So most carbon in the sample is absorbed CO, and the visible light absorption enhancement is attributed to the doping of nitrogen.

3.4. UV-Vis Absorption Enhancement of Nitrogen-Doped Bismuth Titanate Nanocrystalline Films. The UV-Vis absorption spectra of bismuth titanate nano crystalline films with different nitrogen doping ratios were characterized in Figure 4(a). The absorption spectra of undoped bismuth titanate were broadened successfully by urea-assisted sol-gel growth of nitrogen doping as shown in the UV-Vis absorption spectra of $Bi_{12}TiO_{20}$ and $Bi_{12}TiO_{20-x}N_x$ ($x = 0.01$)

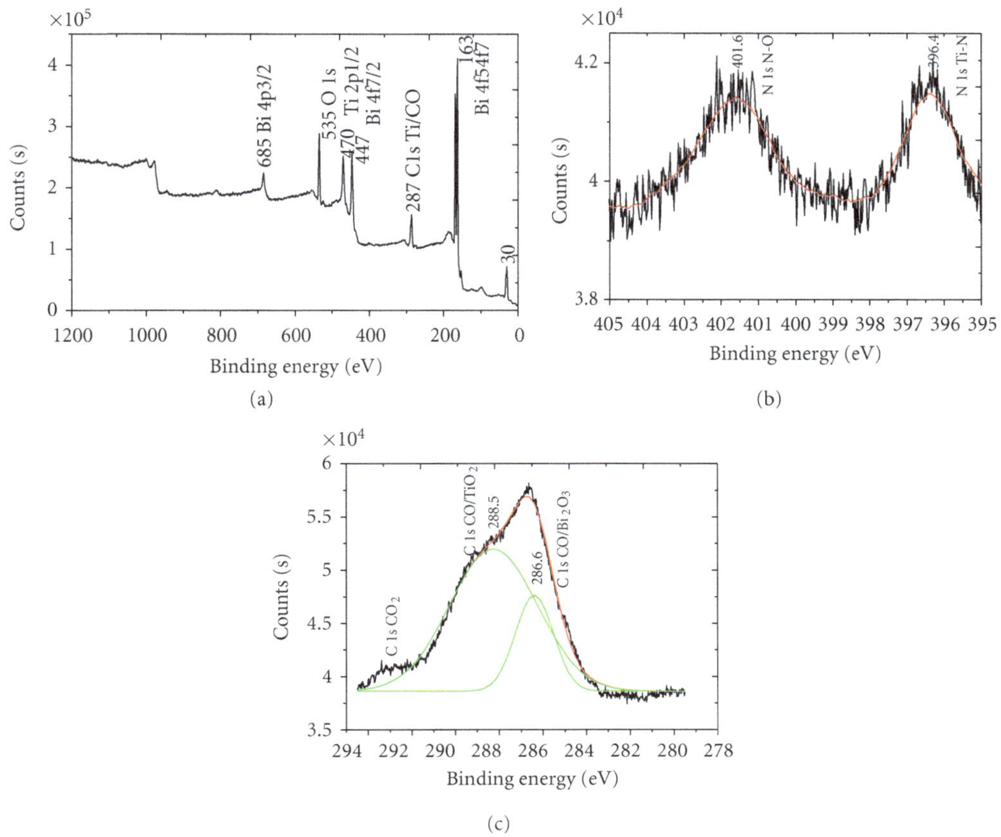

FIGURE 3: XPS absorption spectra of bismuth titanate $Bi_{12}TiO_{20-y}N_y$ ($y = 0.03$): (a) the whole spectrum of $Bi_{12}TiO_{20-y}N_y$ ($y = 0.03$); (b) spectrum of doped nitrogen; (c) spectrum of doped carbon.

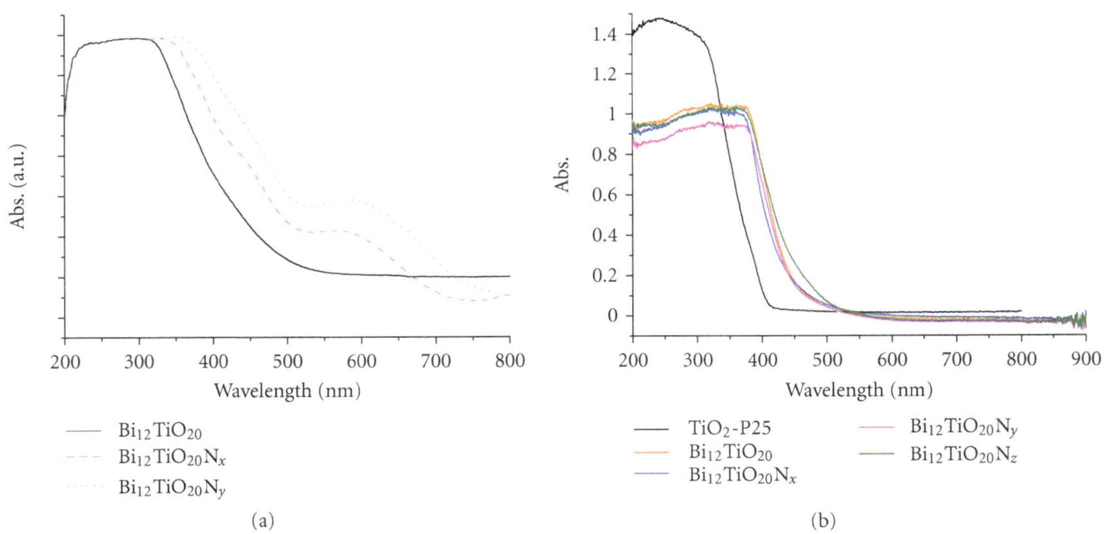

FIGURE 4: (a) Absorption spectra of films of bismuth titanate ($Bi_{12}TiO_{20}$, $Bi_{12}TiO_{20-x}N_x$ ($x = 0.01$) and $Bi_{12}TiO_{20-y}N_y$ ($y = 0.03$)) and (b) UV-Vis diffuse reflection spectra of powders $Bi_{12}TiO_{20}$, $Bi_{12}TiO_{20-x}N_x$ ($x = 0.01$), $Bi_{12}TiO_{20-y}N_y$ ($y = 0.03$), $Bi_{12}TiO_{20-z}N_z$ ($z = 0.06$), and TiO_2-P25.

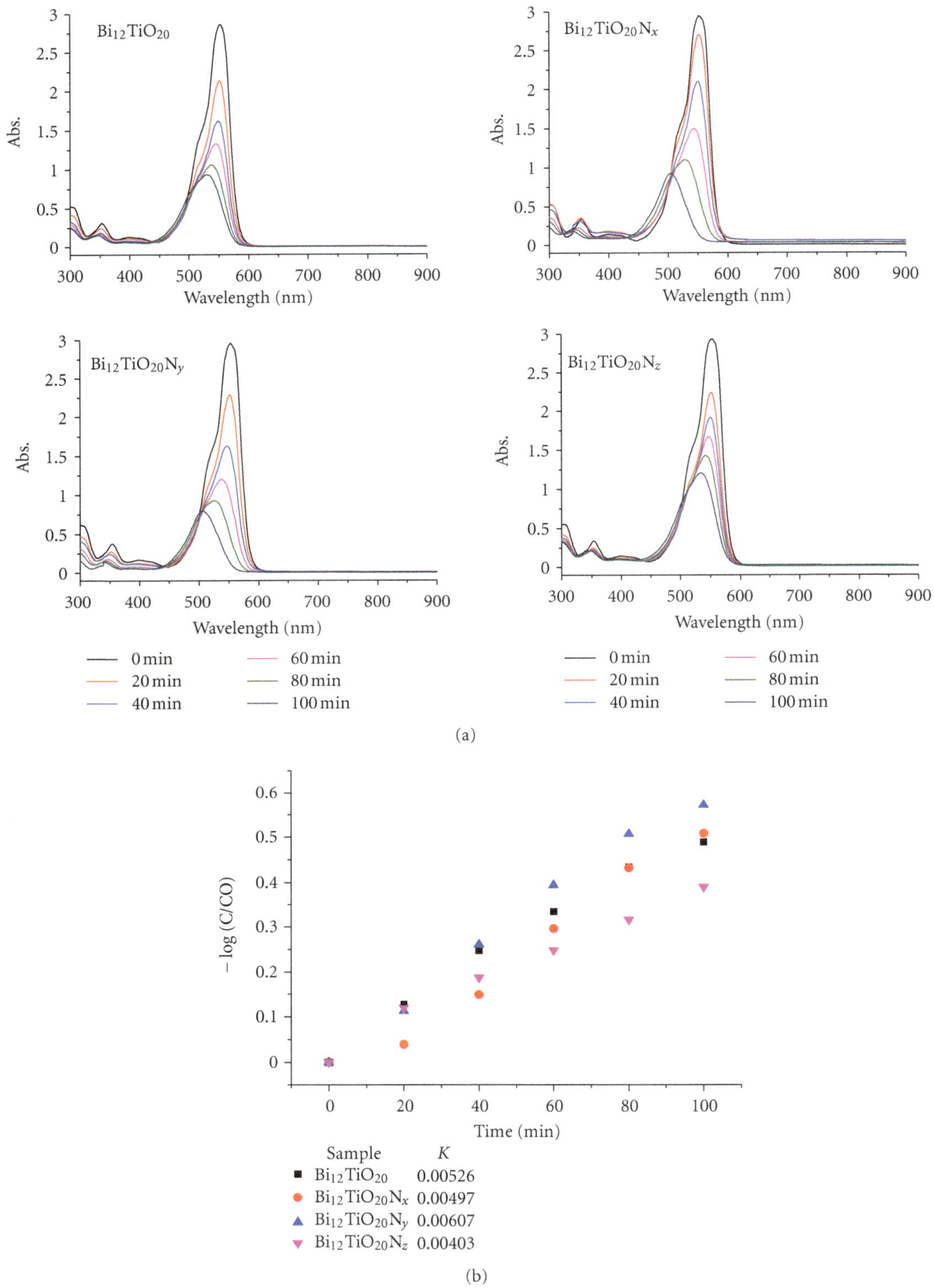

FIGURE 5: Photocatalytic decomposition of RhB by undoped and nitrogen-doped $Bi_{12}TiO_{20}$ photocatalysts: (a) absorption spectra decay of RhB via time; (b) decomposition activity of samples.

and $Bi_{12}TiO_{20-y}N_y$ ($y = 0.03$) films by sol-gel deposition with and without nitrogen doping. The UV-Vis absorption spectra band edge of $Bi_{12}TiO_{20}$ shifted from 400 nm to about 470 nm of nitrogen-doped $Bi_{12}TiO_{20-y}N_y$ ($y = 0.03$ with $NH_2^{\bullet}:Bi^{3+}$ about 1 : 1 solution, and N/(N+O) mole ratios about 3%). The absorption spectra's broadening of nitrogen-doped $Bi_{12}TiO_{20}$ was attributed to the doping of nitrogen's substitution of oxygen site that induced the narrowing of the bandgap. There was an additional absorption peak at 580 nm wavelength that was attributed to the deep-level nitrogen dopant.

The UV-Vis diffuse reflection spectra of nitrogen-doped and undoped $Bi_{12}TiO_{20}$ powders were characterized as shown in Figure 4(b), P25 was also used as a reference. TiO_2-P25 materials have intensive UV absorption and the absorption edge line at about 380 nm, while all $Bi_{12}TiO_{20}$ of nitrogen-doped and undoped have less UV absorption but longer bandgap-related wavelength about 450 nm. The bandgap of nitrogen-doped $Bi_{12}TiO_{20}$ *versus* undoped $Bi_{12}TiO_{20}$ in DRS spectra varied about 20 nm, from 440 nm to 460 nm. The red shift of absorption edge wavelength of nitrogen-doped $Bi_{12}TiO_{20}$ powders in DRS spectra is less than that in absorption spectra of nitrogen-doped $Bi_{12}TiO_{20}$ films, which we attribute to the variation of N/(N+O) ratios from inner to the outer layer. As known, the absorption spectra of films show the absorption of whole bulk materials, while the DRS spectrum more attribute to materials' surfaces. While annealing, there may be more nitrogen dopant in inner part and more oxygen in surfaces, so the absorption spectra edge of whole film materials will show more red shift than the diffuse reflection spectra edge obtained only from powders' surface.

The absorption band edge of $Bi_{12}TiO_{20}$ films and UV-Vis diffuse reflection band edge of $Bi_{12}TiO_{20}$ powders were successfully expanded to visible light region of solar spectra by urea-assisted sol-gel method, which would greatly enhance the utilizing ratio of visible light region of solar spectra in photocatalysis and solar cell researches.

3.5. Influence of Photocatalytic Capability of $Bi_{12}TiO_{20}$ by Nitrogen Doping. To study the relationship of photocatalytic property with nitrogen doping of bismuth titanate, photocatalysis of rhodamine b (RhB) by undoped and nitrogen-doped bismuth titanate was performed. The photocatalytic activity was characterized by the decomposition of 10 mg/L rhodamine b (RhB) solution, with 100 mg of photocatalyst added to 100 mL of rhodamine b (RhB) solution. The reaction system was irradiated by 300 W Xe lamps.

The photocatalytic ability of undoped $Bi_{12}TiO_{20}$ materials had been studied by Yao et al. [24–26] and other groups [28, 29, 31, 33, 34]. Here we are concerned with relationship of photocatalytic ability of undoped and nitrogen-doped $Bi_{12}TiO_{20}$ materials with different amount of urea added in precursors ($NH_2^{\bullet}:Bi^{3+}$ about 0 : 1, 1 : 1, 2 : 1, and 3 : 1). Different amount of urea added in precursor solution deduced to the different amount of nitrogen dopant in the $Bi_{12}TiO_{20}$ photocatalysts. All samples were annealed at 600°C for 30 minutes.

As shown in Figure 5, the performance of catalyst $Bi_{12}TiO_{20}$ and $Bi_{12}TiO_{20-y}N_y$ ($y = 0.03$) have better performance than other catalysts. The undoped and nitrogen-doped $Bi_{12}TiO_{20-y}N_y$ ($y = 0.03$) catalysts with N/(N+O) mole ratio about 0% and 3% were prepared by $NH_2^{\bullet}:Bi^{3+}$ about 0 : 1 and 2 : 1 in the precursor solutions, respectively.

The velocity constants of catalyst $Bi_{12}TiO_{20}$ ($k = 5.26 \times 10^{-3}\,min^{-1}$) and catalyst $Bi_{12}TiO_{20-y}N_y$ ($y = 0.03$; $k = 6.07 \times 10^{-3}\,min^{-1}$) are about 1.5-times the velocity constant of catalyst $Bi_{12}TiO_{20-z}N_z$ ($z = 0.06$; $k = 4.03 \times 10^{-3}\,min^{-1}$). So the nitrogen-doped $Bi_{12}TiO_{20-y}N_y$ ($y = 0.03$) photocatalyst with N/(N+O) mole ratio about 3% have better performance than undoped $Bi_{12}TiO_{20}$, light-doped $Bi_{12}TiO_{20-x}N_x$ ($x = 0.01$), and heavily doped $Bi_{12}TiO_{20-z}N_z$ ($z = 0.06$) photocatalysts.

The variation of whole spectra photocatalytic performance of undoped and nitrogen-doped $Bi_{12}TiO_{20}$ was attribute to the morphologies as observed by SEM in Figure 2. $Bi_{12}TiO_{20-y}N_y$ ($y = 0.03$) have more crystalline nano crystal facets and larger surface areas than other samples, which lead to better photocatalytic performance than other samples. But a high level of nitrogen-doped (above 6% of N/(N+O)) would lead to the formation of more defects in nano crystals of prepared catalysts, and would influence the transportation of photogenerated electrons and holes. Though high level of nitrogen doping can effectively expand the visible light absorption, a high defects density would lead to poor photocatalytic activity.

4. Conclusions

In conclusion, by urea-assisted sol-gel method, undoped and nitrogen-doped $Bi_{12}TiO_{20}$, $Bi_{12}TiO_{20-x}N_x$ ($x = 0.01$), $Bi_{12}TiO_{20-y}N_y$ ($y = 0.03$), and $Bi_{12}TiO_{20-z}N_z$ ($z = 0.06$) powders were prepared. After annealing, the morphologies of as-synthesized samples exhibit sheet-like morphologies. And by controlling the amount of urea added in the precursor solutions, photocatalyst with sheet-like morphologies with good crystalline facet and larger surface area can be synthesized.

The absorption spectra of nitrogen-doped $Bi_{12}TiO_{20-x}N_x$ ($x = 0.01$), $Bi_{12}TiO_{20-y}N_y$ ($y = 0.03$), and $Bi_{12}TiO_{20-z}N_z$ ($z = 0.06$) films were successfully expanded from about 400 nm of $Bi_{12}TiO_{20}$ to about 500 nm, also from DRS spectra the band edge were successfully expanded from about 420 nm to about 460 nm, which would greatly enhance the utilization ratio of visible light in solar spectra. The expanded visible light can be attributed to the doping of nitrogen substituting of oxygen in $Bi_{12}TiO_{20}$ crystal lattice. Photocatalytic performance of nitrogen-doped $Bi_{12}TiO_{20-y}N_y$ ($y = 0.03$) with N/(N+O) mole ratio about 3% is better photocatalytic activity than that of undoped, light-doped (below 1%), and heavily doped $Bi_{12}TiO_{20-z}N_z$ ($z = 0.06$) photocatalysts.

Acknowledgments

This paper was financially supported by research grants from the National Basic Research Program of China (973 Program, no. 2007CB613302), and the National Science Foundation of China (nos. 20973102 and 51021062). The authors also express our thanks for the editor's patience.

References

[1] A. Fujishima, K. Honda, and S. Kikuchi, "Photosensitized electrolytic oxidation on semiconducting n-type TiO_2 electrode," *Kogyo Kagaku Zasshi*, vol. 72, pp. 108–113, 1969.

[2] A. Fujishima and K. Honda, "Electrochemical photolysis of water at a semiconductor electrode," *Nature*, vol. 238, no. 5358, pp. 37–38, 1972.

[3] J. A. Byrne, P. A. Fernandez-Ibañez, P. S. M. Dunlop, D. M. A. Alrousan, and J. W. J. Hamilton, "Photocatalytic enhancement for solar disinfection of water: a review," *International Journal of Photoenergy*, vol. 2011, Article ID 764870, 12 pages, 2011.

[4] J. Gamage and Z. Zhang, "Applications of photocatalytic disinfection," *International Journal of Photoenergy*, vol. 200, Article ID 764870, 11 pages, 2010.

[5] R. Asahi, T. Morikawa, T. Ohwaki, K. Aoki, and Y. Taga, "Visible-light photocatalysis in nitrogen-doped titanium oxides," *Science*, vol. 293, no. 5528, pp. 269–271, 2001.

[6] S. Sato, R. Asahi, T. Morikawa, T. Ohwaki, K. Aoki, and Y. Taga, "Photocatalysts sensitive to visible light—response," *Science*, vol. 295, no. 5555, pp. 626–627, 2002.

[7] K. Nukumizu, J. Nunoshige, T. Takata et al., "TiNxOyFz as a stable photocatalyst for water oxidation in visible light (<570 nm)," *Chemistry Letters*, vol. 32, no. 2, pp. 196–197, 2003.

[8] H. Liu and L. Gao, "(Sulfur, nitrogen)-codoped rutile-titanium dioxide as a visible-light-activated photocatalyst," *Journal of the American Ceramic Society*, vol. 87, no. 8, pp. 1582–1584, 2004.

[9] H. Liu and L. Gao, "Codoped rutile TiO_2 as a new photocatalyst for visible light irradiation," *Chemistry Letters*, vol. 33, no. 6, pp. 730–731, 2004.

[10] M. Mrowetz, W. Balcerski, A. J. Colussi, and M. R. Hoffmann, "Oxidative power of nitrogen-doped TiO_2 photocatalysts under visible illumination," *Journal of Physical Chemistry B*, vol. 108, no. 45, pp. 17269–17273, 2004.

[11] Q. Zhang, J. Wang, S. Yin, T. Sato, and F. Saito, "Synthesis of a visible-light active TiO_2-xSx photocatalyst by means of mechanochemical doping," *Journal of the American Ceramic Society*, vol. 87, no. 6, pp. 1161–1163, 2004.

[12] T. Matsunaga and M. Inagaki, "Carbon-coated anatase for oxidation of methylene blue and NO," *Applied Catalysis B*, vol. 64, no. 1-2, pp. 9–12, 2006.

[13] H. Yamashita, Y. Nishida, O. Chiyoda et al., "Design of efficient $TTiO_2$/SiC photocatalysts from TiC-SiC nano particles for degradation of organic pollutants diluted in water," *Materials Science Forum*, vol. 510-511, pp. 9–12, 2006.

[14] Y. Kesong, D. Ying, H. Baibiao, and H. Shenghao, "Theoretical study of N-doped TiO_2 rutile crystals," *Journal of Physical Chemistry B*, vol. 110, no. 47, pp. 24011–24014, 2006.

[15] H. Yamashita, Y. Nishida, S. Yuan et al., "Design of TiO_2-SiC photocatalyst using TiC-SiC nano-particles for degradation of 2-propanol diluted in water," *Catalysis Today*, vol. 120, no. 2, pp. 163–167, 2007.

[16] K. Yang, Y. Dai, and B. Huang, "Study of the nitrogen concentration influence on N-doped TiO_2 anatase from first-principles calculations," *Journal of Physical Chemistry C*, vol. 111, no. 32, pp. 12086–12090, 2007.

[17] J. A. Rengifo-Herrera, E. Mielczarski, J. Mielczarski, N. C. Castillo, J. Kiwi, and C. Pulgarin, "Escherichia coli inactivation by N, S co-doped commercial TiO_2 powders under UV and visible light," *Applied Catalysis B*, vol. 84, no. 3-4, pp. 448–456, 2008.

[18] G. Shang, H. Fu, S. Yang, and T. Xu, "Mechanistic study of visible-light-induced photodegradation of 4-chlorophenol by $TiO_{2-x}N_x$ (0.021<x<0.049) with low nitrogen concentration," *International Journal of Photoenergy*, vol. 2012, Article ID 759306, 9 pages, 2012.

[19] L. Zhao, J. Ran, Z. Shu, G. Dai, P. Zhai, and S. Wang, "Effects of calcination temperatures on photocatalytic activity of ordered titanate nanoribbon/SnO_2 films fabricated during an EPD process," *International Journal of Photoenergy*, vol. 2012, Article ID 472958, 7 pages, 2012.

[20] N.-H. Lee, H.-J. Oh, S.-C. Jung, W.-J. Lee, D.-H. Kim, and S.-J. Kim, "Photocatalytic properties of nanotubular-shaped TiO_2 powders with anatase phase obtained from titanate nanotube powder through various thermal treatments," *International Journal of Photoenergy*, vol. 2011, Article ID 327821, 7 pages, 2011.

[21] M.-L. Chen and W.-C. Oh, "The improved photocatalytic properties of methylene blue for V_2O_3/CNT/TiO_2 composite under visible light," *International Journal of Photoenergy*, vol. 2010, Article ID 264831, 5 pages, 2010.

[22] A. Kudo and S. Hijii, "H2 or O2 evolution from aqueous solutions on layered oxide photocatalysts consisting of Bi3+ with 6s2 configuration and d0 transition metal ions," *Chemistry Letters*, no. 10, pp. 1103–1104, 1999.

[23] S. Yao, J. Wei, B. Huang et al., "Morphology modulated growth of bismuth tungsten oxide nanocrystals," *Journal of Solid State Chemistry*, vol. 182, no. 2, pp. 236–239, 2009.

[24] W. F. Yao, H. Wang, X. H. Xu et al., "Photocatalytic property of bismuth titanate $Bi_{12}TiO_{20}$ crystals," *Applied Catalysis A*, vol. 243, no. 1, pp. 185–190, 2003.

[25] W. F. Yao, H. Wang, X. H. Xu et al., "Photocatalytic property of Zn-doped $Bi_{12}TiO_{20}$," *Journal of Materials Science Letters*, vol. 22, no. 14, pp. 989–992, 2003.

[26] W. F. Yao, H. Wang, X. H. Xu et al., "Characterization and photocatalytic properties of Ba doped $Bi_{12}TiO_{20}$," *Journal of Molecular Catalysis A*, vol. 202, no. 1-2, pp. 305–311, 2003.

[27] W. F. Yao, X. H. Xu, J. T. Zhou et al., "Photocatalytic property of sillenite $Bi_{24}AIO_{39}$ crystals," *Journal of Molecular Catalysis A*, vol. 212, no. 1-2, pp. 323–328, 2004.

[28] N. Thanabodeekij, E. Gulari, and S. Wongkasemjit, "$Bi_{12}TiO_{20}$ synthesized directly from bismuth (III) nitrate pentahydrate and titanium glycolate and its activity," *Powder Technology*, vol. 160, no. 3, pp. 203–208, 2005.

[29] X. H. Xu, W. F. Yao, Y. Zhang, A. Q. Zhou, Y. Hou, and M. Wang, "Photocatalytic properties of bismuth titanate compounds," *Acta Chimica Sinica*, vol. 63, no. 1, pp. 5–10, 2005.

[30] T. Lin, Z. Pi, M. C. Gong, J. B. Zhong, J. L. Wang, and Y. Q. Chen, "Gas-phase photocatalytic oxidation of benzene over titanium dioxide loaded on $Bi_{12}TiO_{20}$," *Chinese Chemical Letters*, vol. 18, no. 2, pp. 241–243, 2007.

[31] S. Xu, W. Shangguan, J. Yuan, J. Shi, and M. Chen, "Photocatalytic properties of bismuth titanate $Bi_{12}TiO_{20}$ prepared by coprecipitation processing," *Materials Science and Engineering B*, vol. 137, no. 1–3, pp. 108–111, 2007.

[32] S. Xu, W. Shangguan, J. Yuan, J. Shi, and M. Chen, "Preparations and photocatalytic degradation of methyl orange in water on magnetically separable $Bi_{12}TiO_{20}$ supported on nickel ferrite," *Science and Technology of Advanced Materials*, vol. 8, no. 1-2, pp. 40–46, 2007.

[33] J. Zhou, Z. Zou, A. K. Ray, and X. S. Zhao, "Preparation and characterization of polycrystalline bismuth titanate $Bi_{12}TiO_{20}$ and its photocatalytic properties under visible light irradiation," *Industrial and Engineering Chemistry Research*, vol. 46, no. 3, pp. 745–749, 2007.

[34] H. Zhang, M. Lü, S. Liu et al., "Preparation and photocatalytic properties of sillenite $Bi_{12}TiO_{20}$ films," *Surface and Coatings Technology*, vol. 202, no. 20, pp. 4930–4934, 2008.

[35] J. Hou, Y. Qu, D. Krsmanovic, C. Ducati, D. Eder, and R. V. Kumar, "Solution-phase synthesis of single-crystalline $Bi_{12}TiO_{20}$ nanowires with photocatalytic properties," *Chemical Communications*, no. 26, pp. 3937–3939, 2009.

[36] W. Wei, Y. Dai, and B. Huang, "First-principles characterization of Bi-based photocatalysts: $Bi_{12}TiO_{20}$, $Bi_2Ti_2O_7$, and $Bi_4Ti_3O_{12}$," *Journal of Physical Chemistry C*, vol. 113, no. 14, pp. 5658–5663, 2009.

[37] V. Marinova, "Optical properties of $Bi_{12}TiO_{20}$ doped with Al, P, Ag, Cu, Co and co-doped with Al + P single crystals," *Optical Materials*, vol. 15, no. 2, pp. 149–158, 2000.

[38] Z. Shuxian, W. K. Hall, G. Ertl, and H. Knözinger, "X-Ray photoemission study of oxygen and nitric oxide adsorption on MoS_2," *Journal of Catalysis*, vol. 100, no. 1, pp. 167–175, 1986.

[39] S. Badrinarayanan, S. Sinha, and A. B. Mandale, "XPS studies of nitrogen ion implanted zirconium and titanium," *Journal of Electron Spectroscopy and Related Phenomena*, vol. 49, no. 3, pp. 303–309, 1989.

[40] G. Soto, W. De La Cruz, and M. H. Farías, "XPS, AES, and EELS characterization of nitrogen-containing thin films," *Journal of Electron Spectroscopy and Related Phenomena*, vol. 135, no. 1, pp. 27–39, 2004.

[41] C. Ocal and S. Ferrer, "The strong metal-support interaction (SMSI) in Pt-TiO2 model catalysts. A new CO adsorption state on Pt-Ti atoms," *The Journal of Chemical Physics*, vol. 84, no. 11, pp. 6474–6478, 1986.

16

Near-Infrared All-Silicon Photodetectors

M. Casalino, G. Coppola, M. Iodice, I. Rendina, and L. Sirleto

Institute for Microelectronics and Microsystems, National Council of Research, Via P. Castellino 111, 80131 Naples, Italy

Correspondence should be addressed to M. Casalino, maurizio.casalino@na.imm.cnr.it

Academic Editor: Fabio Iacona

We report the fabrication and characterization of all-silicon photodetectors at 1550 nm based on the internal photoemission effect. We investigated two types of structures: bulk and integrated devices. The former are constituted by a Fabry-Perot microcavity incorporating a Schottky diode, and their performance in terms of responsivity, free spectral range, and finesse was experimentally calculated in order to prove an enhancement in responsivity due to the cavity effect. Results show a responsivity peak of about 0.01 mA/W at 1550 nm with a reverse bias of 100 mV. The latter are constituted by a Schottky junction placed transversally to the optical field confined into the waveguide. Preliminary results show a responsivity of about 0.1 mA/W at 1550 nm with a reverse bias of 1 V and an efficient behaviour in both C and L bands. Finally, an estimation of bandwidth for GHz range is deduced for both devices. The technological steps utilized to fabricate the devices allow an efficiently monolithic integration with complementary metal-oxide semiconductor (CMOS) compatible structures.

1. Introduction

Silicon is an extremely attractive platform for large-scale integration of multiple optical and electronic functions on a single silicon chip. This should result in revolutionizing applications in microelectronics, telecommunications, and biological and chemical sensing. Significant advantages exist when using silicon as the base material for such devices: in particular, a vast amount of research is now available to designers and process engineers on all aspects of the material system, as well as the established, global silicon processing industry that has evolved from almost six decades of silicon-based microelectronics fabrication. Silicon wafers have the lowest cost per unit area and the highest crystal quality of any semiconductor material. The industry is able to produce microprocessors with hundreds of millions of components—all integrated onto thumb-sized chips—and offers them at such low prices that they are increasingly used in consumer electronics. Silicon manufacturing represents the most spectacular convergence of technological sophistication and economics of scale [1–3].

For these reasons, in the last two decades, there has been growing interest in photonic devices based on Si-compatible materials [4] in the field of both the optical telecommunications and the optical interconnection. In this contest, tremendous progress in the technological processes based on the use of silicon-on-insulator (SOI) substrates has allowed to obtain reliable and fully effective CMOS-compatible optical components such as: low loss waveguides, high-Q resonators, high-speed modulators, couplers, and optically pumped lasers [5–10]. All these devices have been developed to operate in the wavelength range from C band (1528–1561 nm) to L band (1561–1620 nm) where the defect-free intrinsic bulk silicon has minimal absorption. On the other hand, this transparency window limits silicon applications for photodetection in the infrared range so that the development of high-performance waveguide-integrated photodetectors on Si-CMOS platform has remained an imperative but unaccomplished task so far.

In order to detect light into C and L bands, germanium-based photodetectors have shown impressive advances during the last years. However, the aggressive substrate cleaning processes and thermal mismatch between Ge and Si hinder a seamless integration of Ge-PDs with the standard Si-integrated circuit realization. For these reasons their fabrication and related technologies are much more expensive than silicon-based PDs [11, 12].

In order to develop all-silicon photodetectors and to take advantage of low-cost standard Si-CMOS processing technologies without additional material or process steps, a number of options have been proposed [13], in particular, the two-photon absorption [14], the incorporation of optical dopants/defects with midbandgap energy levels into the Si lattice [15, 16], the internal photoemission effect (IPE) [17], recently adopted also in silicon photodetectors based on surface plasmons [18, 19].

The aim of this work is to investigate photodetectors based on IPE at near-infrared wavelengths. We report the fabrication and characterization of two types of structures: bulk and integrated devices. The performance in terms of responsivity and estimated bandwidth is reported and discussed.

2. Internal Photoemission Theory

Internal photoemission is the optical excitation of electrons into the metal to an energy above the Schottky barrier and then transport of these electrons to the conduction band of the semiconductor (Figure 1).

The standard theory of photoemission from a metal into a vacuum is due to Fowler [20]. Fowler's theory was originally obtained without taking into account the thickness of the Schottky metal layer. In order to study the quantum efficiency for thin metal films, the theory must be further extended, taking into account multiple reflections of the excited electrons from the surfaces of the metal film, in addition to collisions with phonons, imperfections, and cold electrons. Assuming a thin metal film, a phenomenological semiclassical ballistic transport model for the effects of the scattering mechanisms resulting in a multiplicative factor for quantum efficiency has been developed by Vickers [21].

The efficiency in collecting those electrons which have sufficient normal kinetic energy to overcome potential barrier (Φ_B) will depend on their probability of collisions with cold or hot electrons and with the two boundary surfaces of the metal. Supposing that the Schottky barrier is illuminated through the front surface of the metal electrode and that optical absorption is uniform throughout the metal thickness, we indicate with p_t the average probability that photoexcited electrons arrive at the semiconductor surface without collision. Then, we indicate with p_r the accumulated probability of arrival at the reflection boundary of those electrons which have sufficient energy to overcome Φ_B after one collision with cold or hot electrons, which are oriented with equal probability over the entire half-sphere before reflection by the metal surface. Finally, for multiple reflections, we indicate with p the probability that the capturable electrons can bounce from one metal boundary to another. Therefore, the total accumulated probability that the electrons will have sufficient normal kinetic energy to overcome Φ_B is given by [22]

$$P_E = p_t + p_r\left(1 + p + p^2 + p^3 + \cdots\right)\exp\left(-\frac{d}{L}\right) \cong \frac{L}{d}\left[1 - e^{-d/L}\right]^{1/2}, \quad (1)$$

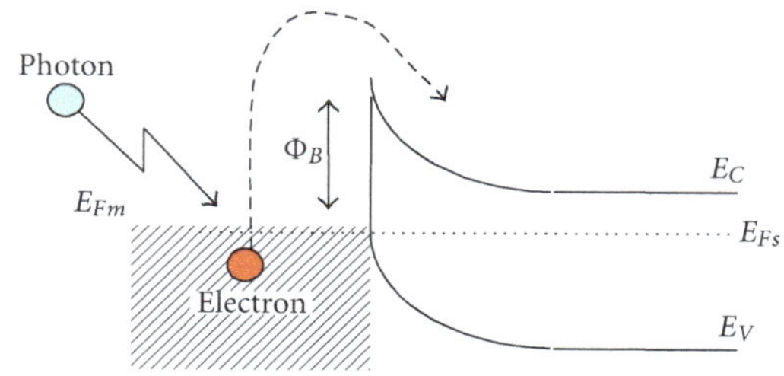

FIGURE 1: Energy band diagram for a metal-semiconductor junction.

where d is the metal thickness and L the mean free path describing the probability of collision with cold or hot electrons [22]. It would be possible to show by plotting (1) that by decreasing metal thickness d P_E increases and that a considerable gain can be obtained when d is much smaller than L (thin film). This gain is due to the increased probability of emission of carriers. In a recent work, Scales and Berini show that a further enhancement of this probability emission can be obtained in structures realized with thin metal film buried in a semiconductor and forming two Schottky barriers [23].

3. IPE-Based Bulk Device

3.1. Proposed Device. The sketch of the proposed device is shown in Figure 2. The resonant cavity is a vertical-to-the-surface Fabry-Perot structure. It is formed by a buried reflector, a top mirror interface and, in the middle, a silicon cavity. The buried reflector is a Bragg mirror, realized by alternating layers of amorphous hydrogenated silicon (a-Si:H) and silicon nitride (Si_3N_4). A thick copper (Cu) Schottky metal layer, working as both top reflector and absorbing medium, is deposited above the silicon substrate. Schottky Cu is surrounded by an aluminum (Al) contact in order to collect the photogenerated carriers.

Due to the multiple reflections of the optical field inside the cavity an improvement in metal absorbance, thus in responsivity, is expected.

3.2. Device Fabrication. The samples were fabricated starting from a slightly doped ($10^{12}\,cm^{-3}$) p-type bi-polished 100 μm thick silicon wafer. On the back, a multilayer Bragg mirror is fabricated by plasma-enhanced chemical vapor deposition technique (PECVD). The mirror is composed by a quarter-wave stack of a-Si:H and Si_3N_4 layers, having nominal refractive index, at 1550 nm, of 3.52 and 1.82, respectively. The reflector is realized with five periods of a-Si:H/Si_3N_4 pairs, whose nominal thickness is 110 nm and 213 nm, respectively. Silicon nitride is deposited at pressure of 1.2 mbar, temperature of 250°C, at 30 W of RF power. In the deposition chamber 10 sccm of NH_3, 88 sccm of SiH_4 (5% in He), and 632 sccm of N_2 are flowed. The deposition rate is 22.93 nm/min, and the suited Si_3N_4 thickness is obtained with a process time of 9 min and 17 sec. Amorphous hydrogenated silicon, instead, is deposited at pressure of 0.8 mbar, temperature of 250°C, power of 2 W, and a SiH_4

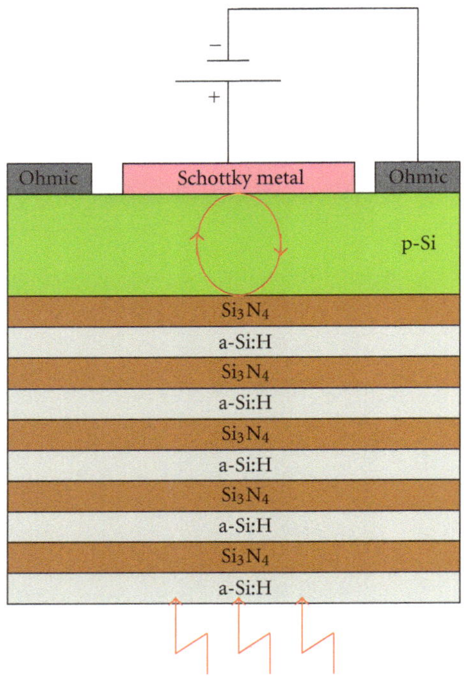

FIGURE 2: Schematic cross-section of the proposed photodetector.

(5% in He) flow of 600 sccm. The a-Si:H deposition rate is 3.15 nm/min, and the desired thickness is obtained with a process time of 34 min and 56 sec. The collecting ohmic contact and the Schottky contact were both realized on the top of the sample. The collecting contact was made by a 200 nm thick aluminum film, thermally evaporated at $3 \cdot 10^{-6}$ mbar and 150°C, patterned by a lift-off process of photoresist Shipley S1813 which, deposited by a spin coater at 4000 rpm, has a thickness of 1.4 μm. Then, an annealing at 475°C in nitrogen for 30 min, in order to get a nonrectifying behavior, was carried out [24]. Finally, the Schottky contact was fabricated. Copper was thermally evaporated and patterned by lift-off, so obtaining a metal film thickness of 200 nm, thicker than field penetration depth. The collecting contact and the Schottky contact are shaped by a ring and a disk (40 μm of radius) with the Schottky metal disk placed inside the ohmic metal ring (Figure 3).

3.3. Responsivity Measurements. The experimental setup for external responsivity measurements is quite simple. The laser beam emitted by a wavelength tunable laser is split by a Y fiber junction. One branch is used to monitor the optical power, while the other one is collimated, chopped, and sent onto the device. The photocurrent produced by our device is measured by a lock-in amplifier. A transimpedance amplifier (TIA) is employed to provide a reverse bias voltage to the photodetector and at the same time for reducing the dark current. The dark current cancellation circuit realised by using a transimpedance amplifier has a limited bandwidth; however it is adequate for our scope, that is dc or quasistatic measurements [25].

FIGURE 3: Top view of the proposed fabricated device.

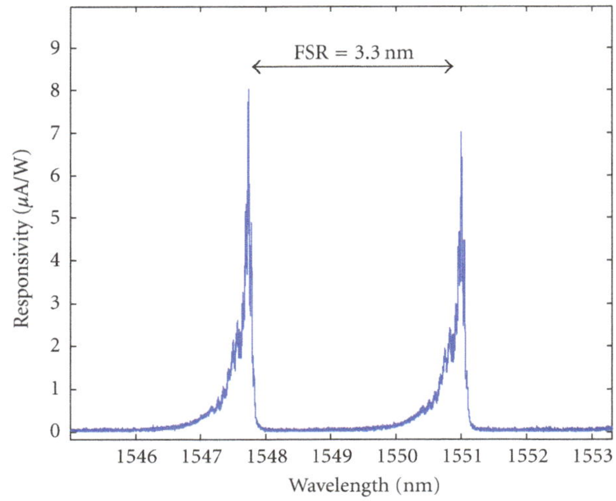

FIGURE 4: Measured external responsivity versus wavelength for the proposed Cu/p-Si photodetector at −100 mV.

Measurements around 1550 nm (step of 0.001 nm) at 100 mV of reverse bias applied were carried out (Figure 4). At 100 mV of the reverse bias applied, device dark current is about 0.4 μA.

We observed two distinct peaks indicating the influence of the cavity. The measured free spectral range of 3.3 nm agrees with the value calculated by the formula $\mathrm{FSR} = \lambda_0^2/(2nL)$ [26], where λ_0 is the central wavelength at the considered range (1550 nm), n is the silicon cavity refractive index (3.48 at 1550 nm [27]), and L is the silicon cavity thickness (100 μm). A peak external responsivity and an experimental finesse of 8.2 μA/W and 33, respectively, are obtained.

In our previous paper [28], a macroscopic device (radius of about 2 mm) was considered and peak external responsivity value of 4.3 μA/W and a finesse value of 4.7 were reported. In this paper, a microscopic device (radius of 40 μm) is investigated and an increase in finesse and responsivity has been obtained. In our opinion, reducing the device dimensions, the imperfection and hence losses are reduced. Therefore, improved responsivity and finesse were expected. Of course, the low responsivity imposes photodetector to be connect to a suitable TIA whose gain can be chosen by taking into

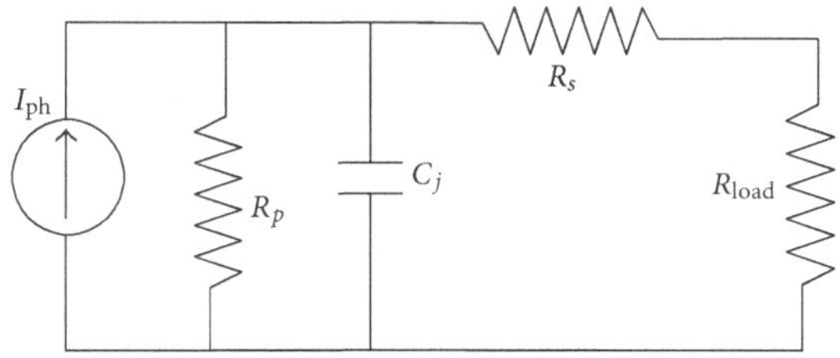

FIGURE 5: Equivalent circuit for small-signal analysis.

account the application field and the related performance requirements in terms of sensitivity and bandwidth.

3.4. Estimated Bandwidth. Figure 5 shows the equivalent circuit for small-signal analysis [25].

In the sketch, the photodetector schematized as a current generator (I_{ph}), the resistance (R_p) and capacitance (C_j) associated to the junction, the series resistance (R_s), and the load resistance (R_L) are reported. R_p (evaluated from the inverse derivative of the current-voltage characteristic) results in being greater than 60 kΩ at any reverse bias and 0.2 MΩ at −100 mV. On the other hand, $R_s = 6.9\,\text{k}\Omega$, extracted by fitting the I-V characteristic with the canonical equation of the Schottky diodes, results in being both much greater than load resistance (typically 50 Ω) and much smaller than parallel resistance R_p.

Hence, the device 3 dB frequency becomes

$$f_{3\,\text{dB}} = \frac{1}{2\pi R_s C_j}. \tag{2}$$

Finally, in order to estimate the bandwidth, the device capacitance has been measured by an LCZ meter (Keithley-3322) which drives the device with a known AC voltage input signal deriving the capacitance by precisely measuring the resultant current. The measured capacitance is smaller than 10 pF at any reverse bias and about 5 pF at −100 mV.

Unfortunately, the series resistance R_s is rather large due to the nonrectifying contact realized without a heavily p-doped region under the contact. For this reason, the estimated device bandwidth is limited to megahertz range but could be easily improved toward GHz operation in an optimized device provided a good ohmic contact.

4. IPE-Based Integrated Device

4.1. Proposed Device. The sketch of the proposed integrated device is shown in Figure 6, it is a rib waveguide fabricated starting from an SOI (silicon-on-insulator) substrate. The Cu Schottky metal is placed on the output facet, thus transversally to the optical field confined into the waveguide, while the collection contacts are placed on the waveguide slabs.

It is noteworthy that the Cu film on the bottom of the trench is deposited on the silicon oxide thick layer, thus the semiconductor/metal contact is only confined on the vertical surface of the trench. This allows to achieve a very narrow active region. On the other hand, the Cu layer on the oxide layer provides a large region for the electrical pad contact without hampering the device capacitance.

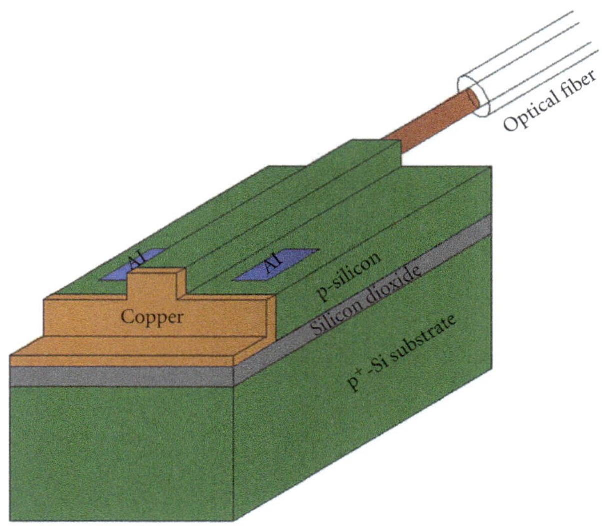

FIGURE 6: Schematic view of the proposed photodetector integrated with SOI waveguide.

4.2. Device Fabrication. In Figure 7, a top view of the proposed integrated photodetector is illustrated.

Rib waveguide was defined on a 1.5 μm thick lightly p-doped top layer of an SOI wafer with a 1.0 μm thick silicon oxide buried layer. The waveguide was formed by dry-etching the Si layer over a 100 nm depth, with a rib width of 2 μm. The rib height was chosen in order to assure a single mode propagation [29]. Alongside of the waveguides collecting ohmic contacts were placed; these contacts were realized by a thermally evaporated aluminium film, annealed at 475°C for 30 minutes. Waveguides were terminated on a deep trench that reaches down the buried oxide layer and formed by reactive ion etching (RIE). The thick photoresist, used as protection mask during trench etching, was not stripped after the RIE process, in this way a self-aligned lift-off process was performed to fabricate the Schottky contact by a 400 nm thick Cu layer, thermally evaporated on the vertical surface of the deep trench.

4.3. Responsivity Measurements. The photodetector responsivity has been measured by launching a light beam of a wavelength tunable laser source from a lensed optical fiber with a spot size of 3 μm into the SOI waveguide. The coupling and waveguide transmission loss have been evaluated by a numeric simulation [30], and a loss of about 6 dB has been estimated. A sourcemeter (Keithley-2410) was employed both to apply a bias voltage across the device and to measure the photogenerated current. Figure 8 illustrates the photogenerated current as function of the wavelength into a range from 1520 nm to 1620 nm at reverse bias of −1 V. At −1 V of the reverse bias applied, device dark current is about 0.01 μA.

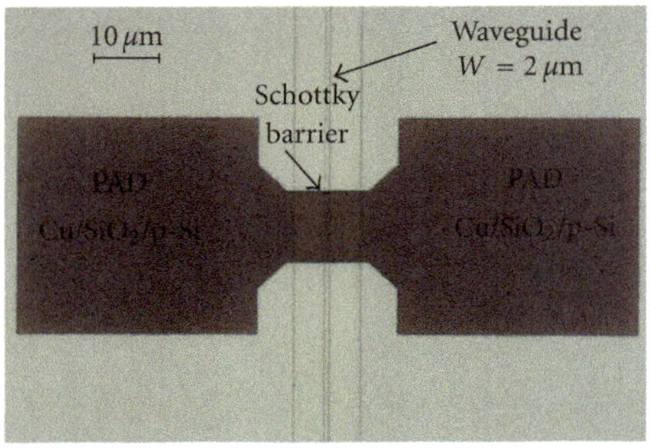

FIGURE 7: Top view of the proposed Si waveguide photodetector.

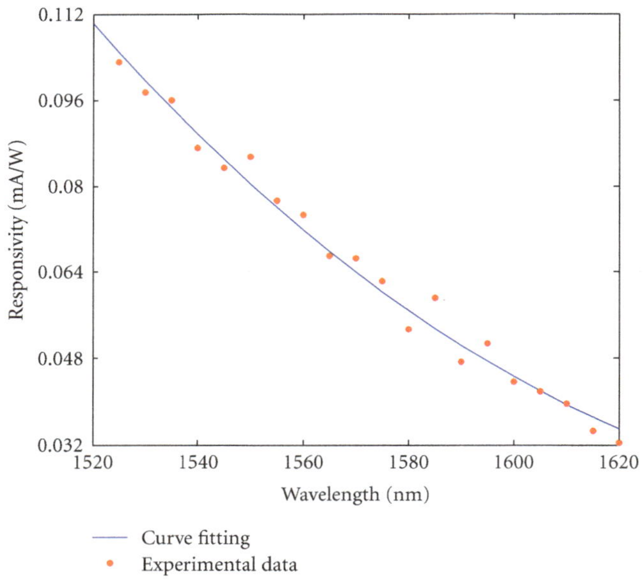

FIGURE 8: Photogenerated current versus wavelength at −1 V of reverse bias applied.

The curve shows a downward trend with the optical wavelengths. This behaviour is typical for photocurrents generated by IPE and can be predicted by theoretical equation recalled in Section 2. Using it as curve fitting of the experimental data, the potential barrier has been extracted obtaining a value of $\Phi_B = 0.74 \pm 0.01$ eV. This confirms absorption due to the IPE and not to other mechanisms generating photocarrires as surface-state absorption or TPA whose behaviour, in terms of photogenerated current, grows by increasing the optical wavelengths [16].

4.4. Estimated Bandwidth. In order to estimate the bandwidth, the device capacitance has been measured by an LCZ meter (Keithley-3322) which drives the device with a known AC voltage input signal deriving the capacitance by precisely measuring the resultant current. The measured capacitance per unit area is about 4 fF/μm^2 at −1 V. Due to the large area of the Cu/p-Si interface of the characterized device (see Figure 7), its bandwidth is limited to megahertz range, in particular this area of 1.5 μm × 5 mm extends in much more of the region around the waveguides, that is, of the active area useful for the photogeneration. However, in an optimized device the bandwidth could be easily improved by confining the Schottky interface strictly around the waveguide. If an active area for an optimized device of 1.5 μm (waveguide height) × 2 μm (waveguide width) is considered, a value of 12 fF for the capacitance can be estimated. It is noteworthy that when the detector area is made sufficiently small, the influence due to the capacitance is reduced and the effect of the transit time dominates [31]. The transit time limited bandwidth is given by

$$f_{\text{tr}} = \frac{0.44 v_{\text{sat}}}{W}, \quad (3)$$

where v_{sat} is the effective Si carrier saturation velocity (10^7 cm s^{-1}) and W is the depletion width. For a depletion width of 1.6 μm at reverse bias of −1 Volt [32], a bandwidth >170 GHz can be estimated.

5. Conclusions

In this paper the fabrication and characterization of all-silicon photodetectors based on IPE at 1550 μm have been reported. In particular, two types of structures, bulk and integrated devices, have been investigated.

Concerning the bulk photodetector, the preliminary device based on resonant structure exhibits a responsivity of about 8.2 μA/W around 1550 nm and the capacitance measurements (pF) encourage pursuing increased bandwidth toward GHz operation. Even if responsivity could be improved by optimizing a number of parameters, first of all the ohmic contact quality, its value could be still suitable for power monitoring applications. By comparing experimental results of this paper with the ones reported in our previous paper [28], we experimentally demonstrate that an improvement in responsivity can be obtained by increasing the cavity finesse. Hence, the proposed photodetector is thus very promising for integration in more complex microcavity having very high finesse (e.g., disk or ring resonator) which could allow the integration on silicon waveguide and a further enhancement in responsivity.

Concerning the integrated devices, the preliminary results exhibit a responsivity of 0.08 mA/W at 1550 nm and an efficient behaviour in both C and L bands.

In a next optimized device its responsivity could be enhanced by a resonant structure defined by etching deep trenches on the silicon rib waveguide. In addition, in order to obtain operation speed in the GHz range, the bandwidth could be improved by realizing a Schottky interface narrowly around the waveguide and by realizing a large pad for electrical contact and testing on the oxide.

References

[1] B. Jalali and S. Fathpour, "Silicon photonics," *Journal of Lightwave Technology*, vol. 24, no. 12, pp. 4600–4615, 2006.

[2] R. Soref, "The past, present, and future of silicon photonics," *IEEE Journal on Selected Topics in Quantum Electronics*, vol. 12, no. 6, pp. 1678–1687, 2006.

[3] L. Pavesi and D. Lockwood, *Silicon Photonics*, vol. 94 of *Topics in Applied Physics*, Springer, Berlin, Germany, 2004.

[4] L. C. Kimerling, L. dal Negro, S. Saini et al., "Monolithic silicon microphotonics," in *Silicon Photonics*, L. Pavesi and D. J. Lockwood, Eds., vol. 94 of *Topics in Applied Physics*, pp. 89–119, Springer, Berlin, Germany, 2004.

[5] L. K. Rowe, M. Elsey, N. G. Tarr, A. P. Knights, and E. Post, "CMOS-compatible optical rib waveguides defined by local oxidation of silicon," *Electronics Letters*, vol. 43, no. 7, pp. 392–393, 2007.

[6] L. Vivien, D. Pascal, S. Lardenois et al., "Light injection in SOI microwaveguides using high-efficiency grating couplers," *Journal of Lightwave Technology*, vol. 24, no. 10, pp. 3810–3815, 2006.

[7] Q. Xu, S. Manipatruni, B. Schmidt, J. Shakya, and M. Lipson, "12.5 Gbit/s carrier-injection-based silicon micro-ring silicon modulators," *Optics Express*, vol. 15, no. 2, pp. 430–436, 2007.

[8] C. P. Michael, M. Borselli, T. J. Johnson, C. Chrystal, and O. Painter, "An optical fiber-taper probe for wafer-scale microphotonic device characterization," *Optics Express*, vol. 15, no. 8, pp. 4745–4752, 2007.

[9] A. Liu, L. Liao, D. Rubin et al., "High-speed optical modulation based on carrier depletion in a silicon waveguide," *Optics Express*, vol. 15, no. 2, pp. 660–668, 2007.

[10] A. Liu, H. Rong, R. Jones, O. Cohen, D. Hak, and M. Paniccia, "Optical amplification and lasing by stimulated Raman scattering in silicon waveguides," *Journal of Lightwave Technology*, vol. 24, no. 3, pp. 1440–1455, 2006.

[11] G. Masini, L. Colace, and G. Assanto, "2.5 Gbit/s polycrystalline germanium-on-silicon photodetector operating from 1.3 to 1.55 μm," *Applied Physics Letters*, vol. 82, no. 15, pp. 2524–2526, 2003.

[12] L. Colace, P. Ferrara, G. Assanto, D. Fulgoni, and L. Nash, "Low dark-current germanium-on-silicon near-infrared detectors," *IEEE Photonics Technology Letters*, vol. 19, no. 22, pp. 1813–1815, 2007.

[13] M. Casalino, G. Coppola, M. Iodice, I. Rendina, and L. Sirleto, "Near-infrared sub-bandgap all-silicon photodetectors: state of the art and perspectives," *Sensors*, vol. 10, no. 12, pp. 10571–10600, 2010.

[14] T. K. Liang, H. K. Tsang, I. E. Day, J. Drake, A. P. Knights, and M. Asghari, "Silicon waveguide two-photon absorption detector at 1.5 μm wavelength for autocorrelation measurements," *Applied Physics Letters*, vol. 81, no. 7, pp. 1323–1325, 2002.

[15] J. D. B. Bradley, P. E. Jessop, and A. P. Knights, "Silicon waveguide-integrated optical power monitor with enhanced sensitivity at 1550 nm," *Applied Physics Letters*, vol. 86, no. 24, Article ID 241103, pp. 1–3, 2005.

[16] H. Chen, X. Luo, and A. W. Poon, "Cavity-enhanced photocurrent generation by 1.55 μm wavelengths linear absorption in a p-i-n diode embedded silicon microring resonator," *Applied Physics Letters*, vol. 81, Article ID 171111, 2009.

[17] S. Zhu, M. B. Yu, G. Q. Lo, and D. L. Kwong, "Near-infrared waveguide-based nickel silicide Schottky-barrier photodetector for optical communications," *Applied Physics Letters*, vol. 92, no. 8, Article ID 081103, 2008.

[18] A. Akbari and P. Berini, "Schottky contact surface-plasmon detector integrated with an asymmetric metal stripe waveguide," *Applied Physics Letters*, vol. 95, no. 2, Article ID 021104, 2009.

[19] Y. Wang, X. Su, Y. Zhu et al., "Photocurrent in Ag-Si photodiodes modulated by plasmonic nanopatterns," *Applied Physics Letters*, vol. 95, no. 24, Article ID 241106, 2009.

[20] R. H. Fowler, "The analysis of photoelectric sensitivity curves for clean metals at various temperatures," *Physical Review*, vol. 38, no. 1, pp. 45–56, 1931.

[21] V. E. Vickers, "Hydrostatic equilibrium and gravitational collapse of relativistic charged fluid balls," *Applied Optics*, vol. 10, pp. 2190–2192, 1971.

[22] E. Y. Chan and H. C. Card, "Near IR interband transitions and optical parameters of metal-germanium contacts," *Applied Optics*, vol. 19, no. 8, pp. 1309–1315, 1980.

[23] C. Scales and P. Berini, "Thin-film schottky barrier photodetector models," *IEEE Journal of Quantum Electronics*, vol. 46, no. 5, Article ID 5485040, pp. 633–643, 2010.

[24] H. C. Card, "Aluminum-silicon schottky barriers and ohmic contacts in integrated circuits," *IEEE Transactions on Electron Devices*, vol. ED-23, no. 6, pp. 538–544, 1976.

[25] S. Donati, *Photodetectors: Devices, Circuits, and Applications*, Prentice Hall PTR, Upper Saddle River, NJ, USA, 1999.

[26] M. S. Ünlü and S. Strite, "Resonant cavity enhanced photonic devices," *Journal of Applied Physics*, vol. 78, no. 2, pp. 607–639, 1995.

[27] E. D. Palik, *Handbook of Optical Constants of Solids*, Academic Press, San Diego, Calif, USA, 1985.

[28] M. Casalino, L. Sirleto, L. Moretti, M. Gioffré, G. Coppola, and I. Rendina, "Silicon resonant cavity enhanced photodetector based on the internal photoemission effect at 1.55 μm: fabrication and characterization," *Applied Physics Letters*, vol. 92, no. 25, Article ID 251104, 2008.

[29] S. P. Pogossian, L. Vescan, and A. Vonsovici, "The single-mode condition for semiconductor rib waveguides with large cross section," *Journal of Lightwave Technology*, vol. 16, no. 10, pp. 1851–1853, 1998.

[30] BeamPROP of RSoftDesignGroup, http://www.rsoftdesign.com/products.php?sub=Component+Design&itm=BeamPROP&det=Product+Overview.

[31] M. Casalino, L. Sirleto, L. Moretti, and I. Rendina, "A silicon compatible resonant cavity enhanced photodetector working at 1.55 vm," *Semiconductor Science and Technology*, vol. 23, no. 7, Article ID 075001, 2008.

[32] S. M. Sze, *Physics of Semiconductor Devices*, John Wiley & Sons, New York, NY, USA, 1981.

17

Preparation, Characterization, and Activity Evaluation of $CuO/F\text{-}TiO_2$ Photocatalyst

Zhang Jinfeng,[1] Yang Yunguang,[2] and Liu Wei[2]

[1] *Department of Physics, Huaibei Normal University, Anhui, Huaibei 235000, China*
[2] *Department of Chemistry, Huaibei Normal University, Anhui, Huaibei 235000, China*

Correspondence should be addressed to Liu Wei, weiliu@chnu.edu.cn

Academic Editor: Baibiao Huang

$CuO/F\text{-}TiO_2$ nanoparticle photocatalyst was prepared by ball milling. The photocatalyst was characterized by X-ray powder diffraction, scanning electron microscopy, transmission electron microscopy, UV-Vis diffuse reflectance spectroscopy, and photoluminescence emission spectroscopy. The photocatalytic activity was evaluated by photocatalytic oxidation of rhodamine B and reduction of $Cr_2O_7^{2-}$. The results showed that, for $F\text{-}TiO_2$ photocatalyst, the photooxidation activity increases remarkably with the increasing amount of NH_4F up to 1.0 g, and the photoreduction activity decreases gradually with the increase in the amounts of NH_4F. For the $CuO/F\text{-}TiO_2$ photocatalyst, the photoreduction activity increases greatly with the increase in the amount of doped p-CuO up to 1.0 wt.%, and the photooxidation activity decreases rapidly with the increase in the amounts of doped p-CuO. Compared with pure TiO_2, the photoabsorption wavelength range of the $CuO/F\text{-}TiO_2$ and $F\text{-}TiO_2$ photocatalysts red shifts and improves the utilization of the total spectrum. The effect of ball milling time on the photocatalytic activity of the photocatalysts was also investigated. The mechanisms of influence on the photocatalytic activity of the photocatalysts were also discussed.

1. Introduction

Since Fujishima and Honda discovered the photocatalytic splitting of water on titanium dioxide (TiO_2) electrodes in 1972 [1], TiO_2 as a photocatalyst has been extensively studied because it has relatively high photocatalytic activity, biological and chemical stability, low cost, nontoxicity, and long-term stability against photocorrosion and chemical corrosion [2–9]. However, the photocatalytic activity of TiO_2 is limited to irradiation wavelengths in the UV region, thus the effective utilization of solar energy is limited to about 3–5% of the total solar spectrum. Some problems still remain to be solved in its application, such as the fast recombination of photogenerated electron-hole pairs. Therefore, improving photocatalytic activity by modification has become a hot topic among the researchers in the near decade [10, 11]. Many investigators have quested for various methods, such as doping transition metals [12–15], doping nonmetallic elements [16–20], and forming composite photocatalysts from different semiconductors [21–24], and so forth, to enhance the photocatalytic activity of TiO_2 and to improve the utilization of visible light.

Recently, fluorinated TiO_2 ($F\text{-}TiO_2$) had been investigated extensively. The results showed that the photocatalytic oxidation activity of the $F\text{-}TiO_2$ is much higher than that of TiO_2 by reason of the extension of the photoabsorption wavelength as a result of doped F element [25–27]. The p-n junction photocatalysts NiO/TiO_2 and $p\text{-}ZnO/n\text{-}TiO_2$ have been studied in our laboratory [21, 28]. The results showed that, compared with pure TiO_2, the p-n junction photocatalysts have higher photocatalytic reduction activity, but lower photocatalytic oxidation activity. It is known that CuO is a p-type semiconductor [29]. When p-type CuO and fluorinated n-type TiO_2 integrates, a p-n heterojunction photocatalyst $CuO/F\text{-}TiO_2$ will be formed, which may improve charge separation and photocatalytic activity of the photocatalyst.

In this paper, $CuO/F\text{-}TiO_2$ powder was prepared by ball milling using NH_4F solution as a disperser. The photocatalytic activity of the photocatalyst was evaluated by photocatalytic oxidation of rhodamine B (RhB) and reduction of $Cr_2O_7^{2-}$. The desired result was obtained. The effect of ball milling time on the photocatalytic activity of the photocatalysts was also investigated. The mechanisms of influence on the photocatalytic activity of the CuO/F-TiO_2photocatalyst were also discussed.

2. Experimental

2.1. Materials. The TiO_2 powder (Anatase 90%, Rutile 10%, with crystallite size of about 50–60 nm) used in the experiments was supplied by Sinopharm Chemical Reagent Co, Ltd. $Cu(NO_3)_2 \cdot 3H_2O$ was supplied by Shanghai Reagent Factory (purity 99.9%). NH_4F, Rhodamine B, and other chemicals used in the experiments were purchased from Shanghai and other China Chemical Reagent Ltd. They were of analytically pure grade. Deionized water was used throughout this study.

2.2. Preparation of F-TiO_2 and CuO/F-TiO_2. Nanoparticle CuO was synthesized by heat treatment of $Cu(NO_3)_2 \cdot 3H_2O$ at 500°C for 6 h in air, and the crystallite size is about 80 nm. The preparation of F-TiO_2 photocatalyst was carried out in a ND2-2L ball mill (made in Tianzun Electronics Co, Ltd., Nanjing University). The procedures for the preparation of F-TiO_2 are as follows: 5.0 g TiO_2 powder and three different sizes of zirconia balls were mixed in the zirconia tank, and then a certain amount of NH_4F (0 g, 0.2 g, 0.5 g, 1.0 g, 1.5 g, 2.0 g, 2.5 g) and H_2O (5 mL) was added. After being milled for a certain time (0–24 h) at the speed of 550 rpm, the wet powder was dried at 110°C in air. The final samples were used for the determination of photocatalytic activity and characterization. CuO/F-TiO_2 photocatalyst was prepared in the same procedure. For each sample, 1.0 g NH_4F, 5.0 g TiO_2, and 5.0 mL H_2O were added in a zirconia tank, varying the weight ratio of CuO(0 wt.%, 0.1 wt.%, 0.3 wt.%, 0.5 wt.%, 1.0 wt.%, 3.0 wt.%, and 5.0 wt.%), and then different CuO/F-TiO_2 powder samples were prepared, respectively. The specific surface area of the different CuO/F-TiO_2 photocatalysts has no obvious change and it is about 24.7 m^2/g. The flow chart of preparation of the CuO/F-TiO_2 is given in Scheme 1.

2.3. Photoreaction Apparatus and Procedure. Experiments were carried out in a photoreaction apparatus [21, 28]. The photoreaction apparatus consists of two parts. The first part is an annular quartz tube. A 375 W medium pressure mercury lamp (Institute of Electric Light Source, Beijing), with a maximum emission at about 365 nm, was used as light sources. The lamp is laid in the empty chamber of the annular tube, and running water passes through an inner thimble of the annular tube. Owing to continuous cooling, the temperature of the reaction solution is maintained at approximately 30°C. The second part is an unsealed beaker with a diameter of 12 cm. At the start of the experiment, the reaction solution (volume, 300 mL) containing reactants and photocatalyst was put in the unsealed beakers, and a magnetic stirring device was used to stir the reaction solution. The distance between the light source and the surface of the reaction solution is 11 cm. In the experiments, the initial pH of the reaction solution was 5.0, the amount of the photocatalyst used was 2.0 g/L, and the initial concentrations of $Cr_2O_7^{2-}$ and RhB were 1.0×10^{-4} and 1.0×10^{-5} mol/L, respectively. In order to disperse the photocatalyst powder, the suspensions were ultrasonically vibrated for 10 or 20 min prior to irradiation. After illumination, the samples taken from the reaction suspension were centrifuged at 7000 rpm for 20 min and filtered through a 0.2 μm millipore filter to remove the particles. The filtrate was then analyzed. In order to determine the reproducibility of the results, at least duplicated runs were carried out for each condition for averaging the results, and the experimental error was found to be within ±4%.

2.4. Characterization. In order to determine the crystal phase composition and the crystallite size of the photocatalysts, X-ray diffraction measurement was carried out at room temperature using a DX-2000 X-ray powder diffractometer with Cu $K\alpha$ radiation and a scanning speed of 3°/min. The accelerating voltage and emission current were 40 kV and 30 mA, respectively. The crystallite size was calculated by X-ray line broadening analysis using the Scherrer equation.

The microcrystalline structure and surface characteristics of the photocatalysts were also investigated by using (X-650 Japan) scanning electron microscope.

Transmission electron microscopy and high-resolution transmission electron microscopy (HR-TEM) images were performed with a JEOL-2010 transmission electron microscope, using an accelerating voltage of 200 kV.

UV-Vis diffuse reflectance spectra measurements were carried out using a Hitachi UV-365 spectrophotometer equipped with an integrating sphere attachment. The analysis range was from 250 to 650 nm, and $BaSO_4$ was used as a reflectance standard.

Photoluminescence emission spectra were recorded on a JASCO FP-6500 type fluorescence spectrophotometer over a wavelength range of 360–500 nm.

2.5. Analysis. The concentration of $Cr_2O_7^{2-}$ in the solution is determined spectrophotometrically using diphenylcarbazide reagent as a developer. The concentration of rhodamine B (RhB) in the solution is determined spectrophotometrically. The photoreduction efficiency of $Cr_2O_7^{2-}$ and the photooxidation efficiency of rhodamine B were calculated from the following expression:

$$\eta = \left[\frac{(C_0 - C_t)}{C_0}\right] \times 100\%, \quad (1)$$

where η is the photocatalytic efficiency; C_0 is the concentration of reactant before illumination; C_t is the concentration of reactant after illumination time t.

3. Results and Discussion

3.1. Characterization of Photocatalysts

3.1.1. XRD Analysis. The fixed ball milling time is 6 h. The XRD patterns of different photocatalysts are shown in Figure 1. It is clear that, when the amount of doped CuO is less than 5.0 wt.%, the diffraction peaks of CuO cannot be

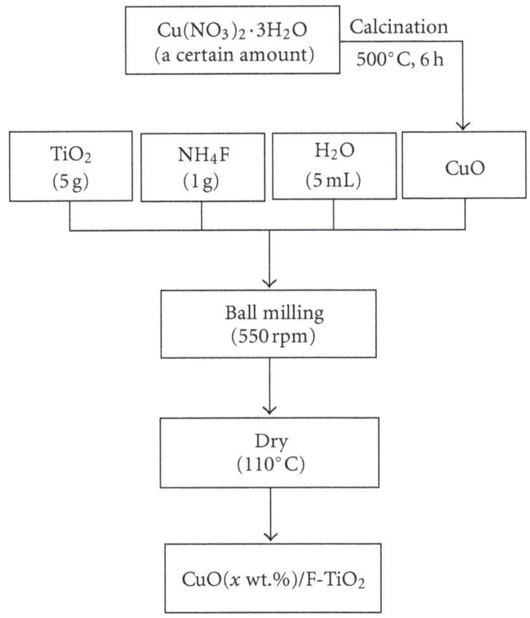

SCHEME 1: The flow chart of preparation of the CuO/F-TiO_2.

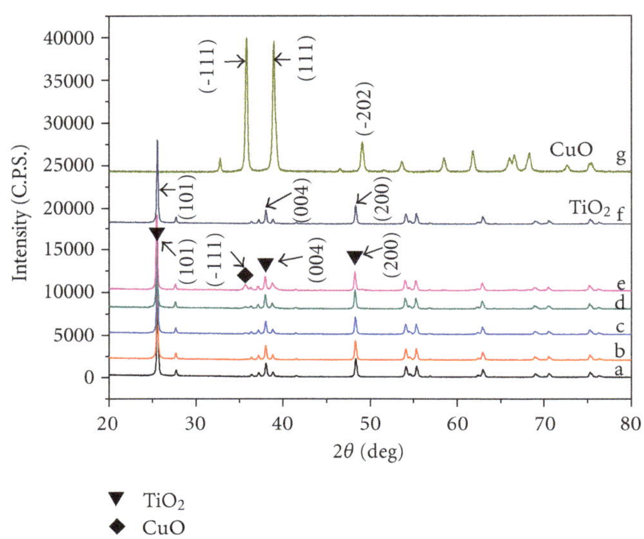

FIGURE 1: XRD patterns of different photocatalysts. (a) F-TiO_2, (b) CuO (0.1 wt.%)/F-TiO_2, (c) CuO (1.0 wt.%)/F-TiO_2, (d) CuO (5.0 wt.%)/F-TiO_2, (e) CuO (10.0 wt.%)/F-TiO_2, (f) TiO_2, and (g) CuO.

found in XRD patterns. This illustrates that CuO is highly dispersed in the bulk phase of the catalyst. When the amount of doped CuO is higher than 5.0 wt.%, the diffraction peaks of CuO can be found in XRD patterns. Since no new crystal phases are found, it can be concluded that a new solid is not formed in the ball milling process of TiO_2, NH_4F, and CuO, or probably because the new materials content are too low to allow detection of their reflection peaks. The similar result was reported in [30]. It is known by the calculation from the Scherrer equation that the diameter of the photocatalyst is not obviously changed. The crystallite size is about 50 nm.

3.1.2. SEM Analysis. SEM was used to investigate the morphology of the samples. Figure 2 shows SEM images of CuO (1.0 wt.%)/F-TiO_2 photocatalysts ball milled for 6 and 24 h, respectively. It can be seen that the appearance is shapeless sheet, and the average diameter of the photocatalyst is about 50–60 nm. The result is the same as that of XRD. From Figure 2, it also can be seen that, when the ball milling time is 6 h, the dispersion degree of the sample is higher than that of the sample ball milled for 24 h. Namely, when the ball milling time is longer than the optimum time, with the increase in the ball milling time, the fresh surfaces formed by

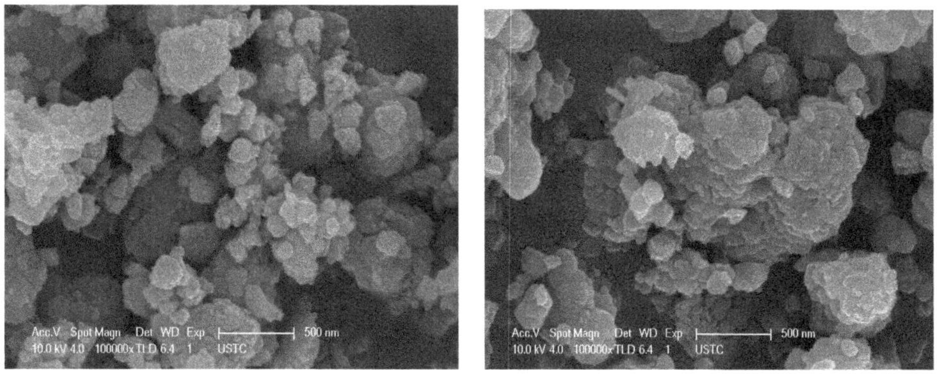

(a) (b)

Figure 2: SEM images of CuO (1.0 wt.%)/F-TiO$_2$ photocatalysts. (a) CuO (1.0 wt.%)/F-TiO$_2$, ball milling time 6 h, (b) CuO (1.0 wt.%)/F-TiO$_2$, ball milling time 24 h.

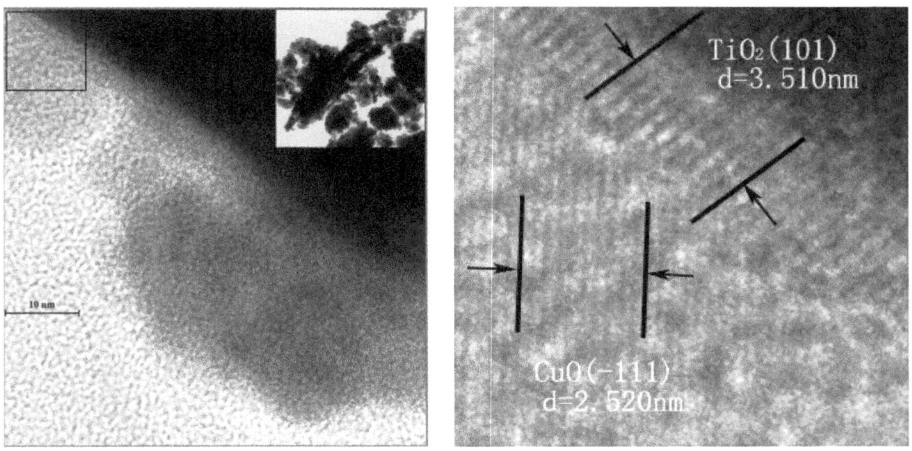

(a) (b)

Figure 3: TEM and HR-TEM images of CuO (1.0 wt.%)/F-TiO$_2$photocatalyst: (a) TEM image, (b) HR-TEM image.

high-energy ball milling possess high surface energy and prefer to agglomerate. The similar result was reported in [30].

3.1.3. TEM Analysis. In order to investigate the interface of the sample, the CuO (1.0 wt.%)/F-TiO$_2$ was chosen for TEM and high-resolution TEM characterization.

Figure 3(a) gives an overview of the typical TEM image of the CuO (1.0 wt.%)/F-TiO$_2$ photocatalyst. It clearly exhibits the existence of CuO nanoparticles with mean sizes of about 50–70 nm dispersing over the particle of TiO$_2$. Figure 3(b) shows the HR-TEM image of the sample corresponding to the rectangle region of the TEM image in Figure 3(a). The upper part depicts the (101) plane of TiO$_2$ with a spacing value of 3.510 nm. The lower part depicts the (−111) plane of CuO with a spacing of 2.520 nm. The good crystalline quality and the clear interface between CuO and TiO$_2$ are advantageous for the separation of the photogenerated charge carriers. Based on the above results, it is suggested that the heterojunction will be formed by ball milling between CuO and TiO$_2$.

3.1.4. UV-Vis Analysis. Figure 4 shows the UV-Vis diffuse reflectance spectra of F-TiO$_2$ and a series of CuO/F-TiO$_2$ photocatalysts. The samples were ball milled for 6 h, respectively. It is known that the bandgap of TiO$_2$ is about 3.2 eV and it can be excited by photons with wavelengths below 387 nm. From Figure 4, it can be seen that, compared with pure TiO$_2$, the absorption wavelength ranges of the F-TiO$_2$ and CuO/F-TiO$_2$ are extended greatly towards visible light. Compared with the F-TiO$_2$, the absorption edge of the photocatalyst CuO/F-TiO$_2$ extends a little to longer wavelength, revealing the good contact between CuO and F-TiO$_2$crystallites as a consequence of the interdispersion of the two phases produced by ball milling process. For F-TiO$_2$, the presence of a strong absorption band at a low wavelength in the spectra near 350 nm indicates that the Ti species are tetrahedral Ti^{4+}. This absorption band is generally

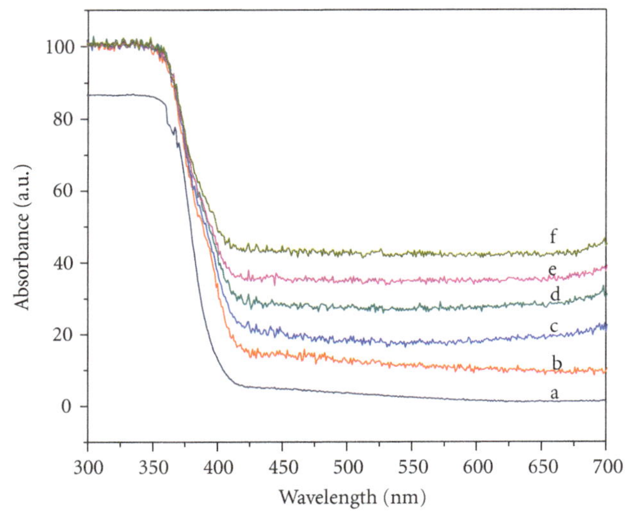

FIGURE 4: UV-Vis diffuse reflectance spectra of different photocatalysts. (a) TiO_2, (b) F-TiO_2, (c) CuO(0.1 wt.%)/F-TiO_2, (d) CuO(0.3 wt.%)/F-TiO_2, (e) CuO(0.5 wt.%)/F-TiO_2, (f) CuO(1 wt.%)/F-TiO_2.

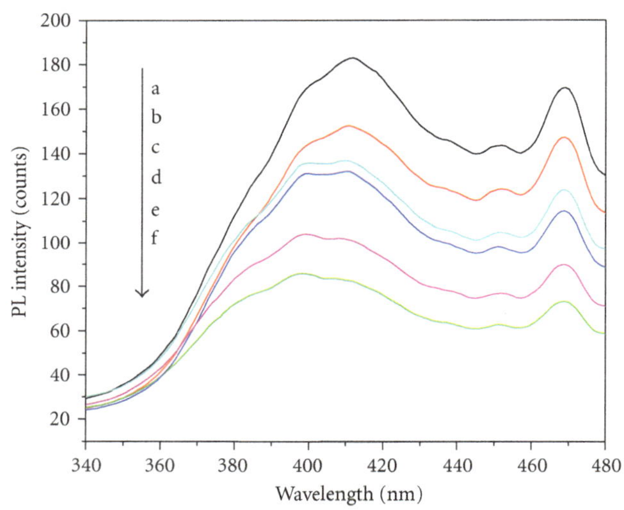

FIGURE 5: Photoluminescence emission spectra of photocatalysts. (a) TiO_2, (b) CuO (0.1 wt.%)/F-TiO_2, (c) CuO (0.3 wt.%)/F-TiO_2, (d) CuO (0.5 wt.%)/F-TiO_2, (e) CuO (3 wt.%)/F-TiO_2, (f) CuO (1 wt.%)/F-TiO_2.

associated with the electronic excitation of the valence band O2p electron to the conduction band Ti3d level. The photoexcited wavelength range of the CuO/F-TiO_2 photocatalyst is connected with the amount of CuO. It increases with the increase in the amount of CuO. In theory, because the absorption wavelength range is extended greatly towards visible light and the absorption intensity increases, the formation rate of electron-hole pairs on the photocatalyst surface also increases greatly, which results in the photocatalyst exhibiting higher photocatalytic activity. From Figure 7, it is clear that the photocatalytic activity of photocatalyst CuO/F-TiO_2 is strongly dependent on the amount of CuO. In the experimental condition, the results of UV-Vis diffuse reflectance spectra are consistent with the evaluation of photoreduction activity. There were similar results in previous reports [28].

3.1.5. Photoluminescence Emission Spectra. The photoluminescence emission spectra have been widely used to investigate the efficiency of charge carrier trapping, immigration, and transfer and to understand the fate of electron-hole pairs in semiconductor particles, since photoluminescence emission is resulted from the recombination of free carriers [31, 32]. In this study, using an ultraviolet light with a 260 nm wavelength as the excitation source, the fluorescence emission spectra of the different samples are shown in Figure 5. It can be seen that the samples have a stronger emission peak at around 410 nm and a weaker emission peak at around 470 nm [21, 33].

From Figure 5, it is clear that the relative intensity of the emission spectra of TiO_2 has the greatest relative intensity, which means that electrons and holes of $TiO_2$are easy to recombine. The relative intensity of the photocatalyst CuO/F-$TiO_2$is lower than that of TiO_2, showing that doping CuO is helpful to inhibit the recombination of electrons and holes and to improve the photocatalytic activity. The amount of CuO can influence the thickness of the superficial space-charge layer of TiO_2. When the amount of CuO is 1.0 wt %, the relative intensity of emission spectra is the lowest, which shows that a 1.0 wt.% doping quantity of CuO can effectively restrain the recombination of photoexcited electrons and holes. When the CuO content is too small, due to the absence of adequate traps, the recombination rate of electron-hole pairs is higher. When the doping quantity is considerably high, the absorption of light and the generation of electrons-holes are both decreased.

3.2. Evaluation of the Photocatalytic Activity

3.2.1. Effect of the Amount of NH_4F on the Photocatalytic Activity of F-TiO_2. The photocatalytic activity was investigated by photocatalytic oxidation of RhB and photocatalytic reduction of $Cr_2O_7^{2-}$. The dark absorption and blank tests are carried out simultaneously. The dark absorption test demonstrated that the adsorption on the catalysts varied depending on the samples, while the blank experiment showed that the substrate with no photocatalyst was hardly degraded after exposure to radiation for 1 h, illustrating that photoinduced self-sensitized photodegradation has little influence on the results of the experiment. All of the data obtained are corrected for absorption after stirring for 30 min in the dark. The detailed procedure was described in [34]. The fixed ball milling time is 3 h, and fixed illumination times for the photocatalytic reduction of $Cr_2O_7^{2-}$ and photocatalytic oxidation of RhB were 20 min and 10 min, respectively. Figure 6 shows the effect of the amount of NH_4F on the photocatalytic reduction of $Cr_2O_7^{2-}$ and photocatalytic oxidation of RhB.

From Figure 6, it can be seen that, without NH_4F present, namely, the pure TiO_2 powder photocatalyst, its photooxidation activity is the lowest, and the photooxidation efficiency for RhB is 15.3%. The photooxidation activity of

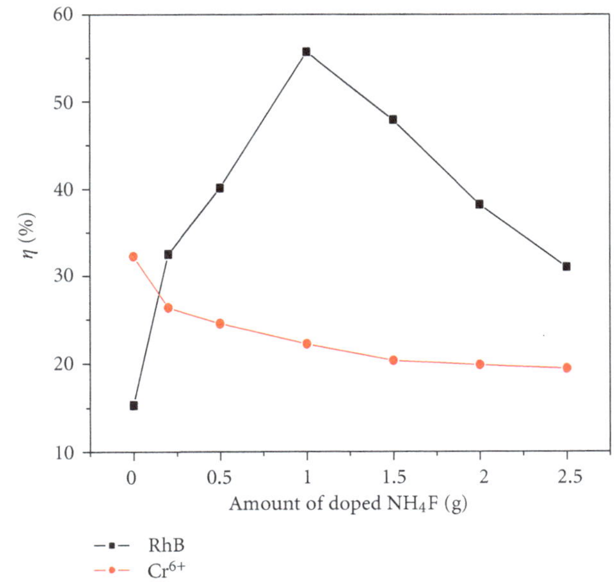

FIGURE 6: Effect of the amount of NH$_4$F on the photocatalytic reduction of $Cr_2O_7^{2-}$ and photocatalytic oxidation of RhB.

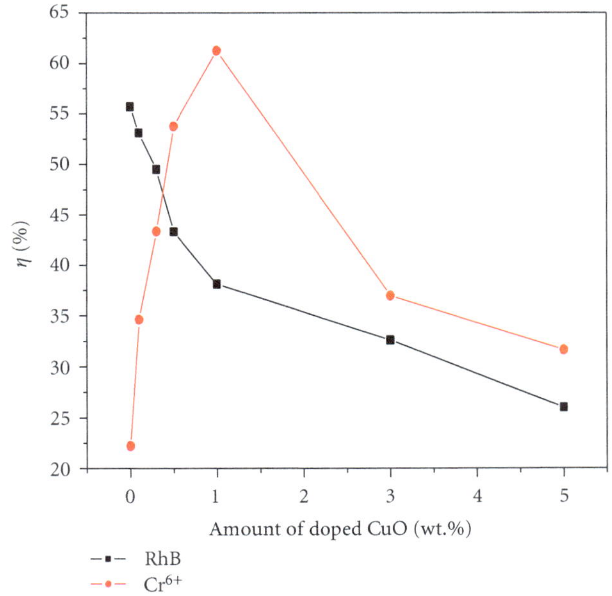

FIGURE 7: Effect of the amount of doped CuO on the photocatalytic reduction of $Cr_2O_7^{2-}$ and photocatalytic oxidation of RhB.

F-TiO$_2$ increases remarkably with the increasing amount of NH$_4$F up to 1.0 g. When the amount of NH$_4$F is 1.0 g, the photocatalytic activity of the F-TiO$_2$ photocatalyst is at its peak, and the photooxidation efficiency for RhB is 55.7%. When the amount of doped NH$_4$F is higher than optimal amount, the photooxidation activity of F-TiO$_2$ decreases gradually. At the same time, it is clear that the photooxidation activity of F-TiO$_2$ is higher than that of pure TiO$_2$ photocatalyst in the experimental condition. F-TiO$_2$ enhanced the photodegradation of RhB in aqueous solutions. It was proposed that the fluorinated surface favored the generation of free OH radicals, which was responsible for the enhanced photocatalytic oxidation activity. The result has been reported in earlier literature [27].

However, under the same condition, for the photocatalytic reduction of $Cr_2O_7^{2-}$, the photocatalytic reduction activity of F-TiO$_2$ decreases gradually with the increase in the amounts of doped NH$_4$F. For the pure TiO$_2$ powder photocatalyst, its photocatalytic reduction activity is at its peak, and the photocatalytic reduction efficiency is 32.2%. When the amounts of NH$_4$F are 0.2 and 2.0 g, the photoreduction efficiencies are 26.3 and 19.8%, respectively.

3.2.2. Effect of Amount of Doped CuO on the Photocatalytic Activity. In order to obtain the optimum concentration of CuO, the effect of the amount of CuO on the photocatalytic activity of CuO/F-TiO$_2$photocatalyst was studied. The experiments were carried out with different concentrations of CuO varying from 0% to 5 wt.%. Figure 7 shows the effects of the amount of doped CuO on the photocatalytic reduction of $Cr_2O_7^{2-}$ and photocatalytic oxidation of RhB. The fixed ball milling time for each sample was 3 h, and the fixed illumination times for the photocatalytic reduction of $Cr_2O_7^{2-}$ and photocatalytic oxidation of RhB were 20 min and 10 min, respectively. The amounts of TiO$_2$ and NH$_4$F were 5.0 g and 1.0 g, respectively. It can be seen that, for the photocatalytic reduction of $Cr_2O_7^{2-}$, the photoreduction activity of CuO/F-TiO$_2$ increases remarkably with the increase in the amount of doped CuO up to 1.0 wt.%. The optimum amount of doped CuO is 1.0 wt.%. When the amount of doped is higher than the optimal amount, the photoreduction activity of the CuO/F-TiO$_2$ photocatalyst decreases gradually as the amount of doped p-CuO increases. The results also show that, without CuO present, namely, the F-TiO$_2$ powder photocatalyst, its photoreduction activity is the lowest, and the photoreduction efficiency is 22.2%. When the amount of doped p-CuO is 1.0 wt.%, the photoreduction activity of CuO/F-TiO$_2$ is at its peak, and the photoreduction efficiency is 61.2%. From Figures 6 and 7, it is clear that the photoreduction activity of CuO/F-TiO$_2$is higher than that of pure TiO$_2$ and F-TiO$_2$ photocatalyst. It is proposed that, when the amount of CuO is lower than the optimum amount, the increase of the amount of CuO can increase trapping sites of the carriers, which prolongs the lifetime of the carriers, thereby improving the photocatalytic activity. The other important reason is that CuO is a p-type semiconductor. When the F-TiO$_2$ doped with suitable amount of p-CuO, both p-CuO and F-TiO$_2$ can form the p-n junction photocatalyst by ball milling. Therefore, the photoreduction activity increases [33]. But when the amount of p-CuO is higher than the optimum amount of doping, the high concentration dopant ions act as recombination centers of electrons and holes, decrease the thickness of the space-charge layer on the surface of TiO$_2$ particle, and reduce the photon absorption [21]. However, under the same condition, for the photocatalytic oxidation of RhB, the photooxidation activity of p-CuO/F-TiO$_2$ decreases rapidly with the increase

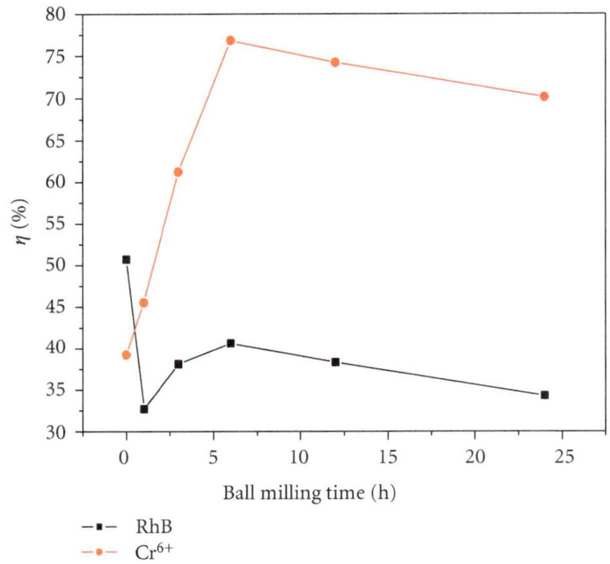

FIGURE 8: Effect of ball milling time on the photocatalytic activity.

in the amounts of doped p-CuO. For the pure F-TiO_2 photocatalyst, its photooxidation activity is at its peak, and the photooxidation efficiency of F-TiO_2 is 55.7%. When the amounts of doped p-CuO are 0.3 and 1.0 wt.%, the photooxidation efficiencies are 49.5 and 38.1%, respectively.

The photocatalyst CuO/F-TiO_2 has higher photocatalytic reduction activity than that of TiO_2 and F-TiO_2. The possible reasons are as follows.

First, it can be explained by the p-n junction principle [21, 28]. It is known that CuO is p-type semiconductor, and TiO_2 is n-type semiconductor. When p-type CuO and n-type TiO_2 integrate p-n junctions will be formed between CuO and TiO_2, and the inner electric field will be also formed at the same time in the interface. So a number of micro p-n heterojunction CuO/F-TiO_2photocatalysts will be formed after doping CuO powder into TiO_2 granule. At the equilibrium, the inner electric field formed made p-type semiconductor CuO region have the negative charge while TiO_2 region have the positive charge. The electron-hole pairs will be created under UV light illumination. With the effect of the inner electric field, the holes flow into the negative field and the electrons move to the positive field. Therefore, the photogenerated electron-hole pairs will be separated effectively by the p-n heterojunction formed in CuO/F-TiO_2.

Second, compared with F-TiO_2, the photoabsorption wavelength range of the CuO/F-TiO_2 photocatalyst red shifts and extends the wavelength range of photoexcitation and enhances the utilization of the total spectrum. The results also show that the photoluminescence emission intensity of the CuO/F-TiO_2 photocatalyst is lower than that of the pure F-TiO_2. It indicates that doping CuO is helpful to inhibit the recombination of electrons and holes and improve the photocatalytic activity. At the same time, it was reported that p-type semiconductor is a collector of photoexcited holes [35]. Therefore, there are enrichment electrons in the interface to react with $Cr_2O_7^{2-}$ adsorbed on the photocatalyst surface, so the p-n heterojunction photocatalyst CuO/F-TiO_2 has higher photocatalytic reduction activity, but lower photocatalytic oxidation activity.

3.2.3. Effect of Ball Milling Time on the Photocatalytic Activity. The effect of ball milling time on the photocatalytic activity of CuO (1.0 wt.%)/F-TiO_2 photocatalyst is shown in Figure 8. It can be seen that the ball milling time influences the photocatalytic activity strongly. Without ball milling, the photoreduction efficiency is 39.2%, and the photooxidation efficiency is 51.3%. The photoreduction efficiency of $Cr_2O_7^{2-}$ increases gradually with the increase in ball milling time up to 6 h. When the ball milling time is 1, 3, 6, 12, and 24 h, the photoreduction efficiency is 45.5, 61.2, 76.8, 74.2, and 70.1%, respectively. For the photocatalytic oxidation of RhB, the photooxidation efficiency decreases rapidly with the increase in the ball milling time. When the ball milling time is 1 h, the photooxidation efficiency is the lowest, and it is 32.7%. However, when the ball milling time is longer than 1 h, the photooxidation efficiency increases gradually with the increase in ball milling time up to 6 h. When the ball milling time is 6 h, the photooxidation efficiency of RhB is 40.6%. When the ball milling time is longer than 6 h, the photooxidation efficiency decreases gradually. The reason is that, without ball milling, F-TiO_2 and p-CuO only play their own photocatalytic role, and the p-n junction photocatalysts are not formed. However, after ball milling, TiO_2 and p-CuO can form p-n junction photocatalyst, which results in the increase of the photoreduction activity and the decrease of the photooxidation activity rapidly. Another reason is that, with the increase in the ball milling time, the specific surface area of the photocatalyst increases. Correspondingly, the number of active sites per unit weight of photocatalyst also increases [21]. But when the ball milling time is longer than the optimum time, it is proposed that, with the increase in the ball milling time, the fresh surface formed by high-energy ball milling possesses high surface energy and prefers to agglomerate [36, 37], resulting in the decrease of the photocatalytic activity. The assumption is proved by the results of SEM.

3.2.4. Hydroxyl Radical Analysis. The formation of hydroxyl radicals (•OH) on the surface of CuO (1.0 wt.%)/F-TiO_2 photocatalyst is detected by a photoluminescence (PL) technique with terephthalic acid as a probe molecule. The method is rapid, sensitive, and specific, needing only simple standard PL instrumentation. Terephthalic acid readily reacts with •OH to produce a highly fluorescent product, 2-hydroxyterephthalic acid, whose PL peak intensity is in proportion to the amount of OH radicals produced in water. Experimental procedures were reported in earlier reports [27]. After UV irradiation for 10 min, the reaction solution was filtrated to measure the PL intensity at 425 nm excited by 315 nm light of 2-hydroxyterephthalic acid.

Figure 9 shows the changes of PL spectra of different samples from 5×10^{-4} mol/L terephthalic acid solution in 2×10^{-3} mol/L NaOH. From Figure 9, it can be seen that, when the ball milling time is 6 h, the PL peak intensity of the sample is the highest. It means that the formation rate of

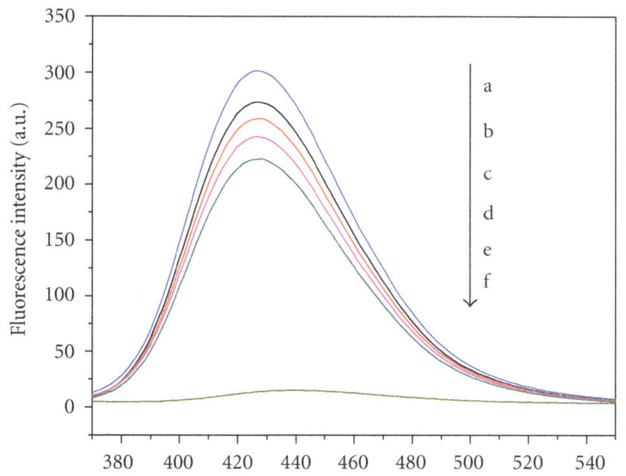

FIGURE 9: Effect of ball milling time on the changes of PL spectra. (a) 6 h, (b) 3 h, (c) 12 h, (d) 24 h, (e) 1 h, (f) pure terephthalic acid.

OH radicals on its surface is much larger than that of other samples. The PL peak intensities of the samples decrease are as follows: ball milling time 6 h > 3 h > 12 h > 24 h > 1 h. The result of PL peak intensity is consistent with that of the photooxidation activity.

4. Conclusions

The CuO/F-TiO_2 photocatalyst was prepared by ball milling method. The photooxidation activity of F-TiO_2 increases greatly with the increase in the amount of NH_4F up to 1.0 g. The photoreduction activity of CuO/F-TiO_2 increases remarkably with the increase in the amount of doped p-CuO up to 1.0 wt.%. The ball milling time has a significant influence on the photocatalytic activity of the photocatalyst. The optimum ball milling time is 6 h. Compared with pure TiO_2, the photoabsorption wavelength range of the F-TiO_2 and CuO/F-TiO_2 photocatalysts red shifts and improves the utilization of the total spectrum. As the formation of the p-n heterojunction and p-type CuO species acts as holes traps and collector, the photogenerated electron-hole pairs of the CuO/F-TiO_2 photocatalyst are separated by the inner electric field and the photocatalytic reduction activity is enhanced greatly.

Acknowledgments

This study was supported by the Natural Science Foundation of China (nos. 20973071, 51172086, and 21103060) and the Key Project of Science and Technology Research of Ministry of Education of China (208062).

References

[1] A. Fujishima and K. Honda, "Electrochemical photolysis of water at a semiconductor electrode," *Nature*, vol. 238, no. 5358, pp. 37–38, 1972.

[2] T. Yan, J. Long, X. Shi, D. Wang, Z. Li, and X. Wang, "Efficient photocatalytic degradation of volatile organic compounds by porous indium hydroxide nanocrystals," *Environmental Science and Technology*, vol. 44, no. 4, pp. 1380–1385, 2010.

[3] X. Xin, M. Scheiner, M. Ye, and Z. Lin, "Surface-treated TiO_2 nanoparticles for dye-sensitized solar cells with remarkably enhanced performance," *Langmuir*, vol. 27, no. 23, pp. 14594–14598, 2011.

[4] X. Xu, X. Li, P. Lin et al., "A general templated method to homogeneous and composition-tunable hybrid TiO_2 nanocomposite fibers," *Chemical Communications*, vol. 47, no. 9, pp. 2538–2540, 2011.

[5] Z. J. Zhou, J. Q. Fan, X. Wang, W. H. Zhou, Z. L. Du, and S. X. Wu, "Effect of highly ordered single-crystalline TiO_2 nanowire length on the photovoltaic performance of dye-sensitized solar cells," *ACS Applied Materials & Interfaces*, vol. 3, pp. 4349–4353, 2011.

[6] K. Lv, J. Yu, J. Fan, and M. Jaroniec, "Rugby-like anatase titania hollow nanoparticles with enhanced photocatalytic activity," *CrystEngCommunity*, vol. 13, no. 23, pp. 7044–7048, 2011.

[7] D. Yang, C. Chen, Z. Zheng et al., "Grafting silica species on anatase surface for visible light photocatalytic activity," *Energy and Environmental Science*, vol. 4, no. 6, pp. 2279–2287, 2011.

[8] Y. Wang, D. Zhao, H. Ji et al., "Sonochemical hydrogen production efficiently catalyzed by Au/TiO_2," *Journal of Physical Chemistry C*, vol. 114, no. 41, pp. 17728–17733, 2010.

[9] R. Shi, J. Lin, Y. Wang, J. Xu, and Y. Zhu, "Visible-light photocatalytic degradation of $BiTaO_4$ photocatalyst and mechanism of photocorrosion suppression," *Journal of Physical Chemistry C*, vol. 114, no. 14, pp. 6472–6477, 2010.

[10] R. J. Wang, G. H. Jiang, Y. W. Ding et al., "Photocatalytic activity of heterostructures based on TiO_2 and halloysite nanotubes," *ACS Applied Materials & Interfaces*, vol. 3, pp. 4154–4158, 2011.

[11] D. Y. Wu and M. Long, "Realizing visible-light-induced self-cleaning property of cotton through coating N-TiO_2 film and loading AgI particles," *ACS Applied Materials & Interfaces*, vol. 3, pp. 4770–4774, 2011.

[12] M. Liu, X. Qiu, M. Miyauchi, and K. Hashimoto, "Cu(II) oxide amorphous nanoclusters grafted Ti^{3+} self-doped TiO_2: an efficient visible light photocatalyst," *Chemistry of Materials*, vol. 23, no. 23, pp. 5282–5286, 2011.

[13] N. Zhang, S. Liu, X. Fu, and Y.-J. Xu, "Synthesis of M@TiO_2 (M = Au, Pd, Pt) core-shell nanocomposites with tunable photoreactivity," *Journal of Physical Chemistry C*, vol. 115, no. 18, pp. 9136–9145, 2011.

[14] T. A. Kandiel, A. A. Ismail, and D. W. Bahnemann, "Mesoporous TiO_2 nanostructures: a route to minimize Pt loading on titania photocatalysts for hydrogen production," *Physical Chemistry Chemical Physics*, vol. 13, no. 45, pp. 20155–20161, 2011.

[15] P. Sangpour, F. Hashemi, and A. Z. Moshfegh, "Photoenhanced degradation of methylene blue on cosputtered M:TiO_2 (M = Au, Ag, Cu) nanocomposite systems: a comparative study," *Journal of Physical Chemistry C*, vol. 114, no. 33, pp. 13955–13961, 2010.

[16] K. Yang, Y. Dai, and B. Huang, "Density functional study of boron-doped anatase TiO_2," *Journal of Physical Chemistry C*, vol. 114, no. 46, pp. 19830–19834, 2010.

[17] P. Xu, J. Lu, T. Xu, S. Gao, B. Huang, and Y. Dai, "I2-hydrosol-seeded growth of $(I_2)_n$-C-codoped meso/nanoporous TiO_2 for visible light-driven photocatalysis," *Journal of Physical Chemistry C*, vol. 114, no. 20, pp. 9510–9517, 2010.

[18] W. Guo, Y. Shen, L. Wu, Y. Gao, and T. Ma, "Effect of N dopant amount on the performance of dye-sensitized solar cells based on N-Doped TiO_2 electrodes," *Journal of Physical Chemistry C*, vol. 115, no. 43, pp. 21494–21499, 2011.

[19] J. Lin, J. Lin, and Y. Zhu, "Controlled synthesis of the $ZnWO_4$ nanostructure and effects on the photocatalytic performance," *Inorganic Chemistry*, vol. 46, no. 20, pp. 8372–8378, 2007.
[20] A. T. Kuvarega, R. W. M. Krause, and B. B. Mamba, "Nitrogen/palladium-codoped TiO_2 for efficient visible light photocatalytic dye degradation," *Journal of Physical Chemistry C*, vol. 115, no. 45, pp. 22110–22120, 2011.
[21] S. Chen, W. Zhao, W. Liu, and S. Zhang, "Preparation, characterization and activity evaluation of p-n junction photocatalyst p-ZnO/n-TiO_2," *Applied Surface Science*, vol. 255, no. 5, pp. 2478–2484, 2008.
[22] Z.-R. Tang, F. Li, Y. Zhang, X. Fu, and Y.-J. Xu, "Composites of titanate nanotube and carbon nanotube as photocatalyst with high mineralization ratio for gas-phase degradation of volatile aromatic pollutant," *Journal of Physical Chemistry C*, vol. 115, no. 16, pp. 7880–7886, 2011.
[23] V. Štengl, D. Popelková, and P. Vláčil, "TiO_2-graphene nanocomposite as high performace photocatalysts," *Journal of Physical Chemistry C*, vol. 115, no. 51, pp. 25209–25218, 2011.
[24] J. Zhang, J. Yu, Y. Zhang, Q. Li, and J. R. Gong, "Visible light photocatalytic H_2-production activity of CuS/ZnS porous nanosheets based on photoinduced interfacial charge transfer," *Nano Letters*, vol. 11, no. 11, pp. 4774–4779, 2011.
[25] H. Park and W. Choi, "Effects of TiO_2 surface fluorination on photocatalytic reactions and photoelectrochemical behaviors," *Journal of Physical Chemistry B*, vol. 108, no. 13, pp. 4086–4093, 2004.
[26] W. Choi, "Pure and modified TiO_2 photocatalysts and their environmental applications," *Catalysis Surveys from Asia*, vol. 10, no. 1, pp. 16–28, 2006.
[27] J. Yu, W. Wang, B. Cheng, and B. L. Su, "Enhancement of photocatalytic activity of Mesporous TiO_2 powders by hydrothermal surface fluorination treatment," *Journal of Physical Chemistry C*, vol. 113, no. 16, pp. 6743–6750, 2009.
[28] S. F. Chen, S. J. Zhang, W. Liu, and W. Zhao, "Preparation and activity evaluation of p-n junction photocatalyst NiO/TiO_2," *Journal of Hazardous Materials*, vol. 155, no. 1-2, pp. 320–326, 2008.
[29] H. Yu, J. Yu, S. Liu, and S. Mann, "Template-free hydrothermal synthesis of CuO/Cu_2O composite hollow microspheres," *Chemistry of Materials*, vol. 19, no. 17, pp. 4327–4334, 2007.
[30] S. F. Chen, L. Chen, S. Gao, and G. Y. Cao, "The preparation of nitrogen-doped photocatalyst TiO_2—XN x by ball milling," *Chemical Physics Letters*, vol. 413, no. 4-6, pp. 404–409, 2005.
[31] T. Cai, M. Yue, X. Wang, Q. Deng, Z. Peng, and W. Zhou, "Preparation, characterization, and photocatalytic performance of $NdPW_{12}O_{40}/TiO_2$ composite catalyst," *Chinese Journal of Catalysis*, vol. 28, no. 1, pp. 10–16, 2007.
[32] J. W. Tang, Z. G. Zou, and J. H. Ye, "Photophysical and photocatalytic properties of $AgInW_2O_8$," *The Journal of Physical Chemistry B*, vol. 107, pp. 14265–14269, 2003.
[33] S. Kang, K. Shin, K. Prabakar, and C. Lee, "Optical and electrical properties of ZnO doped with nitrogen," *Physica Status Solidi B*, vol. 241, no. 12, pp. 2830–2834, 2004.
[34] S. F. Chen, X. L. Yu, H. Y. Zhang, and W. Liu, "Preparation and photocatalytic activity evaluation of composite Fe-TiO_2/TiO_2 photocatalyst," *Journal of the Electrochemical Society*, vol. 157, no. 5, pp. K96–K102, 2010.
[35] A. Lohner, M. Woerner, T. Elsaesser, and W. Kaiser, "Picosecond capture of photoexcited holes by shallow acceptors in p-type GaAs," *Physical Review Letters*, vol. 68, no. 26, pp. 3920–3923, 1992.
[36] S. F. Chen, L. Chen, G. Shen, and C. Gengyu, "The preparation of coupled WO3/TiO_2 photocatalyst by ball milling," *Powder Technology*, vol. 160, no. 3, pp. 198–202, 2005.
[37] J. Wang, S. Yin, M. Komatsu, Q. Zhang, F. Saito, and T. Sato, "Preparation and characterization of nitrogen doped $SrTiO_3$ photocatalyst," *Journal of Photochemistry and Photobiology A*, vol. 165, no. 1–3, pp. 149–156, 2004.

18

Combined Application of UV Photolysis and Ozonation with Biological Aerating Filter in Tertiary Wastewater Treatment

Zhaoqian Jing and Shiwei Cao

College of Civil Engineering, Nanjing Forestry University, Nanjing 210037, China

Correspondence should be addressed to Zhaoqian Jing, carljing@163.com

Academic Editor: Meenakshisundaram Swaminathan

To enhance the biodegradability of residual organic pollutants in secondary effluent of wastewater treatment plants, UV photolysis and ozonation were used in combination as pretreatment before a biological aerating filter (BAF). The results indicated that UV photolysis could not remove much COD (chemical oxygen demand), and the performance of ozonation was better than the former. With UV photolysis combined with ozonation (UV/O_3), COD removal was much higher than the sum of that with UV photolysis and ozonation alone, which indicated that UV photolysis could efficiently promote COD removal during ozonation. This pretreatment also improved molecular weight distribution (MWD) and biodegradability greatly. Proportion of organic compounds with molecular weight (MW) <3 kDalton was increased from 51.9% to 85.9%. COD removal rates with BAF and O_3/BAF were only about 25% and 38%, respectively. When UV/O_3 oxidation was combined with BAF, the average COD removal rate reached above 61%, which was about 2.5 times of that with BAF alone. With influent COD ranging from 65 to 84 mg/L, the effluent COD was stably in the scope of 23–31 mg/L. The combination of UV/O_3 oxidation with BAF was quite efficient in organic pollutants removal for tertiary wastewater treatment.

1. Introduction

In recent years, contamination of surface and subsurface water by organic pollutants has received much attention. Wastewater treatment plants (WWTPs) play important roles in such pollutants removal. In China, more and more WWTPs have been built not only in large cities, but also in small towns and counties. To meet the national discharge standard for municipal WWTPs, most of them have been built with secondary treatment processes, which usually comprise physical interception, precipitation, and biological treatment. Although the secondary treatment processes are efficient in bulk and soluble pollutants removal, there are still some residual organic compounds, nitrogen, phosphorus, and pathogenic bacteria in the effluent. Some researchers have found that there are more than 200 organic micropollutants in the secondary effluent of municipal WWTPs [1, 2]. Most of these organic pollutants are refractory and difficult to be further degraded, and therefore their discharge in water may cause long-term environmental effects to aquatic and terrestrial organisms [3, 4].

In late May, 2007, a drinking water crisis took place in Wuxi, Jiangsu Province, China, owing to the large scale growth of algae bloom in Wuxi's drinking water source—Taihu Lake. This crisis has made nearly 170 WWTPs being required to be upgraded with deep treatment or tertiary treatment to relieve the pollution load to Taihu Lake. Coagulation and filtration have been usually applied in tertiary treatment for residual suspended solids and partial colloid substances removal. However, most refractory and volatile organic compounds cannot be eliminated by these processes [5]. Owing to the low biodegradability of residual organic compounds, it is difficult to treat the secondary effluent directly with further biological processes. Appropriate processes should be found for refractory pollutants removal in tertiary wastewater treatment.

Many advanced oxidation processes (AOPs) can be used for destruction of refractory and high molecular organic compounds in wastewater, such as ozonation, ultraviolet (UV) photolysis, hydrogen peroxide (H_2O_2) oxidation, and Fenton oxidation. These AOPs can be used alone or in combination. During these processes, hydroxyl radicals with redox potential of 2.8 V are usually generated as the basic oxidants for pollutants removal [6]. Hydroxyl radicals are highly reactive and consequently short-lived. In the application of these AOPs, it is vital to provide appropriate reaction conditions and produce enough hydroxyl radicals. Fenton oxidation is usually accomplished in acid conditions, resulting in large quantities of acid and base consumption [7]. It can be used in small scale industrial wastewater treatment or special pollutants removal. Ozone is a strong oxidant that decomposes in water to form hydroxyl radicals which are stronger oxidizing agents than ozone itself. The use of ozonation for refractory and complex organic pollutants oxidation is very popular in water and wastewater treatment [8, 9]. During ozonation, the high molecular pollutants are decomposed into small molecular compounds, which are readily to be biodegraded [10]. Ozonation can also reduce the toxic effects of some micropollutants, but does not decrease the viability of bacteria [11]. Therefore, it is feasible to apply ozonation with biological processes in wastewater treatment. UV light is often used in water and wastewater disinfection. UV photolysis can selectively reduce some organic compounds, but it alone is not efficient enough for pollutants removal. However, many researchers have found UV photolysis can enhance the oxidation potential of some oxidation processes. When UV is used with H_2O_2, a strong oxidant whose photolytic dissociation yields hydroxyl radicals [12, 13], many recalcitrant pollutants such as pharmaceuticals and pesticides can be oxidized [14]. When UV photolysis is combined with ozonation (UV/O_3 oxidation), more hydroxyl radicals are produced via the photolysis of reaction intermediates such as H_2O_2 [15], and the organic pollutants are decomposed more completely. UV/O_3 oxidation has been used in pharmaceuticals, drinking water and industrial wastewater treatment [16–18]. In WWTPs, UV/O_3 oxidation can be used as pretreatment for biological process, in which complex pollutants are decomposed into more biodegradable substances. Therefore, more organic compounds can be removed in the subsequent biological treatment process.

As described above, although there are many studies on UV/O_3 oxidation in organic pollutants removal from water and wastewater, there are few studies in the application of UV/O_3 oxidation with biological processes in tertiary wastewater treatment. In this study, UV/O_3 oxidation was combined with a biological aerating filter (BAF) to treat the secondary effluent from a WWTP. The application of UV/O_3 oxidation in combination with BAF could achieve high organic pollutants removal with low cost [19]. The characteristics of the secondary effluent were evaluated. COD (chemical oxygen demand) removal performance and molecular weight distribution (MWD) changes during UV photolysis, ozonation, and their combination were investigated. Contrast experiments were also carried out to study the performances of COD removal with BAF, O_3/BAF and UV/O_3/BAF.

2. Materials and Methods

2.1. Reactors. A pilot system shown in Figure 1 was built in a WWTP in Nanjing, Jiangsu Province of China. UV/O_3 oxidation was carried out in a cylinder glass reactor with diameter of 300 mm and effective volume of 4 L. A 10 W low-pressure mercury UV lamp with main wavelength of 254 nm was installed in the center of the reactor. Ozone was produced with an ozone generator (CF-G-3-2.5 g, Qingdao Guolin Industry Co., Ltd, China) with dry air. Ozone dosage was controlled with a flowmeter.

The wastewater after pretreatment with UV photolysis and ozonation was pumped into a BAF, an organic glass cylinder reactor with diameter of 100 mm. Clay ceramisites with diameter of 10 mm were used as filling material with depth of 1.5 m. This BAF was operated with hydraulic load of 1.5 m^3/m^2 h and gas/water ratio of 5 : 1.

2.2. Analytical Methods. The samples from the influent, oxidation reactor, and final effluent of BAF were taken regularly and the concentrations of COD, BOD_5 (5-day biological oxygen demand) and NH_4^+-N (ammonia) were analyzed according to the standard methods (APHA, 1999). MWD was measured using ultrafiltration membrane filters with cutoff sizes of 1–100 kDalton. DOC (dissolved organic carbon) after membrane filtration with different cutoff sizes was measured using a Shimadzu TOC-V instrument. pH, dissolved oxygen, ozone concentration, and temperature were monitored regularly.

2.3. Influent Characteristics. The effluent from the secondary clarifiers in the WWTP was directly used as influent of the combined process. This WWTP was run with an A^2/O (anaerobic/anoxic and oxic) process, which had removed most soluble organic compounds and nitrogen. The main indexes of influent during this study were as follows: COD 63.2–91.4 mg/L, BOD_5 9–23 mg/L, NH_4^+-N 5.3–8.6 mg/L, pH 7.1–7.4, and water temperature 19–25°C.

3. Results and Discussion

3.1. Evaluation of Secondary Effluent. COD and BOD_5 in the effluent from the secondary clarifiers of the WWTP in 2009 are shown in Figure 2. COD in the effluent ranged from 63.2 mg/L to 91.4 mg/L while BOD_5 was in the scope of 9–23 mg/L. According to the national discharge standard for municipal wastewater treatment plant in China, when the effluent from a WWTP is discharged into an enclosed watershed, COD and BOD_5 in the effluent are required below 50 mg/L and 10 mg/L, respectively. It is obvious that the effluent could not meet the national discharge standard. This WWTP was run with A^2/O process, in which most biodegradable organic compounds had been degraded. The ratio of BOD_5/COD (B/C) was low, ranging from 0.17 to 0.25, which indicated that most of the residual organic substances in the effluent were difficult to be biodegraded.

3.2. COD Removal with UV Photolysis, Ozonation, and UV/O_3 Oxidation. Figure 3 shows COD variation in the oxidation reactor with time under UV photolysis, ozonation, and UV/O_3 oxidation. It can be seen that COD removal with UV photolysis alone was low. In 30 min contacting

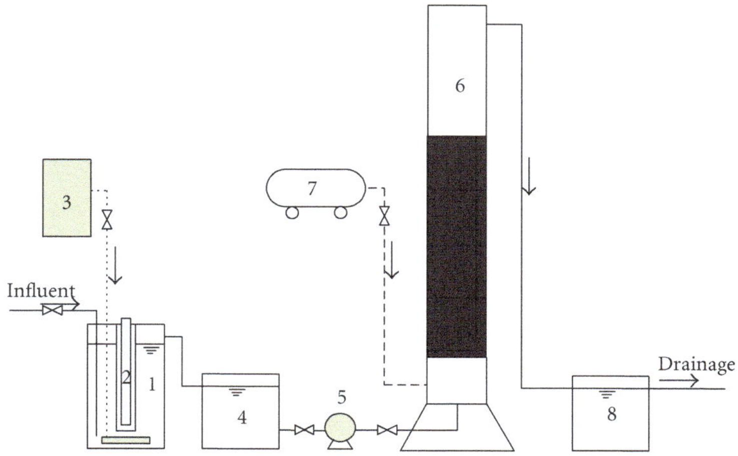

FIGURE 1: Schematic diagram for UV/O_3 oxidation combined with BAF, (1) UV/O_3 oxidation reactor, (2) UV lamp, (3) ozone generator, (4) medium water tank, (5) pump, (6) BAF, (7) aerator, (8) final water tank.

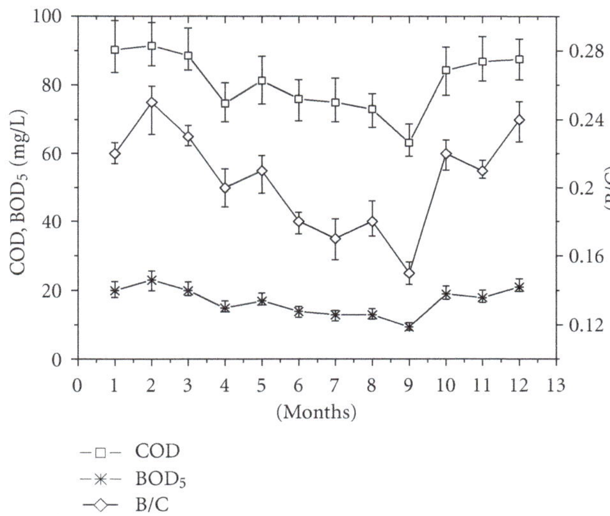

FIGURE 2: COD, BOD_5, and B/C changes in the effluent from the secondary clarifiers.

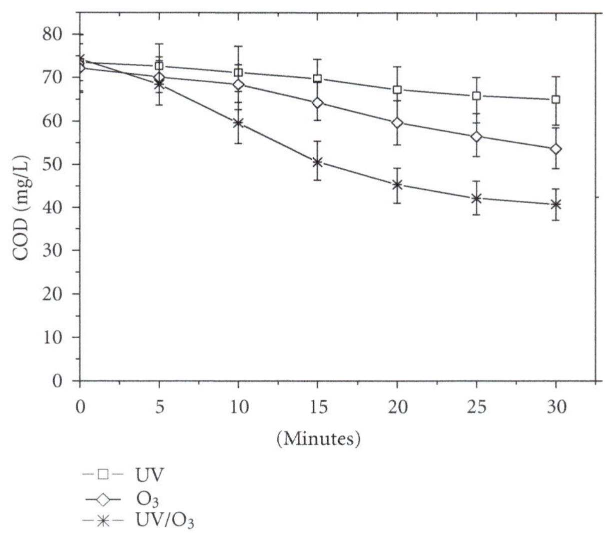

FIGURE 3: COD changes with time under UV photolysis, ozonation, and UV/O_3 oxidation (O_3 dosage 8 mg/L, temperature 21°C, pH 7.2).

time, COD only declined from 73.5 mg/L to 65.1 mg/L. The performance of ozonation was much better compared with UV photolysis. After 30 min ozonation, COD decreased from 72.2 mg/L to 53.7 mg/L. More than 25% of COD was removed during ozonation. The combination of UV photolysis and ozonation led to significant improvement of COD removal compared with UV photolysis and ozonation alone. In 30 min, COD reduced from 74.3 mg/L to 40.9 mg/L. The removal rate of COD attained about 45%, which was much higher than the sum of COD removal rates with UV photolysis and ozonation alone. The results indicated that UV photolysis could greatly improve COD removal during ozonation [20].

It can also be observed that COD concentration decreased with contact time rise. During UV/O_3 oxidation, COD decreased from the original 74.3 mg/L to 59.7, 45.4, and 40.9 mg/L at contact time of 10, 20, and 30 min, respectively. The extension of contact time was beneficial for COD removal. Nevertheless, the reactor volume needed to be amplified with contact time extension. More construction investment and operation cost would be required. It can be seen that contact time influence on COD removal was relieved with contact time exceeding 20 min. Therefore the subsequent experiments were done with contact time of 20 min.

3.3. Influence of O_3 Dosages on UV/O_3 Oxidation. A series of experiments of UV/O_3 oxidation was done at different ozone dosages in the range of 0–12 mg/L. Results of COD variation with ozone dosages are shown in Figure 4. It can

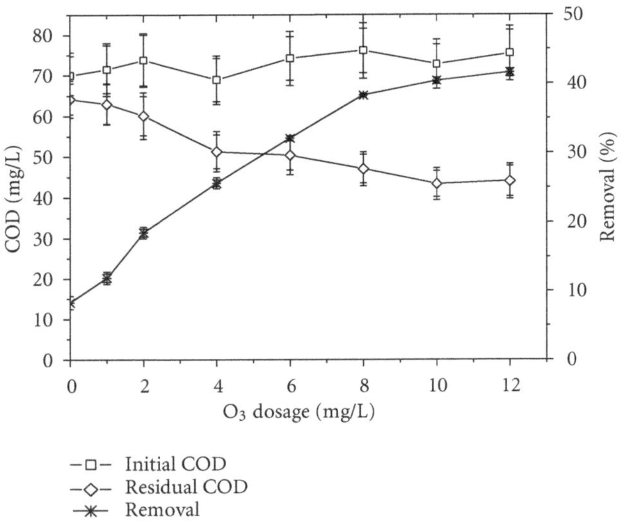

Figure 4: Influence of O_3 dosage on COD removal in UV/O_3 oxidation (contact time 20 min, temperature 20°C, pH 7.2).

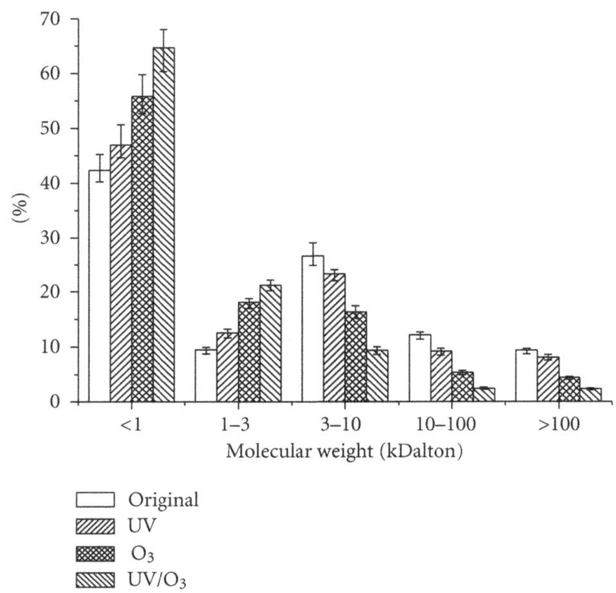

Figure 5: Molecular weight distribution under UV photolysis, ozonation, and UV/O_3 oxidation (contact time 20 min, O_3 dosage 8 mg/L, temperature 21°C, pH 7.2).

be seen that COD removal rose with ozone dosage increase. COD removal increased from 8.3% to 41.6% with dosage ranging from 0 to 12 mg/L, while residual COD decreased from 64.2 mg/L to 44 mg/L. It can also be noticed that when ozone dosage was under 2 mg/L, COD removal was low. When ozone dosage was increased from 2 to 8 mg/L, COD removal was enhanced greatly. These results indicated that ozone dosage was an important factor of UV/O_3 oxidation. However, COD changed little when ozone dosage was above 8 mg/L. This indicated that there were still some organic compounds which were difficult to be degraded with UV/O_3

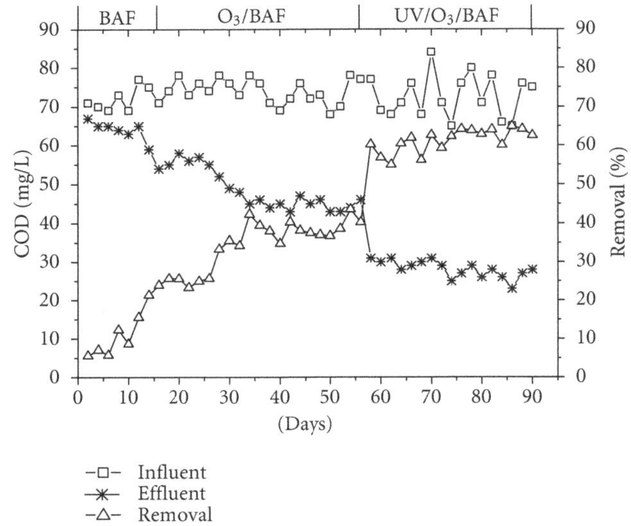

Figure 6: COD removal contrast with BAF, O_3/BAF, and UV/O_3/BAF (oxidation contact time 20 min, O_3 dosage 8 mg/L).

oxidation, and excessive ozone dosage could not make further COD removal.

3.4. Molecular Weight Distributions under UV Photolysis, Ozonation, and UV/O_3 Oxidation. MWD under different operations is shown in Figure 5. In the effluent from the secondary clarifiers, organics with molecular weight (MW) <1 kDalton were 42.4%. After UV photolysis, there was a minor increase of organics with MW <3 kDalton. After ozonation, the proportions of organics with MW <1 kDalton and 1–3 kDalton were increased obviously. Organics with MW <3 kDalton attained about 74%, which was much higher than 51.9% in the original influent. The combination of UV photolysis and ozonation changed MWD more greatly. Organics with MW <1 kDalton and 1–3 kDalton reached 64.7% and 21.2%, respectively. The proportion of organics >10 kDalton decreased from the original 21.5% to 4.7%. UV/O_3 oxidation improved MWD of organics greatly and produced more biodegradable substances [21]. Consequently, the wastewater after UV/O_3 oxidation was more appropriate for further biological treatment.

3.5. Performance of UV/O_3 Oxidation Combined with BAF. Figure 6 shows COD removal with BAF alone (1–26 d), O_3/BAF (27–56 d), and UV/O_3/BAF (57–90 d). At the beginning, the BAF was started with the activated sludge of this WWTP as inoculum. COD removal increased gradually with time as more biofilm grew on the ceramisites. At the end of Day 16, COD removal rate attained about 24%. Subsequently, COD removal rate was maintained around 25%. Owing to the low biodegradability of organics in the influent, BAF alone could not remove more COD. COD in the effluent from the BAF was still above 50 mg/L, exceeding the standard limit for WWTPs in China.

In the second phase (27–56 d), ozonation was combined with BAF, and there was an obvious increase in

COD removal. The combination of ozonation and BAF could remove 33%–44% COD. The average removal rate reached 38%. Ozonation had decomposed partial organic compounds into biodegradable substances, which could be readily biodegraded in subsequent BAF.

When UV photolysis and ozonation were combined with BAF (UV/O_3/BAF, from Day 57 to Day 90), there was further improvement of COD removal. The average COD removal rate was above 61%, which was 2.5 and 1.6 times of that in BAF and O_3/BAF, respectively. With influent COD ranging from 65 mg/L to 84 mg/L, the effluent COD was stably maintained in the scope of 23–31 mg/L. These results indicated that the combination of UV/O_3 oxidation with BAF was an appropriate process for low biodegradable wastewater treatment. UV/O_3 oxidation could not only remove partial organic pollutants, but also efficiently enhance wastewater biodegradability [22]. With UV/O_3 oxidation combined with BAF, most organics were removed from the wastewater with the comprehensive activities of physicochemical decomposition and biodegradation.

4. Conclusion

The effluent from the secondary clarifiers of the WWTP comprises many refractory organic pollutants, most of which are difficult to be biodegraded. Direct treatment with further biological processes cannot make satisfactory performance. AOPs are usually efficient in refractory pollutants removal, and can be combined with biological processes in low biodegradable wastewater treatment. In this study, UV/O_3 oxidation was combined with BAF in tertiary treatment. The results indicated that although UV photolysis alone was not quite efficient for COD removal, it could improve the performance of ozonation. When UV photolysis was combined with ozonation, 45% of COD in the wastewater from the secondary effluent was removed, and more biodegradable organic compounds were produced. It was really feasible to combine UV/O_3 oxidation with BAF to improve COD removal in tertiary wastewater treatment. The combination of UV/O_3 oxidation with BAF could remove more than 61% of COD in the wastewater from the second clarifiers, which was nearly 2.5 and 1.6 times of that in BAF and O_3/BAF, respectively.

Acknowledgments

This research was supported by the project of the priority academic program development of Jiangsu Higher Education Institutions (2011-6), and science and technology project of China Housing and Urban-Rural Development Ministry (2010-K7-10).

References

[1] J. Hollender, S. G. Zimmermann, S. Koepke et al., "Elimination of organic micropollutants in a municipal wastewater treatment plant upgraded with a full-scale post-ozonation followed by sand filtration," *Environmental Science and Technology*, vol. 43, no. 20, pp. 7862–7869, 2009.

[2] V. J. P. Vilar, A. I. E. Gomes, V. M. Ramos, M. I. Maldonado, and R. A. R. Boaventura, "Solar photocatalysis of a recalcitrant coloured effluent from a wastewater treatment plant," *Photochemical and Photobiological Sciences*, vol. 8, no. 5, pp. 691–698, 2009.

[3] J. Gong, Y. Liu, and X. Sun, "O_3 and UV/O_3 oxidation of organic constituents of biotreated municipal wastewater," *Water Research*, vol. 42, no. 4-5, pp. 1238–1244, 2008.

[4] M. Bundschuh, R. Pierstorf, W. H. Schreiber, and R. Schulz, "Positive effects of wastewater ozonation displayed by in situ bioassays in the receiving stream," *Environmental Science and Technology*, vol. 45, no. 8, pp. 3774–3780, 2011.

[5] M. Petala, V. Tsiridis, P. Samaras, A. Zouboulis, and G. P. Sakellaropoulos, "Wastewater reclamation by advanced treatment of secondary effluents," *Desalination*, vol. 195, no. 1–3, pp. 109–118, 2006.

[6] F. J. Beltrán, M. González, and J. F. González, "Industrial wastewater advanced oxidation. part 1. UV radiation in the presence and absence of hydrogen peroxide," *Water Research*, vol. 31, no. 10, pp. 2405–2414, 1997.

[7] S. Kenfack, V. Sarria, J. Wéthé et al., "From laboratory studies to the field applications of advanced oxidation processes: a case study of technology transfer from Switzerland to burkina faso on the field of photochemical detoxification of biorecalcitrant chemical pollutants in water," *International Journal of Photoenergy*, vol. 2009, Article ID 104281, 8 pages, 2009.

[8] R. G. Rice, "Applications of ozone for industrial wastewater treatment—a review," *Ozone*, vol. 18, no. 6, pp. 477–515, 1997.

[9] K. Ikehata and M. G. El-Din, "Aqueous pesticide degradation by ozonation and ozone-based advanced oxidation processes: a review (part II)," *Ozone*, vol. 27, no. 3, pp. 173–202, 2005.

[10] F. J. Beltràn, J. F. García-Araya, and P. M. Alvarez, "Wine distillery wastewater degradation. 1. Oxidative treatment using ozone and its effect on the wastewater biodegradability," *Journal of Agricultural and Food Chemistry*, vol. 47, no. 9, pp. 3911–3918, 1999.

[11] M. Mišík, S. Knasmueller, F. Ferk et al., "Impact of ozonation on the genotoxic activity of tertiary treated municipal wastewater," *Water Research*, vol. 45, no. 12, pp. 3681–3691, 2011.

[12] A. Mandal, K. Ojha, A. K. De, and S. Bhattacharjee, "Removal of catechol from aqueous solution by advanced photo-oxidation process," *Chemical Engineering Journal*, vol. 102, no. 2, pp. 203–208, 2004.

[13] K. J. Friesen, T. M. El-Morsi, and A. S. Abd-El-Aziz, "Photochemical oxidation of short-chain polychlorinated n-alkane mixtures using H_2O_2/UV and the photo-fenton reaction," *International Journal of Photoenergy*, vol. 6, no. 2, pp. 81–88, 2004.

[14] M. Klavarioti, D. Mantzavinos, and D. Kassinos, "Removal of residual pharmaceuticals from aqueous systems by advanced oxidation processes," *Environment International*, vol. 35, no. 2, pp. 402–417, 2009.

[15] W. Zhao, Q. Liao, J. Zhang, Y. Yang, J. Dai, and D. Zhao, "Oxidation of cationic red X-GRL by ozonation combined with UV radiation in aqueous solution: degradation, kinetics, and modeling," *Chemical Engineering Journal*, vol. 171, no. 2, pp. 628–639, 2011.

[16] F. J. Beltrán, J. M. Encinar, and J. F. González, "Industrial wastewater advanced oxidation. Part 2. Ozone combined with hydrogen peroxide or UV radiation," *Water Research*, vol. 31, no. 10, pp. 2415–2428, 1997.

[17] C. Hofl, G. Sigl, O. Specht, I. Wurdack, and D. Wabner, "Oxidative degradation of AOX and COD by different

advanced oxidation processes: a comparative study with two samples of a pharmaceutical wastewater," *Water Science and Technology*, vol. 35, no. 4, pp. 257–264, 1997.

[18] W. K. Lafi, M. Al-Anber, Z. A. Al-Anber, M. Al-Shannag, and A. Khalil, "Coagulation and advanced oxidation processes in the treatment of olive mill wastewater (OMW)," *Desalination and Water Treatment*, vol. 24, no. 1–3, pp. 251–256, 2010.

[19] A. S. Stasinakis, "Use of selected advanced oxidation processes (AOPs) for wastewater treatment—a mini review," *Global NEST Journal*, vol. 10, no. 3, pp. 376–385, 2008.

[20] W. S. Kuo, "Synergistic effects of combination of photolysis and ozonation on destruction of chlorophenols in water," *Chemosphere*, vol. 39, no. 11, pp. 1853–1860, 1999.

[21] J. Chung and J. O. Kim, "Application of advanced oxidation processes to remove refractory compounds from dye wastewater," *Desalination and Water Treatment*, vol. 25, no. 1–3, pp. 233–240, 2011.

[22] K. Ito, W. Jian, W. Nishijima, A. U. Baes, E. Shoto, and M. Okada, "Comparison of ozonation and AOPs combined with biodegradation for removal of THM precursors in treated sewage effluents," *Water Science and Technology*, vol. 38, no. 7, pp. 179–186, 1998.

19

Structural and Optoelectronic Properties of CdSe Tetrapod Nanocrystals for Bulk Heterojunction Solar Cell Applications

Nguyen Tam Nguyen Truong, Thao Phuong Ngoc Nguyen, and Chinho Park

School of Chemical Engineering, Yeungnam University, 280 Daehak-Ro, Gyeongsan, Gyeongbuk 712-749, Republic of Korea

Correspondence should be addressed to Chinho Park; chpark@ynu.ac.kr

Academic Editor: Peter Rupnowski

Semiconducting CdSe tetrapod nanoparticles were prepared, and their structural and optical properties were examined. The surface capping molecule, octylphosphonic acid, was replaced with butylamine after the particle synthesis. The exchange of surface ligands changed the physical properties of the nanocrystals, which resulted in a slight decrease in the nanoparticles size. The effects of changing surface ligands of CdSe tetrapod nanocrystals on the structural and optoelectronic properties were investigated, and it was found that the surfactant of nanoparticles could affect the device performance by enhancing the charge carrier separation at the active layer interfaces. Power conversion efficiency of the bulk heterojunction solar cells having the structure of glass/ITO/PEDOT:PSS/(CdSe + PCPDTBT)/Al was improved from 1.21% to 1.52% with the use of ligand-exchanged nanoparticles.

1. Introduction

Bulk heterojunction solar cells (BHJ) based on blends of semiconducting nanocrystals (NCs) and conducting polymers have attracted considerable attention owing to their low cost, simple processing, and ability to dissolve each component to obtain composite layers optimized for solar energy conversion [1, 2]. Currently, polymer-fullerene solar cells and polymer solar cells have reached power conversion efficiencies (PCEs) of approximately 7.4% [3] and 8.62% [4], respectively. Solar cells with the fullerenes replaced with semiconducting inorganic nanoparticles (NPs) as the electron acceptors are called hybrid BHJ solar cells [5]. Although the PCEs of hybrid BHJ solar cells are lower than those of fullerene-based BHJ cells, they have the potential to show better performance, because NPs have a high absorption coefficient, high intrinsic charge mobility, and can be synthesized into various shapes (sphere, rod, elongated sphere, tetrapod, etc.) with size of 3 to 100 nm, desirable exciton dissociation characteristics, percolation pathway development, and charge transport properties [6–11]. CdSe NPs are one the most selected materials owing to their well-established synthesis method [12, 13]. CdSe NPs also act as good electron acceptors in combination with the commonly used semiconducting polymers [14]. The effects of the various processing conditions, shapes of the nanocrystals, and surfactants used during synthesis as well as the semiconducting polymers selected were investigated intensively [15].

The first hybrid BHJ solar cells using mixtures of semiconducting polymer and inorganic nanocrystals were reported in 1996 by Greenham et al. [1], who found that the addition of CdSe nanoparticles to polymers in devices relied on the ability of the nanoparticles to disperse within the polymer to make larger interfacial surface area for carrier transfer between the two materials and to improve continuous pathways to the contacts. Hybrid BHJ solar cells based on blends of CdSe tetrapods and polymers have been reported to show the efficiency of 3.13% (Dayal et al.) [14], and their devices contained a low-band gap polymer, poly[2,6-(4,4-bis-(2-ethylhexyl)-4H-cyclopenta[2,1-b;3,4-b′]-dithiophene)-alt-4,7-(2,1,3-benzothiadiazole)] (PCPDTBT), and CdSe nanocrystals. The improved cell efficiency of this world's best CdSe hybrid BHJ solar cell was attributed to the enhanced sunlight absorption from the PCPDTBT polymer and improved charge collection by the CdSe tetrapod nanostructures. The effects of the surface ligand of CdSe nanoparticles on the device performance have also been reported [16], which have been investigated to engineer the nanoparticle/polymer blend's morphology and phase separation behavior. Surface ligands on the nanoparticles can act as traps for charge carrier transport or can dominate the nanoparticles' electrical properties by occupying surface electronic states or acting as insulators that reduce charge transport [17]. The properties of surfactant could affect the cosolubility of CdSe/polymer blends in solvents and also affect the optoelectronic properties of the active layer.

This work discusses mainly the structural and optoelectronic properties of synthesized CdSe tetrapod nanocrystals. Furthermore, the effects of surface ligand modification on the physical properties of CdSe NPs and surface morphology of the CdSe/polymer active layer were investigated, and the improvement of the bulk heterojunction solar cell performance has been demonstrated as a result of the nanoparticles' surface ligand exchange.

2. Experimental

CdSe tetrapods were synthesized using the method reported elsewhere [14]. In detail, 0.7 g of cadmium oxide (CdO), 5.2 g of trioctylphosphonic acid (TOP), and 2.14 g of octylphosphonic acid (OPA) were heated to 300°C in a two-necked flask under a nitrogen (N_2) flow. When the color of the cadmium solution turned colorless, the heat was removed and the solution was kept under a nitrogen flow. The selenium precursor was prepared by mixing 0.84 g of selenium (Se) with 2.54 g of tributylphosphine (TBP) in 0.6 mL of toluene and cooled in a refrigerator for 20 min prior to the injection. After 48 hrs, the cadmium precursor was reheated to 300°C, and the selenium precursor was injected rapidly at 300°C. The tetrapods were grown at 250°C for 50 min. The synthesized CdSe tetrapods were washed 5 times with a mixture of toluene and ethanol to remove the excess capping ligands and centrifuged 5 times. After the synthesis, the OPA ligands of as-synthesized OPA-capped CdSe tetrapods were replaced with butylamine (BA) ligands using a liquid-liquid exchange procedure.

The ultraviolet-visible spectroscopy (UV-vis), photoluminescence (PL), X-ray diffraction (XRD), and transmission electron microscopy (TEM) were used to characterize the structural and optical properties of the CdSe nanotetrapod particles. FT-IR spectra (450–4000 cm^{-1}) were recorded on an Excalibur Series FTS 3000 spectrometer (Bio-Rad) at a resolution of 16 with 32 scans using KBr pellets. The UV-vis (Cary 5000) spectra of ligand-capped CdSe tetrapod nanoparticles were obtained using 1.0 cm path length quartz cells. The samples were prepared by dispersing the nanoparticles in ethanol before the measurements. The photoluminescence (PL) was measured using a custom-made PL system at room temperature using 488 nm (argon laser) as the excitation source. The nanocrystal structure was measured using powder X-ray diffraction (XRD) on an MPD PANalytical, using CuKα radiation. The size and shape of the nanocrystals were estimated with a high-resolution transmission electron microscope (HR-TEM, H-7600).

Active layer for BHJ solar cells was formed from the blends of tetrapod NCs and semiconducting polymers. ITO-coated glass substrates were cleaned sequentially with trichloroethylene (TCE), acetone, and methanol for 10 min in an ultrasonic bath. They were then treated in a nitrogen plasma chamber at 50 W for 10 min to improve the surface properties. A thin (~70 nm) layer of poly(ethylenedioxythiophene) doped with poly(styrene sulfonic acid) (PEDOT:PSS, Sigma Aldrich) was spin-coated on the cleaned ITO-glass substrates at 4000 rpm for 30 s and dried at 100°C for 30 min. The CdSe nanoparticles (with different surface ligands) and PCPDTBT blends were first prepared in chlorobenzene. Approximately 150 ± 5 nm thick photoactive CdSe/PCPDTBT layers were deposited by dropping the blends and spin coating them onto the substrates at 4000 rpm for 30 s. The films of (ITO/PEDOT:PSS/(CdSe+PCPDTBT)) were treated thermally for 10 min on a hot plate in a nitrogen glove box at 150°C. The structural properties and surface morphology of the active layers were investigated systematically using a range of techniques including atomic force microscopy (AFM) and scanning electron microscopy (SEM). The bulk heterojunction solar cells having glass/ITO/PEDOT:PSS/(CdSe+PCPDTBT)/Al structure were fabricated. A 100 nm thick Al electrode was added to the top of the photoactive layers by thermal evaporation to complete the device structure. The current density-voltage (J-V) characteristics were investigated under AM1.5G illumination by a solar simulator (Keithley 69911).

3. Results and Discussion

Figures 1(a) and 1(b) show TEM and HR-TEM images of the as-synthesized CdSe nanocrystals (NCs). A large fraction of branches was obtained without selective precipitation, which shows that the tetrapods had been formed. The tetrapod limbs were approximately ~26 nm long and ~7 nm thick. Figure 1(b) shows the lattice fringes of the CdSe tetrapods with fringe spacing of 0.34 nm, corresponding to the (002) planes for the wurtzite structure of CdSe [18].

OPA-capped CdSe nanoparticles (NPs) were first synthesized, and the NPs' ligands were replaced from OPA to butylamine. The various cappings of the CdSe NPs with different surface ligands were confirmed by FT-IR spectroscopy as in Figure 2. The intense symmetric C–H stretching vibrations of the $-CH_2-$ groups at 2948 and 2848 cm^{-1} and a shoulder band of asymmetric stretching of the terminal $-CH_3$ group at 2956 cm^{-1} can be associated with the alkyl chain of OPA as illustrated in Figure 2(a). The IR bands at 1456 and 1402 cm^{-1} can be assigned to the $-CH_2-$ bending vibration and the C–O–H bending vibration [19, 20]. The band at 1017 cm^{-1} is indicated as the P–O–H bond, which is similar to the FT-IR analysis of OPA ligand [21]. However, the peaks at 2956 and 1017 cm^{-1} of the OPA ligand were non existent in Figure 2(b), meaning that the OPA ligand was replaced by the BA ligand after the ligand exchange process. The presence of butylamine is shown by the peak at ~3700–3781 cm^{-1}, which corresponds to the N–H stretching vibration modes [8, 22].

The structure of CdSe tetrapods (Tpods) was confirmed by X-ray diffraction as shown in Figure 3. Comparison of the XRD patterns of as-synthesized nanocrystals with those of

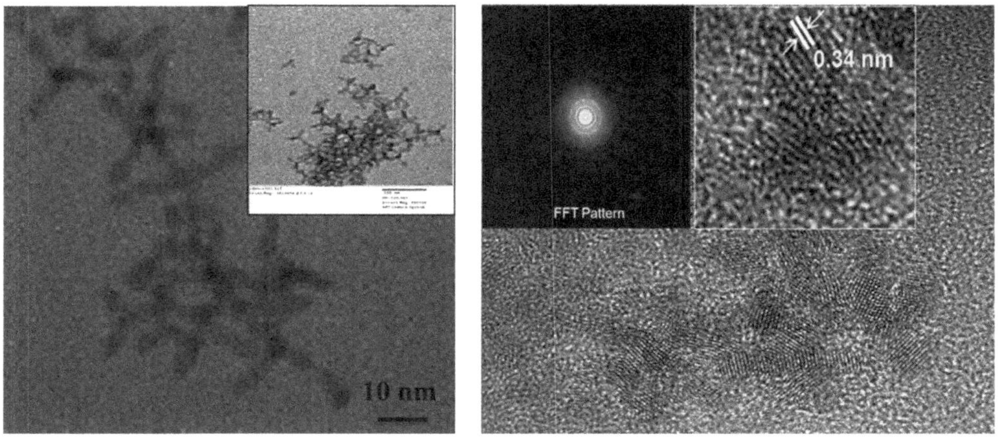

(a) (b)

FIGURE 1: (a) Transmission electron microscopic image and (b) high-resolution transmission electron microscopic image of CdSe tetrapod nanocrystals.

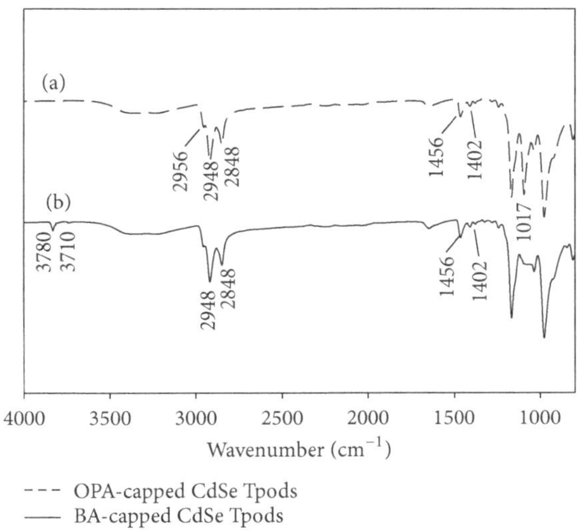

FIGURE 2: FT-IR spectra of (a) OPA-capped CdSe tetrapods (Tpods) and (b) BA-capped CdSe Tpods.

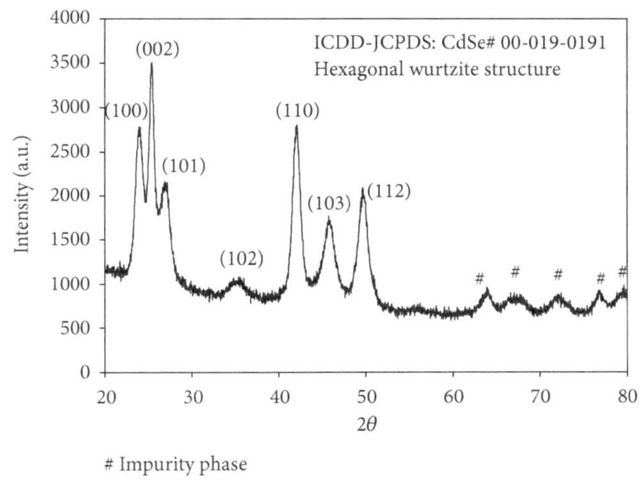

FIGURE 3: The X-ray diffraction patterns of CdSe tetrapods with the reference peak positions of bulk Wurtzite CdSe (ICDD-JCPDS: CdSe # 00-019-0191) displayed.

bulk CdSe crystals showed that the as-synthesized CdSe had a hexagonal wurtzite structure (ICDD-JCPDS: CdSe #00-019-0191) [23] with the arms elongated in the (002) direction. The diffraction peaks due to the reflections from (103), (100), and (101) were assigned as wurtzite phase related peaks. The (103) diffraction peak is significantly broader than the (100) and (101) peaks. This is consistent with the XRD data from the bulk tetrapod nanocrystals [24–27], which showed a more intense and narrow diffraction peak at the 2θ values of ~24 and ~27°, that is, consistent with the tetrapod limb in the c-axis direction of the wurtzite phase.

Figures 4(a) and 4(b) show the optical properties of the OPA-capped CdSe tetrapods determined by UV-vis and PL measurements. The CdSe nanocrystals were dispersed in ethanol prior to the spectroscopic measurements. The absorption edge of the UV-vis spectra was ~547 nm, and the emission peak of the PL spectra was ~593 nm (band gap energy ~2.09 eV). The effects of ligand exchange on the optical properties of the nanoparticles were investigated by UV-vis and PL, and the results are shown in Figures 5(a) and 5(b). After the ligand exchange from OPA to butylamine, the absorption edge of the UV-Vis spectra was shifted to a shorter wavelength from 547 to 542 nm, as shown in Figure 5(a), and the intensity of the emission peak of the PL spectra was decreased and shifted towards a higher energy from 2.09 to 2.14 eV, as shown in Figure 5(b). The shifting of the absorption edge and emission peak toward a shorter wavelength is often called a "blue-shift," which may be due to the quantum confinement effect caused by the decrease in nanocrystal s' size [28, 29]. The luminescence intensity from the smaller nanocrystals is likely to be much stronger, because smaller-sized nanocrystals possess a larger

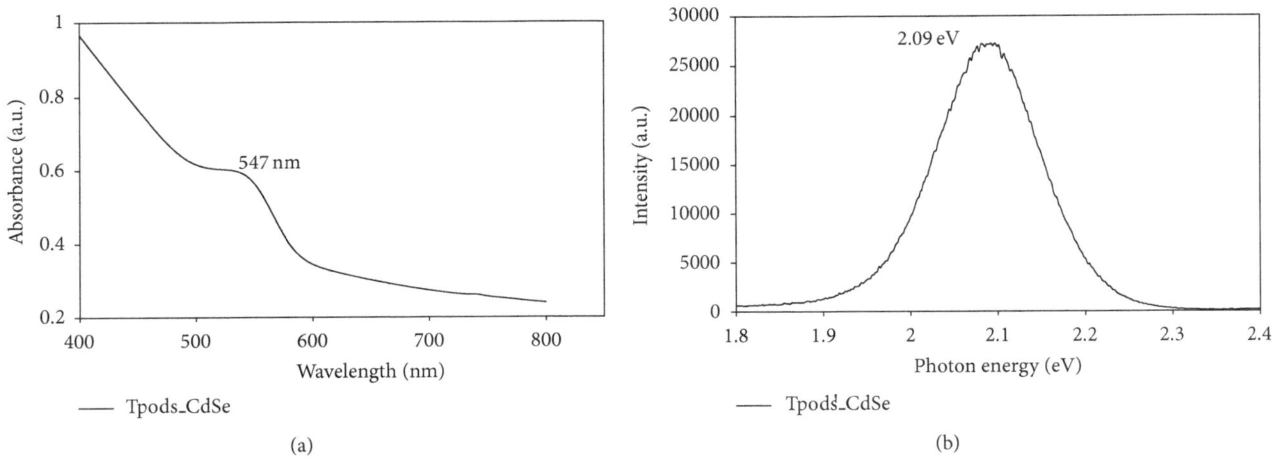

FIGURE 4: (a) Absorption and (b) photoluminescence spectra of the CdSe tetrapods capped with the OPA ligand.

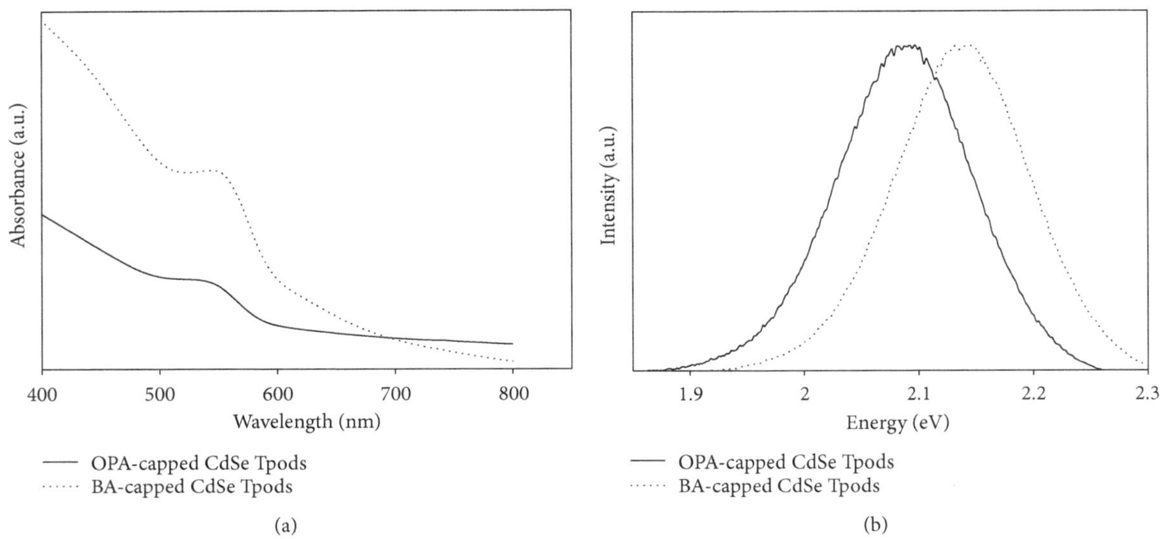

FIGURE 5: (a) Absorption and (b) photoluminescence spectra of CdSe tetrapods capped with OPA and BA ligand.

surface-to-volume ratio than larger-sized nanocrystals, and most of the photo-generated charge carriers undergo recombination at the surface vacancies. The sizes of the nanoparticles before and after the ligand exchange were measured by HR-TEM, and it was found that the size of the BA-capped CdSe tetrapods was smaller than the OPA-capped ones (data not shown here). This result is consistent with the peak shifts of UV-vis and PL spectra.

The effects of the surface ligand exchange of the CdSe tetrapods on the morphological properties of the CdSe/polymer film for bulk heterojunction solar cell applications were also investigated. The poly[2,6-(4,4-bis-(2-ethylhexyl)-4H-cyclopenta[2,1-b;3,4-b′]-dithiophene)-alt-4,7-(2,1,3-benzothiadiazole)] (PCPDTBT) was used as a light-absorbing and electron-donating material. The semiconducting OPA- or BA-capped CdSe nanoparticles were used as electron acceptors. Figures 6(a) and 6(b) show the SEM images of the surface of a composite film of OPA- or BA-capped CdSe and PCPDTBT with a loading of 5 mg/mL and the CdSe tetrapod concentration of 70 wt.%. The surface of the film with a BA ligand was much smoother than that with the OPA ligand. Figures 6(c) and 6(d) show the AFM images of the surface morphology of the composite films. The surface roughness of the composite films was both smooth and homogeneous, and the surface roughness of the film with BA-capped CdSe (~1.69 nm) was smoother than that of the OPA-capped CdSe (~2.52 nm). A smoother surface of the film with the BA-capped CdSe tetrapods was believed to be caused by the smaller particle size of the BA-capped CdSe tetrapods and the positive effects of butylamine in the dispersion characteristics of the nanoparticles in the solvents [30]. The rough surface was considered to be the self-organization signature of the polymer, and the solar cell devices generally showed higher efficiency on a smooth surface than on a rough surface [31].

The *J*-*V* curves of BHJ solar cells fabricated by using the CdSe nanoparticles capped with various surface ligands

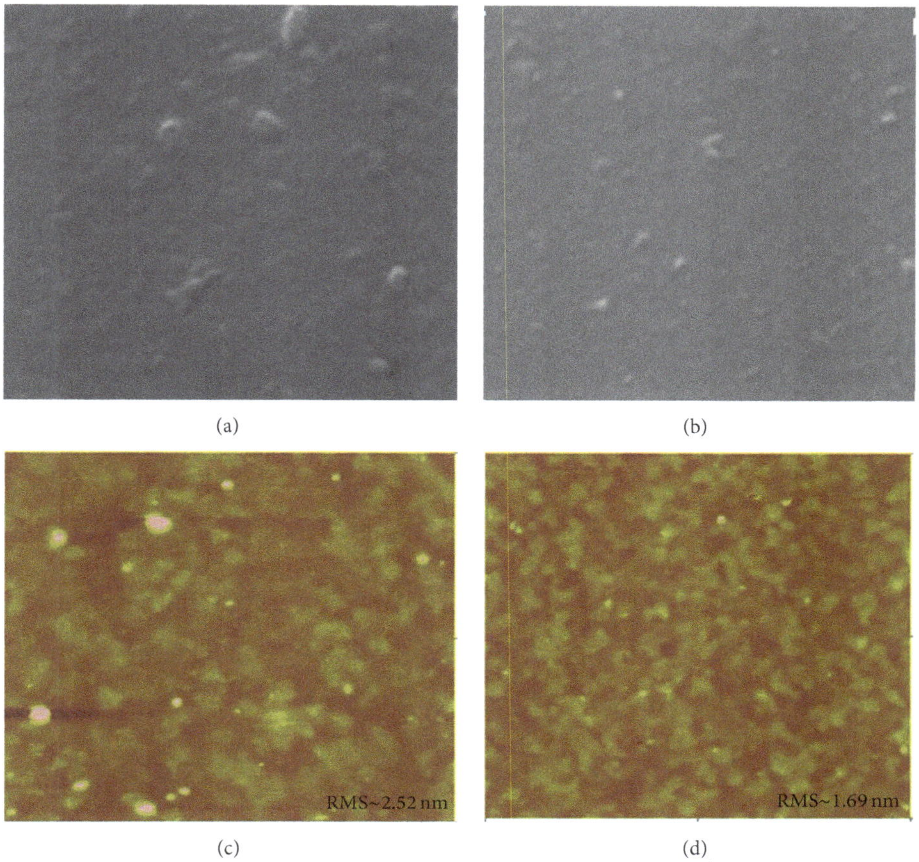

FIGURE 6: Scanning electron microscopic (SEM) image and surface roughness (RMS) of the (CdSe+PCPDTBT) (7 : 3) active layer annealed at 150°C with (a, c) OPA-capped and (b, d) BA-capped CdSe.

with the optimal composition of 70 wt.% CdSe in the CdSe/PCPDTBT composite layers were recorded and shown in Figure 7 along with the device parameters. Power conversion efficiency (η) was improved by changing the surfactant from OPA (η ~ 1.21%) to BA (η ~ 1.52%) mainly due to the increase of current density (J_{sc}: 2.73 (OPA) to 3.41 (BA)), improvement of the surface morphology (RMS) from OPA-sample (~2.52 nm) to BA-sample (~1.69 nm), and increase of charge carrier separation at the polymer/nanoparticle interface which is observed in the quenching of photoluminescence intensity [8, 32]. Efficient separation of charge carriers prevents radiative recombination and thus quenches the photoluminescence of the nanoparticles and polymer components. The blue-shifting of the photoluminescence was also recorded when OPA was replaced by butylamine, which could be a potentially useful method of increasing charge carrier separation at polymer/nanoparticle interface in the BHJ solar cells. The BHJ solar cells made of CdSe with BA-capping ligand showed the best performance (J_{sc} ~ 3.41 mA/cm^2, V_{oc} ~ 0.72 V, FF ~ 61.8%, and PCE ~ 1.52%). The power conversion efficiency of BHJ solar cells fabricated in this study was, however, still relatively low compared with that of world record BHJ solar cells made with tetrapod-shaped CdSe nanocrystals and PCPDTBT polymer [14]. This is possibly attributed to the poorer charge collection due to unoptimized surface morphology and thermal treatment, which might lead to poorer interfacial properties between the polymer and CdSe nanocrystals, creating a phase separation within the active layers. We thus believe that the findings of this study with the PCE values obtained open a new route to improving the BHJ solar cell performance by utilizing another kind of ligands for the CdSe nanoparticles.

4. Conclusions

High-quality CdSe tetrapod nanoparticles were synthesized in this study. A large fraction of branches was obtained without selective precipitation, and the hexagonal wurtzite structure was confirmed. After the ligand exchange from OPA to the BA ligand, the size of the nanoparticles decreased, and the band gap energy increased from 2.09 up to 2.14 eV. The surface roughness of the BA-capped CdSe/PCPDTBT active layer was smoother than that of the OPA-capped CdSe one. The BHJ solar cell performance was significantly improved, showing the power conversion efficiencies of up to 1.52%

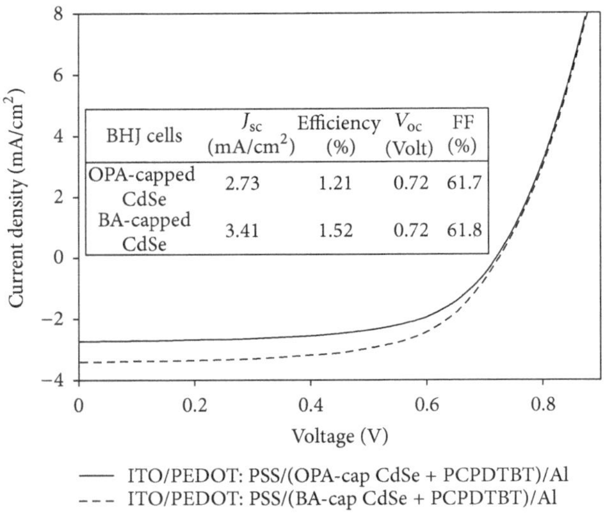

Figure 7: *J-V* curves for the devices with OPA-capped and butylamine-capped CdSe tetrapod nanocrystals (CdSe/[CdSe+PCPDTBT] = 0.7).

(J_{sc} ~ 3.41 mA/cm^2) under AM1.5G simulated solar irradiation.

Acknowledgments

This research was supported by the Basic Science Research Program through the National Research Foundation of Korea (NRF) funded by the Ministry of Education, Science and Technology (2010-0023839) and the Human Resources Development Program of Korea Institute of Energy Technology Evaluation and Planning (KETEP) Grant (no. 20104010100580) funded by the Korean Ministry of Trade, Industry and Energy.

References

[1] N. C. Greenham, X. Peng, and A. P. Alivisatos, "Charge separation and transport in conjugated-polymer/semiconductor-nanocrystal composites studied by photoluminescence quenching and photoconductivity," *Physical Review B*, vol. 54, no. 24, pp. 17628–17637, 1996.

[2] B. R. Saunders and M. L. Turner, "Nanoparticle-polymer photovoltaic cells," *Advances in Colloid and Interface Science*, vol. 138, no. 1, pp. 1–23, 2008.

[3] Y. Liang, Z. Xu, J. Xia et al., "For the bright future-bulk heterojunction polymer solar cells with power conversion efficiency of 7.4%," *Advanced Materials*, vol. 22, no. 20, pp. E135–E138, 2010.

[4] L. Dou, J. You, J. Yang et al., "Tandem polymer solar cells featuring a spectrally matched low-bandgap polymer," *Nature Photonics*, vol. 6, no. 3, pp. 180–185, 2012.

[5] W. U. Huynh, J. J. Dittmer, and A. P. Alivisatos, "Hybrid nanorod-polymer solar cells," *Science*, vol. 295, no. 5564, pp. 2425–2427, 2002.

[6] H. Kim, J. Y. Moon, and H. S. Lee, "Growth of ZnO nanorods on various substrates by electrodeposition," *Electronic Materials Letters*, vol. 5, no. 3, pp. 135–138, 2009.

[7] P. A. Chate, S. S. Patil, J. S. Patil, D. J. Sathe, and P. P. Hankare, "Nanocrystaline CdSe: structural and photoelectrochemical characterization," *Electronic Materials Letters*, vol. 8, no. 6, pp. 553–558, 2012.

[8] N. T. N. Truong and C. Park, "Enhancement of CdSe/poly (3-hexylthiophene) bulk hetero junction solar cell efficiency by surface ligand exchange and thermal treatment," *Japanese Journal of Applied Physics*, vol. 51, article 10NE27, 6 pages, 2012.

[9] H. Chen, J. Yoo, Y. Liu, and G. Zhao, "Green synthesis and characterization of se nanoparticles and nanorods," *Electronic Materials Letters*, vol. 7, no. 4, pp. 333–336, 2011.

[10] D. Kim, C. Hwang, D. Gwoo et al., "Synthesis and characterization of CdS nanocrystals in a novel phosphate glass," *Electronic Materials Letters*, vol. 7, no. 4, pp. 309–312, 2011.

[11] S. K. Mishra, R. K. Srivastava, S. G. Prakash, R. S. Yadav, and A. C. Panday, "Structural, photoconductivity and photoluminescence characterization of cadmium sulfide quantum dots prepared by a co-precipitation method," *Electronic Materials Letters*, vol. 7, no. 1, pp. 31–38, 2011.

[12] X. Luo, P. Liu, N. T. N. Truong, U. Farva, and C. Park, "Photoluminescence blue-shift of CdSe nanoparticles caused by exchange of surface capping layer," *Journal of Physical Chemistry C*, vol. 115, no. 43, pp. 20817–20823, 2011.

[13] C. S. Stan, M. S. Secula, and T. H. Oh, "Highly luminescent polystyrene embedded CdSe quantum dots obtained through a modified colloidal synthesis route," *Electronic Materials Letters*, vol. 8, pp. 275–281, 2012.

[14] S. Dayal, N. Kopidakis, D. C. Olson, D. S. Ginley, and G. Rumbles, "Photovoltaic devices with a low band gap polymer and CdSe nanostructures exceeding 3% efficiency," *Nano Letters*, vol. 10, no. 1, pp. 239–242, 2010.

[15] B. Sun, E. Marx, and N. C. Greenham, "Photovoltaic devices using blends of branched CdSe nanoparticles and conjugated polymers," *Nano Letters*, vol. 3, no. 7, pp. 961–963, 2003.

[16] C. Querner, P. Reiss, S. Sadki, M. Zagorska, and A. Pron, "Size and ligand effects on the electrochemical and spectroelectrochemical responses of CdSe nanocrystals," *Physical Chemistry Chemical Physics*, vol. 7, no. 17, pp. 3204–3209, 2005.

[17] G. Kalyuzhny and R. W. Murray, "Ligand effects on optical properties of CdSe nanocrystals," *Journal of Physical Chemistry B*, vol. 109, no. 15, pp. 7012–7021, 2005.

[18] H. Lee, S. W. Yoon, J. P. Ahn et al., "Synthesis of type II CdTe/CdSe heterostructure tetrapod nanocrystals for PV applications," *Solar Energy Materials and Solar Cells*, vol. 93, no. 6-7, pp. 779–782, 2009.

[19] J. Liu, W. Wang, H. Yu, Z. Wu, J. Peng, and Y. Cao, "Surface ligand effects in MEH-PPV/TiO_2 hybrid solar cells," *Solar Energy Materials and Solar Cells*, vol. 92, no. 11, pp. 1403–1409, 2008.

[20] G. Guerrero, P. H. Mutin, and A. Vioux, "Anchoring of phosphonate and phosphinate coupling molecules on titania particles," *Chemistry of Materials*, vol. 13, no. 11, pp. 4367–4373, 2001.

[21] W. Zhang, S. Ge, Y. Wang et al., "Use of functionalized WS2 nanotubes to produce new polystyrene/polymethylmethacrylate nanocomposites," *Polymer*, vol. 44, no. 7, pp. 2109–2115, 2003.

[22] G. Ramis and G. Busca, "FTIR spectra of adsorbed n-butylamine," *Journal of Molecular Structure*, vol. 193, pp. 93–100, 1989.

[23] P. Chou, C. Chen, C. Cheng et al., "Spectroscopy and femtosecond dynamics of type-II CdTe/CdSe core-shell quantum dots," *ChemPhysChem*, vol. 7, no. 1, pp. 222–228, 2006.

[24] W. W. Yu, Y. A. Wang, and X. Peng, "Formation and stability of size-, shape-, and structure-controlled CdTe nanocrystals: ligand effects on monomers and nanocrystals," *Chemistry of Materials*, vol. 15, no. 22, pp. 4300–4308, 2003.

[25] Q. Pang, L. Zhao, Y. Cai et al., "CdSe nano-tetrapods: controllable synthesis, structure analysis, and electronic and optical properties," *Chemistry of Materials*, vol. 17, no. 21, pp. 5263–5267, 2005.

[26] Y. Jun, S. Lee, N. Kang, and J. Cheon, "Controlled synthesis of multi-armed CdS nanorod architectures using monosurfactant system," *Journal of the American Chemical Society*, vol. 123, no. 21, pp. 5150–5151, 2001.

[27] L. Carbone, S. Kudera, E. Carlino et al., "Multiple wurtzite twinning in CdTe nanocrystals induced by methylphosphonic acid," *Journal of the American Chemical Society*, vol. 128, no. 3, pp. 748–755, 2006.

[28] N. Chandrasekharan and P. V. Kamat, "Tuning the properties of CdSe nanoparticles in reverse micelles," *Research on Chemical Intermediates*, vol. 28, no. 7–9, pp. 847–856, 2002.

[29] J. H. Adair, T. Li, T. Kido et al., "Recent developments in the preparation and properties of nanometer-size spherical and platelet-shaped particles and composite particles," *Materials Science and Engineering R*, vol. 23, no. 4-5, pp. 139–242, 1998.

[30] M. L. Monroe, Y. W. Kim, N. T. N. Truong et al., "Effect of surface modification by solvent exchange on hybrid bulk heterojunction solar cell performance," *Proceeding of Materials Research Society*, vol. 1013, 1013-Z07-10, 2007.

[31] C. N. Hoth, P. Schilinsky, S. A. Choulis, and C. J. Brabec, "Printing highly efficient organic solar cells," *Nano Letters*, vol. 8, no. 9, pp. 2806–2813, 2008.

[32] J. D. Olson, G. P. Gray, and S. A. Carter, "Optimizing hybrid photovoltaics through annealing and ligand choice," *Solar Energy Materials and Solar Cells*, vol. 93, no. 4, pp. 519–523, 2009.

Effect of Processing Parameters on Thickness of Columnar Structured Silicon Wafers Directly Grown from Silicon Melts

Jin-Seok Lee, Bo-Yun Jang, and Young-Soo Ahn

Korea Institute of Energy Research, Daejeon 305-343, Republic of Korea

Correspondence should be addressed to Jin-Seok Lee, jslee@kier.re.kr

Academic Editor: C. W. Lan

In order to obtain optimum growth conditions for desired thickness and more effective silicon feedstock usage, effects of processing parameters such as preheated substrate temperatures, time intervals, moving velocity of substrates, and Ar gas blowing rates on silicon ribbon thickness were investigated in the horizontal growth process. Most of the parameters strongly affected in the control of ribbon thickness with columnar grain structure depended on the solidification rate. The thickness of the silicon ribbon decreased with an increasing substrate temperature, decreasing time interval, and increasing moving velocity of the substrate. However, the blowing of Ar gas onto a liquid layer existing on the surface of solidified ribbon contributed to achieving smooth surface roughness but did not closely affect the change of ribbon thickness in the case of a blowing rate of $\geq 0.65\,\mathrm{Nm^3/h}$ because the thickness of the solidified layer was already determined by the exit height of the reservoir.

1. Introduction

In the photovoltaic industry growing at rates exceeding 30% per year, crystalline silicon (Si) is the dominant photovoltaic material and accounts for approximately 90% of the total solar cell production [1]. However, the current bottleneck in crystalline Si photovoltaics due to the shortage of Si feedstock increases Si prices and puts additional pressure on the fraction of wafer cost in the module [2].

The direct growth technology of Si ribbon wafer from the molten Si is one of the several routes for responding to increasing Si wafer consumption with almost 100% silicon usage. Technology with no kerf loss and high productivity can significantly lower manufacturing costs for solar cells [3]. The ribbon Si technologies are generally classified in two ways by the transport direction of the solidified ribbon with respect to the movement of the liquid-solid interface during crystallization: (i) liquid-solid interface moves in line with ribbon transport direction (also called vertical growth, e.g., edge-defined film-fed growth (EFG) and string ribbon (SR)); (ii) liquid-solid interface moves almost perpendicular to the ribbon transport direction (also called horizontal growth, e.g., ribbon growth on substrate (RGS)) [4]. Horizontal growth has much faster ribbon growth rate than vertical growth, finally leading to the effective production of silicon wafers with low manufacturing costs and high production rates, because thermal conduction through the directly contacted substrate is more efficient at heat removal from the ribbon.

In addition, one option to make a more efficient use of the Si materials in horizontal growth system is the use of thinner Si wafers. For instance, the total amount of Si used per W_p decreases by about 20% when using $200\,\mu$m wafers instead of $300\,\mu$m wafers [5]. However, the effect of processing parameters such as preheated substrate temperatures, time intervals, moving velocity of substrates, and gas blowing rates on Si ribbon thickness has been rarely reported in order to achieve more effective Si usage in the horizontal growth process so far. The present investigation reports the experimental findings for processing parameters on the control of Si ribbon thickness in the horizontal growth process. These considerations will meet the demands for next-generation solar cell production with the optimum growth conditions for the desired thickness of the Si wafer.

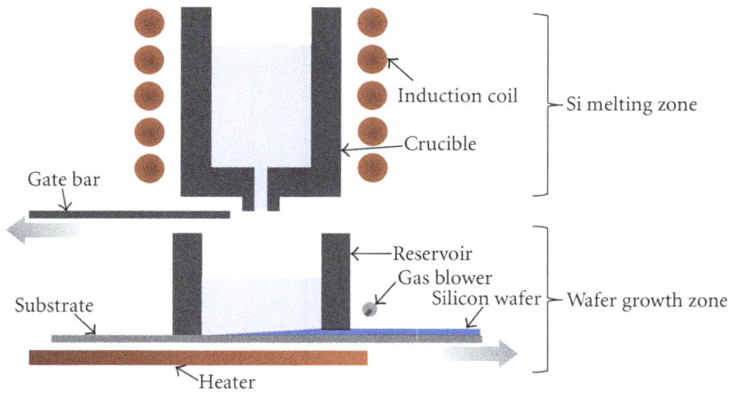

FIGURE 1: Schematic diagram of a horizontal growth system for ribbon-typed Si wafers.

2. Experimental Procedure

The system for the growth of columnar structured Si wafers shown in Figure 1 is separated into Si melting zones by induction coil and ribbon-type wafer growth zones in a vacuum chamber. After the Si (purity > 9N) in the melting zone is completely melted by heating to 1450°C at a pressure of 10^{-4} Torr, liquid Si is poured into the SiC-coated graphite reservoir positioned at the growth zone. As soon as liquid Si is solidified by releasing the latent heat through the SiC-coated graphite substrate preheated at a range of 700–1000°C and moved underneath the reservoir with a velocity of 315 ~ 926 cm/min after a certain amount of time (hereafter called time interval) at a range of 0.5–2.5 sec, then the solidified Si on the substrate moves away from the exit of the melt reservoir of 1 mm. At the moment, the blowing of Ar gas on a solidified wafer surface removed liquid Si residue at a flow rate of 0.5–1.1 Nm^3/h, which improved the surface roughness of the Si wafer as a consequence. Dense SiC layer with thickness of around 50 μm and surface roughness of around 4 μm was coated on graphite substrate by commercially available chemical-vapor-reaction (CVR) method at 1700°C in vacuum, where the surface modification provided the advantage for preventing the carbon contamination from graphite remaining high thermal conductivity. The dimension of the Si wafer, especially the thickness, could be determined by the control of processing parameters such as substrate temperatures, time intervals, moving velocity of substrates, and gas blowing rates. Horizontal growth systems and experimental procedures of Si wafers are described elsewhere in detail [6].

The thickness of as-grown Si wafers was measured by a sharp-tipped micrometer to be approximately 1mm without an additional polishing step, in which five measurements were carried out and averaged. The Si wafer was etched using a 10% KOH solution to measure the microstructures and defects of the wafer using a differential interference contrast optical microscope (BX60MF, Olympus, Japan).

3. Results and Discussion

The thickness of the ribbon-type silicon wafer grown in a horizontal growth system can be controlled by the shape and height of the reservoir exit if it is possible to perfectly engineer the liquid-solid interface in the reservoir. However, fine control of the liquid-solid interface only with the reservoir structure is very difficult in nature without considering processing parameters such as substrate temperatures, time intervals, moving velocity of substrates, and gas blowing rates. Therefore, this study aims to present a manner to control the thickness of columnar structured-silicon wafers horizontally and directly grown from silicon melts. The area for the growth interface of liquid-solid is now very large compared to the wafer thickness in the horizontal growth system. The latent heat of the silicon melt is extracted mainly by conduction through the substrate and also partially by radiation through the surface of the silicon melt. One of the reasons for the high productivity during the horizontal growth system is the ability of the substrate to dissipate heat through the bottom side that allows for better control of solidification, through the moving velocity and heat extraction rate [7]. Therefore, pull speed of the substrate and heat extraction on the substrate are important variables in optimum Si wafer production. Growth rates of 4–9 m/min in horizontal growth systems have been previously demonstrated [8].

Figure 2 shows the thickness variation of ribbon-type Si wafers depending on the substrate temperature preheated at the range of 700–1000°C at the constant time interval of 2.5 sec, moving velocity of the substrate of 485 cm/min, and blowing rate of Ar gas of 0.8 Nm^3/h. Substrate temperature strongly relates to behaviors of heat extraction of solidified Si wafers. Even though the substrate is preheated up to 1000°C, the onset of solidification starts due to temperature differences between Si melt (~1414°C) and substrates. Solidification rate increases with decreasing substrate temperatures because of larger temperature differences. At the moment, the temperature of the melt can be more rapidly decreased by the heat extraction through the cold ceramic substrate. Therefore, higher substrate temperatures were

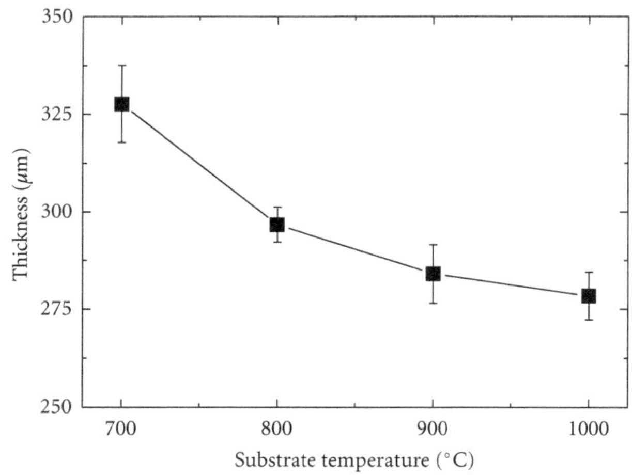

FIGURE 2: Thickness variation of ribbon-type Si wafers depending on the substrate temperature preheated at the range of 700–1000°C at the constant time interval of 2.5 sec, moving velocity of the substrate of 485 cm/min, and blowing rate of Ar gas of 0.8 Nm^3/h.

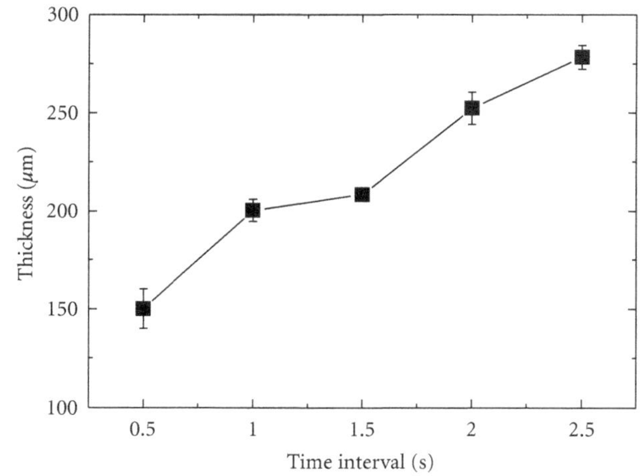

FIGURE 3: Thickness variation of ribbon-type Si wafers depending on the time interval at the range of 0.5–2.5 sec at the constant substrate temperature of 1000°C, moving velocity of substrates of 485 cm/min, and blowing rate of Ar gas of 0.8 Nm^3/h.

beneficial to obtain thinner Si wafers, which could contribute to reduce the Si consumption.

Assuming steady-state conditions, consideration of thermal flux during the growth of ribbon-type silicon wafers in the horizontal growth system is used to calculate the dependence of wafer thickness on the moving velocity of the substrate at a given temperature gradient between the silicon melt and substrate with (1) as follows [9]:

$$V_p = \frac{4\alpha\lambda_s s}{(2\lambda_s - \alpha t) d L \rho_s} \Delta T, \quad (1)$$

where V_p is the moving velocity of the substrate (pulling speed), α the effective coefficient of heat transfer, s the length of the liquid-solid interface, λ_s the thermal conductivity of solid silicon, d the wafer thickness, ρ_s the density of solid silicon, L the crystallization heat of silicon, t the crystallization time, and ΔT the temperature gradient between the melt and substrate. The substrate temperature determines the ΔT and a lower substrate temperature increases the growth velocity according to the above equation. In this study, the temperature of liquid Si was > 1412°C when the melt was poured. Temperature differences, ΔT, were about > 400°C due to the substrate temperature pre-heated in the range of 700–1000°C. Note that wafer thickness, d, can be easily controlled by operating ΔT if V_p ($= ds/dt$) is set at a constant moving velocity and the rest of the parameters such as α, s, λ_s, ρ_s, L, t in the equation are fixed. Therefore, thickness, d, decreases with the decreasing temperature gradient, ΔT, by fast solidification time.

In equipment for the growth of Si wafers in this study, the zone for the melting of Si materials and the zone for the growth of Si wafer are separated for the fabrication of more homogeneous wafer quality. Figure 3 shows the thickness variation of ribbon-type Si wafers depending on the time interval in the range of 0.5–2.5 sec at the constant substrate temperature of 1000°C, moving velocity of substrate of 485 cm/min and blowing rate of Ar gas of 0.8 Nm^3/h. The thickness of the wafer increased with an increased time interval because a longer time interval allows for the increase of residence time in a reservoir filled with high-temperature melt, which allows for longer and slower growth time by increasing the solidification time. This is also supported by (1) which indicates that a longer time interval induces a longer crystallization time, t, resulting in a thicker wafer. Slower growth occurring with increasing thickness of the wafer is due to the additional heat transport through the solidified silicon. As soon as silicon melts in the melting zone are poured into the reservoir, crystal growth begins on a cold (below silicon melting temperature) substrate underneath the reservoir. Then, the substrate, pre-heated up to a desired temperature, is continuously moved underneath the reservoir filled with silicon melts after a certain time interval. Once the thickness of the initial wafer is determined during the crystal growth with a certain time interval, the thickness of the following wafer by the movement of the substrate tends to depend on the thickness of the initial wafer. Therefore, the time interval is one of the important factors to control the thickness of ribbon-type silicon wafers especially in a continuous growth system.

Figure 4 shows the thickness variation of ribbon-type Si wafers depending on the moving velocity of the substrate (pulling speed) at the range of 315–926 cm/min at a constant substrate temperature of 1000°C, time interval of 2.5 sec, and blowing rate of Ar gas of 0.8 Nm^3/h. The lowest moving velocity of 315 cm/min in this study induced an excessively increased wafer thickness of 753 μm, but increasing moving velocity up to 926 cm/min drastically reduced the thickness of the wafer up to 145 μm due to decreasing the solidification time. However, defects such as pores were detected on the wafer surface in cases where the moving velocity exceeded above 926 cm/min. Compared to other silicon crystallization methods, the silicon crystal growth rate is very high, which is applicable to enhance a competitive price in the

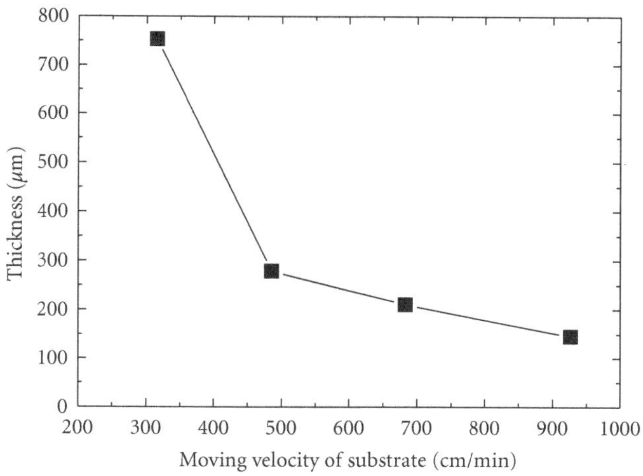

FIGURE 4: Thickness variation of ribbon-type Si wafers depending on the moving velocity of substrates (pulling speed) at the range of 315–926 cm/min at the constant substrate temperature of 1000°C, time interval of 2.5 sec, and blowing rate of Ar gas of 0.8 Nm^3/h.

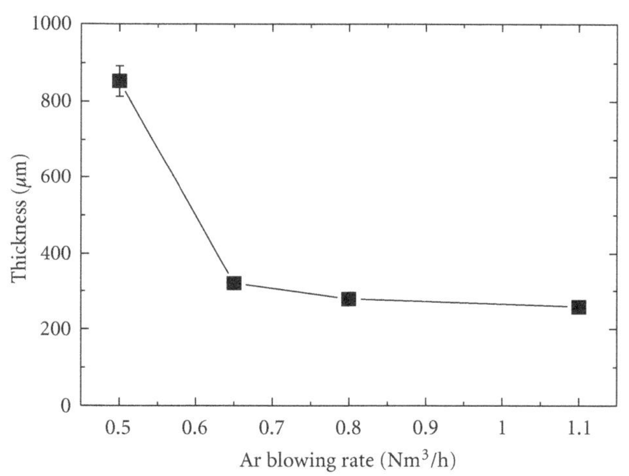

FIGURE 5: Thickness variation of ribbon-type Si wafers depending on the blowing rate of Ar gas at the range of 0.5–1.1 Nm^3/h at the constant substrate temperature of 1000°C, time interval of 2.5 sec, and moving velocity of substrates of 485 cm/min.

solar cell industry. The moving velocity essentially affects the residence time of the melt mass subject to the heat extraction rate through the substrate [7]. At low moving velocity, the longer residence time affords nearly complete solidification of the upper part of the silicon film. However, high moving velocity with a higher productivity encourages that a substantial portion of the upper layer of silicon wafer is still in liquid phase, which means that wafer thickness should be controllable by following the processing step such as the blowing of Ar gas onto the surface of liquid silicon film on solid silicon film.

From the results by Schönecker et al. [10], the initial crystal growth velocity is high in the beginning of the process but decreases rapidly. The reason is caused by the facts that (i) the substrate temperature is not constant during crystal growth, (ii) there is a finite heat transfer between liquid silicon and the substrate, which slows down the crystallization speed, and (iii) the silicon melt is most likely in turbulent flow, transporting heat from the hotter top surface of the melt to the liquid-solid interface. Therefore, the crystallization speed affecting wafer thickness is lower in reality, depending on system properties such as silicon-substrate properties, melt temperature, and melt flux behavior. It is very difficult to fabricate continuously grown wafers with uniform thickness with high reliability in the industrial field.

In order to solve the problem, the blowing system of Ar gas is employed in this study. Figure 5 shows the thickness variation of ribbon-type Si wafers depending on the blowing rate of Ar gas at a range of 0.5–1.1 Nm^3/h at the constant substrate temperature of 1000°C, time interval of 2.5 sec, and moving velocity of substrates of 485 cm/min. Thickness was not reduced at the lowest blowing rate of 0.5 Nm^3/h in this study because a low blowing rate is not sufficient to flatten the surface with a liquid portion. However, a blowing rate of more than 0.65 Nm^3/h could considerably reduce the thickness compared to wafer thickness in the case of a blowing rate of 0.5 Nm^3/h. Although there was an effect on the decrease of wafer thickness above 0.65 Nm^3/h, thickness variation is not as large at the range of 0.65–1.1 Nm^3/h. The solidification height (thickness of solid film) is determined at a certain time interval in the reservoir filled with silicon melts, in which solidification rapidly begins as soon as silicon melts make contact with cold substrate. The thickness of casted film (liquid film on solid film) is strongly dependent on the exit height of the reservoir. If the thickness of solid film is determined in the reservoir at a certain time interval before the casted film on the substrate moves to the blowing zone of Ar gas, thickness is no longer reduced to below the thickness of solid film in casted film even with a high Ar gas blowing rate. However, the step of Ar blowing was considerably effective on processing for thin wafers with smooth surface roughness if the melt layer remained on the surface of solidified wafer.

Figure 6 shows optical microscope images for the surface and cross-section of as-grown ribbon-type silicon wafers at a substrate temperature of 1000°C, time interval of 2.5 sec, moving velocity of the substrate of 485 cm/min, and a blowing rate of Ar gas of 0.8 Nm^3/h. While some grains are almost dislocation-free, others contain a high density of dislocations. Silicon wafers grown in a horizontal growth system usually have a high density of grain boundaries, dislocations, twins, and other intragrain defects because large thermal stresses are generated during rapid solidification and subsequent cooling [1]. The inhomogeneous distribution of dislocations is due to the fact that the thermal stress exceeds the critical shear stress only in some particular orientated grains. Grains with none or a low density of dislocations usually contain twin boundaries, while those with highly dislocated grains usually do not have twins. This means that twin planes act as barriers for dislocation propagation and contribute to the inhomogeneous distribution of dislocations across various grains [1]. A typical surface image (Figure 6(a)) of a wafer shows a mean grain size of around 150 μm and rather low densities of

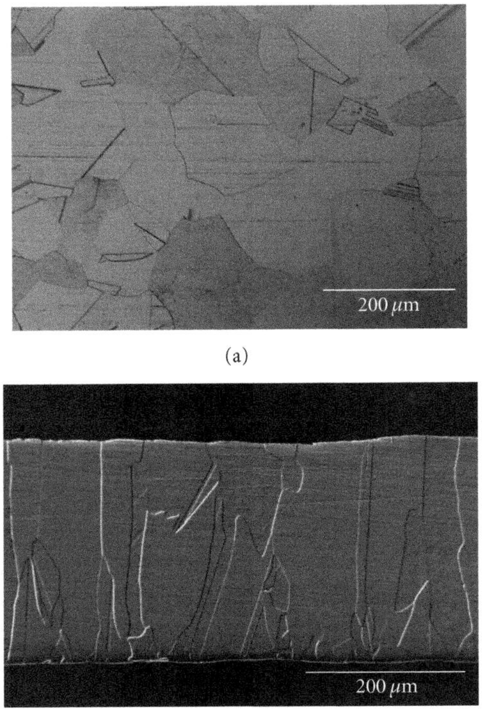

(b)

Figure 6: Optical microscope images for (a) surface and (b) cross section of as-grown ribbon-type silicon wafers at a substrate temperature of 1000°C, time interval of 2.5 sec, moving velocity of substrates of 485 cm/min, and blowing rate of Ar gas of 0.8 Nm^3/h.

dislocation and twin through the consideration of processing parameters compared to previous reports [1]. However, subsequent steps for increasing grain size and stress relief in wafers are presumably necessary for improved efficiency of solar cells. A cross-sectional image (Figure 6(b)) typically indicates a columnar grain structure and smooth surface roughness by the blowing effect of Ar gas. Although there are relatively few grain boundaries in the up-and-down direction, it is expected that those columnar structures will induce an excellent result through current flows along the grain boundaries in the cells [11].

4. Conclusion

Processing parameters such as a pre-heated substrate temperature, time interval, moving velocity of substrates, and the gas blowing rate to control the thickness of the Si ribbon thickness were investigated for achieving more effective Si usage in the horizontal growth process. The substrate temperature strongly related to the change of ribbon thickness depending on the behavior of heat extraction of solidified the Si wafer. Higher substrate temperatures were beneficial to obtain thinner Si wafers, which could contribute to reduce the Si consumption. A longer time interval allows the increase of residence time in the reservoir, so that ribbon thickness increased by increasing the solidification time. The highest moving velocity of 926 cm/min in this study significantly reduced the thickness of wafers up to 145 μm, because the moving velocity essentially affects the residence time of the melt mass subject to the heat extraction rate through the substrate. The effect of Ar blowing was negligible if the thickness of solid film is determined in the reservoir. However, the step of Ar blowing was considerably effective on processing for thin wafers with smooth surface roughness. Processing conditions were established to grow the desired thickness of the Si ribbon with the columnar grain structure and smooth surface roughness in a horizontal growth process for low-cost solar cell application.

Acknowledgment

This work was supported by the New and Renewable Energy Technology Development Program of the Korea Institute of Energy Technology Evaluation and Planning (KETEP) Grant funded by the Korea Government Ministry of Knowledge Economy (*no. 20103020010060*).

References

[1] J. Lu, G. A. Rozgonyi, and A. Schönecker, "Carrier lifetime limitation defects in polycrystalline silicon ribbons grown on substrate (RGS)," *Materials Science in Semiconductor Processing*, vol. 11, no. 1, pp. 20–24, 2008.

[2] G. Hahn, S. Seren, M. Kaes et al., "Review on ribbon silicon techniques for cost reduction in PV," in *Proceedings of the IEEE 4th World Conference on Photovoltaic Energy Conversion*, pp. 972–975, Hawaii, USA, May 2006.

[3] J. P. Kalejs, "Silicon ribbons and foils—state of the art," *Solar Energy Materials and Solar Cells*, vol. 72, no. 1-4, pp. 139–153, 2002.

[4] K. Nakajima and N. Usami, *Crystal Growth of Si for Solar Cells*, Springer, 2009.

[5] C. J. J. Tool, A. R. Burgers, P. Manshanden, and A. W. Weeber, "Effect of wafer thickness on the performance of mc-Si solar cells," in *Proceedings of the 17th European Photovoltaic Solar Energy Conference*, Munich, Germany, 2001.

[6] S. J. Ko, B. Y. Jang, J. S. Kim, Y. S. Ahn, and S. Nahm, "Polycrystalline silicon wafer fabricated by direct growth from liquid silicon," *Journal of the Korean Physical Society*, vol. 57, no. 1, pp. 97–102, 2010.

[7] H. M. Jeong, H. S. Chung, and T. W. Lee, "Computational simulations of Ribbon-Growth on substrate for photovoltaic silicon wafer," *Journal of Crystal Growth*, vol. 312, no. 4, pp. 555–562, 2010.

[8] R. O. Bell and J. P. Kalejs, "Growth of silicon sheets for photovoltaic applications," *Journal of Materials Research*, vol. 13, no. 10, pp. 2732–2739, 1998.

[9] H. Lange and I. A. Schwirtlich, "Ribbon Growth on Substrate (RGS)—a new approach to high speed growth of silicon ribbons for photovoltaics," *Journal of Crystal Growth*, vol. 104, no. 1, pp. 108–112, 1990.

[10] A. Schönecker, L. Laas, A. Gutjahr, P. Wyers, A. Reinink, and B. Wiersma, "Ribbon-growth-on-substrate: progress in high-speed crystalline silicon wafer manufacturing," in *Proceedings of the 29th IEEE Photovoltaic Specialists Conference*, pp. 316–319, New Orlens, La, USA, May 2002.

[11] H. Yamatsugu, H. Mitsuyasu, T. Takakura et al., "Crystallization on dipped substrate wafer technology for crystalline silicon solar cells reduces wafer costs," *Photovoltaic International*, 2008.

Preparation and Characterization of $Cu(In,Ga)Se_2$ Thin Films by Selenization of $Cu_{0.8}Ga_{0.2}$ and In_2Se_3 Precursor Films

Jiang Liu, Da-Ming Zhuang, Ming-Jie Cao, Chen-Yue Wang, Min Xie, and Xiao-Long Li

Department of Mechanical Engineering, Tsinghua University, Beijing 100084, China

Correspondence should be addressed to Da-Ming Zhuang, dmzhuang@tsinghua.edu.cn

Academic Editor: Roel van De Krol

Se-containing precursor films with two different compositions were prepared by magnetron sputtering from $Cu_{0.8}Ga_{0.2}$ and In_2Se_3 targets, and then were selenized using Se vapor. The effects of precursor composition and selenization temperature on the film properties were investigated. The results show that $Cu_{2-x}Se$ phase plays a critical role in film growth and electrical properties of CIGS films. The Cu-rich films exhibit different surface morphology and better crystallinity, as compared to the Cu-poor films. All the CIGS films exhibit p-type conductivity. The resistivity of the Cu-rich films is about three orders of magnitude lower than that of the Cu-poor films, which is attributed to the presence of p-type highly conductive $Cu_{2-x}Se$ phase.

1. Introduction

$Cu(In_{1-x}Ga_x)Se_2$ (CIGS) thin films have received considerable attention in recent years because of their application in solar cells, which have many advantages in high conversion efficiency [1, 2], the possibility of low-cost production [3, 4], and long-term stability [5, 6]. The world-record efficiency for small-area CIGS-based thin film solar cells has recently surpassed 20% [7]. One of the special qualities of the CIGS material is its variable band-gap [8]. The bandgap of chalcopyrite CIGS absorber could vary from 1.04 eV in $CuInSe_2$ to 1.68 eV in $CuGaSe_2$ by substituting indium for gallium. The addition of small amounts of Ga could not only raise the bandgap to more suitably match the AM1.5 solar spectrum, more importantly, but could also improve the electrical properties of CIGS films [9–11]. However, the CIGS films with higher gallium concentration show a significant loss of efficiency relative to lower gallium content films [12]. The optimal [Ga]/[In + Ga] ratio of CIGS absorber for high efficiency CIGS solar cell is considered to be 0.2~0.3 from a number of experimental results [13–15].

A variety of processing techniques have been developed to prepare high quality CIGS thin films. The two most reported processes are the coevaporation from elemental sources and the selenization of metallic precursors. The coevaporation process consists of three stages, which all need precise control over the deposition rate of each element. The selenization of metallic precursors is a two-stage process, which involves the deposition of Cu-In-Ga metallic precursors in the first step followed by their selenization using H_2Se gas or Se vapor [16]. The presence of low-melting metal indium in Cu-In-Ga metallic precursors could always result in rough morphologies of the CIGS films and loss of indium from films during the ramp-up and selenization stages [17, 18]. Moreover, the CIGS films by selenization of metallic precursors have a poor adhesion to substrate, due to a threefold volume expansion caused by the incorporation of Se into the metallic precursors during selenization [18]. Few works have been done using the selenium-containing precursors to prepare CIGS films. In this study, we introduce $Cu_{0.8}Ga_{0.2}$ and In_2Se_3 as precursor films and avoid the direct addition of elemental indium. The precursors were then treated by selenization using Se vapor. The effects of precursor compositions and selenization temperature on the film properties were investigated.

2. Experimental

High purity In_2Se_3 ceramic target and CuGa alloy target with Ga content of 20 at% were used. The size of the two targets is 360 mm × 80 mm. In the first step the CuGa and In_2Se_3

films were deposited on Mo-coated and bare soda-lime glass substrate by middle frequency (MF) magnetron sputtering of CuGa and In_2Se_3 targets. By adjusting the sputtering time and current of the two targets, the precursor films with different compositions could be prepared. Generally, relative proportion of metallic elements in CIGS films lies on that of the precursor films. The copper content of the final film and during the deposition process has a strong influence on film growth and CIGS solar cell properties [19, 20]. For the purpose of comparison, two sets of as-sputtered films were prepared. One was Cu-rich composition ([Cu]/[In + Ga] > 1) and another was Cu-poor composition ([Cu]/[In + Ga] < 1). The substrate holder inside the sputtering equipment could move back and forth along the short transverse direction of targets to keep the homogeneity of the as-sputtered films. The substrate temperature and working pressure during sputtering was kept constant at 200°C and 0.5 Pa, respectively. In the second step, the as-sputtered films were selenized with Se vapor in a three-chamber furnace which can be easily scaled up to provide a large-area in-line sequential CIGS process. Prior to heating, the chambers were evacuated and flushed two times with high purity nitrogen. The selenization time was held constant at 30 min and the selenization temperature varied from 400°C to 600°C.

Crystalline structures of the samples were examined by X-ray diffraction carried out with D/max-RB diffractometer (Rigaku) using Cu Ka radiation (0.1541 nm). The compositions of the films were tested by X-ray fluorescence (LAB CENTER XRF-1800). The surface and cross-section micrographs of the films were obtained by thermal field emission scanning electron microscope (TFE-SEM, LEO-1530). Raman spectra were recorded using Renishaw Raman microscope (model RM2000) with an argon ion laser wavelength of 514.5 nm. Hall-effect measurements were used on films grown on bare soda-lime glass to determine the resistivity and carrier concentration of CIGS films.

3. Results and Discussion

The XRD patterns of the precursor films deposited on bare soda-lime glass with two different compositions are presented in Figure 1. The corresponding composition data are shown in Table 1. To investigate the relative compositional change, the corresponding atomic ratios of Cu/(In + Ga), Ga/(In + Ga), and 2∗Se/(Cu+3∗(In + Ga)) were also calculated. The films were intentionally deposited into two groups, Cu-rich and Cu-poor precursor films, which have [Cu]/[In + Ga] of 1.277 and 0.935, respectively. Since Ga is incorporated in CuGa alloy target, the Cu-rich precursor film deserves to have a slightly higher [Ga]/[In + Ga] than the Cu-poor precursor film. The 2∗[Se]/([Cu]+3∗[In + Ga]) of the two precursor films are far less than that of near-stoichiometric $Cu(In_{1-x}Ga_x)Se_2$ compound and need be improved during selenization. Figure 1 indicates that XRD patterns of the two precursor films are almost the same. The diffraction peaks at 2θ = 26.6° corresponds to (112) plane of CIGS chalcopyrite structure. Two broad peaks at around 2θ = 44° are attributed to Cu (111) and CIGS (204/220). No In-Se

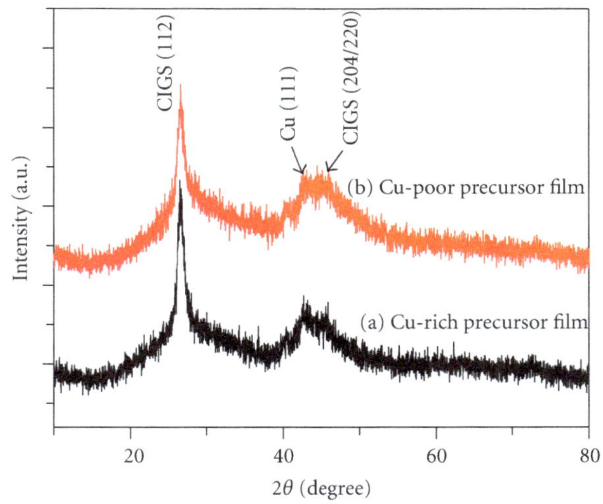

FIGURE 1: XRD patterns of the precursor films of two different compositions: (a) Cu-rich precursor film, (b) Cu-poor precursor film.

related peaks or other characteristic peaks are found, which could be explained by the low crystallinity of the films. After selenization at 550°C for 30 min, XRF measurements (Table 1) show that the CIGS films selenized from Cu-rich and Cu-poor precursor films still conserve their overall Cu-rich and Cu-poor compositions, respectively, with corresponding [Cu]/[In + Ga] of 1.293 and 0.938. The [Ga]/[In + Ga] ratios of the CIGS films are nearly equal to these of their respective precursor films, which could draw the conclusion that the element concentrations of indium and gallium kept almost constant during selenization. Additionally, the 2∗[Se]/([Cu]+3∗[In + Ga]) ratios improved noticeably after selenization, which of Cu-rich film is higher than that of Cu-poor film due to the monovalence chemical state of copper.

Figure 2 shows the XRD patterns of CIGS films selenized from Cu-rich precursor films at different selenization temperatures for 30 min. All samples show characteristic peaks of the chalcopyrite-type structure and are oriented along the (112) direction parallel to the substrate. The full width at half maximum (FWHM) of (112) peaks of the films decreases with the increase of selenization temperature, which may be attributed to the crystalline quality improvement and grain growth. For the CIGS film selenized at 550°C, two additional peaks at 2θ = 31.4° and 55.9° could be found, which correspond to $MoSe_2$ (100) and $MoSe_2$ (110), respectively, according to Joint Committee on Powder Diffraction Standards (JCPDS) Number 29-0914. The intensity of the two peaks of $MoSe_2$ gets stronger when the selenization temperature increases from 550°C to 600°C, which is indicative of higher degree of selenization. Note that when the selenization temperature was 600°C, the soda-lime glass substrate warped but the film deposited on it still did not peel off and could be used for XRD analysis and other tests.

Figure 2(b) shows the detailed XRD patterns of CIGS films from 2θ = 42° to 47° where (204/220) peak of chalcopyrite structure lies. Overall, the position of the (204/220) peak seems to shift to higher diffraction angle as

Table 1: Chemical compositions (measured by XRF) of the precursor films and subsequent selenized films at 550°C for 30 min.

Sample	Cu (at%)	In (at%)	Ga (at%)	Se (at%)	[Cu]/[In + Ga]	[Ga]/[In + Ga]	2∗[Se]/([Cu] + 3∗[In + Ga])
Cu-rich precursor film	39.80	21.87	9.29	29.05	1.277	0.298	0.436
Cu-poor precursor film	32.27	27.08	7.45	33.21	0.935	0.216	0.489
CIGS film selenized from Cu-rich precursor film	29.14	15.80	6.74	48.32	1.293	0.299	0.999
CIGS film selenized from Cu-poor precursor film	23.94	20.00	5.52	50.55	0.938	0.216	1.006

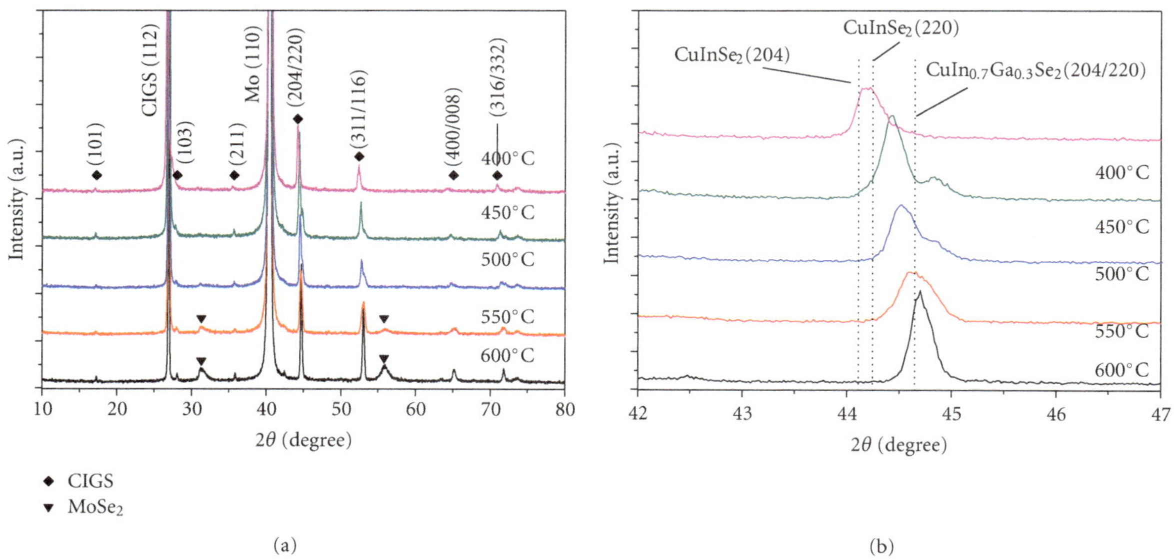

Figure 2: (a) XRD patterns of CIGS films selenized from Cu-rich precursor films at different selenization temperatures for 30 min. (b) Detailed XRD plots of the (204/220) peaks. JCPDS (Joint Committee on Powder Diffraction Standards) Numbers 42-1120, 29-0914, 40-1487, and 35-1102 are referred to for Mo, $MoSe_2$, $CuInSe_2$, and $CuIn_{0.7}Ga_{0.3}Se_2$, respectively.

the selenization temperature increases. The $Cu(In_{1-x}Ga_x)Se_2$ solid solution is based on $CuInSe_2$. Gallium is added to obtain CIGS with a wider bandgap; accordingly, the lattice constants, a and c, decrease. With the increasing of [Ga]/[In + Ga] ratio, the diffraction peaks of chalcopyrite phase shift to higher 2θ side. Theoretically, if the lattice constant c/a ratio does not equal two, the (204) and (220) peaks would uncouple. For the CIGS film selenized at 400°C, the splitting of the (204) and (220) peaks is observed, which is in very good agreement with the $CuInSe_2$ phase (JCPDS Number 40-1487). For the CIGS film selenized at 550°C or 600°C, the (204) and (220) peaks almost overlap, the positions of which match well with the $CuIn_{0.7}Ga_{0.3}Se_2$ solid solution (JCPDS Number 35-1102). Based on the above experimental observations, it could be concluded that $CuInSe_2$ may form first at a relatively low selenization temperature of 400°C, or the formation of $CuInSe_2$ occurs faster than the formation of $CuGaSe_2$, which is also observed in other literatures [21–23]. It could be easily understood that the selenization temperature has a strong influence on the reaction rate and atomic diffusion coefficient. With the increasing of selenization temperature, the gallium diffuses into $CuInSe_2$, thus resulting in the formation of $CuIn_{0.7}Ga_{0.3}Se_2$ solid solution.

To further investigate the structural properties of CIGS thin films, Raman spectroscopy was adopted. Raman spectra of CIGS films selenized at different selenization temperatures for 30 min from Cu-rich precursor films and Cu-poor precursor films are presented in Figure 3. All spectra are normalized to the peak at around 172–174 cm^{-1}, which corresponds to the dominant A_1 mode of CIGS films resulting from the vibration of the Se anions in the x-y plane with the cations at rest [24, 25]. And mixed B_2/E modes are also observed at around 213 cm^{-1}, which represent vibrations of anions and cations together [24]. For all the Cu-rich films (Figure 3(a)), an additional mode appears at around 258–260 cm^{-1}, which is assigned to the A_1 mode of Cu-Se compounds like CuSe or Cu_2Se and is labeled with $Cu_{2-x}Se$ in the following literatures [19, 26]. For the Cu-poor films (Figure 3(b)), the peak of $Cu_{2-x}Se$ A_1 mode exists only when the selenization temperature was 400°C. From the above observation, it is clear that inadequately reaction for the Cu-poor films at low selenization temperature could result in phase segregation. For the Cu-rich films, even at high selenization of up to 600°C, $Cu_{2-x}Se$ segregation is always present, which could be explained with the Cu_2Se-In_2Se_3 pseudo-binary phase diagram [27], where excess Cu couldn't be incorporated in the chalcopyrite phase.

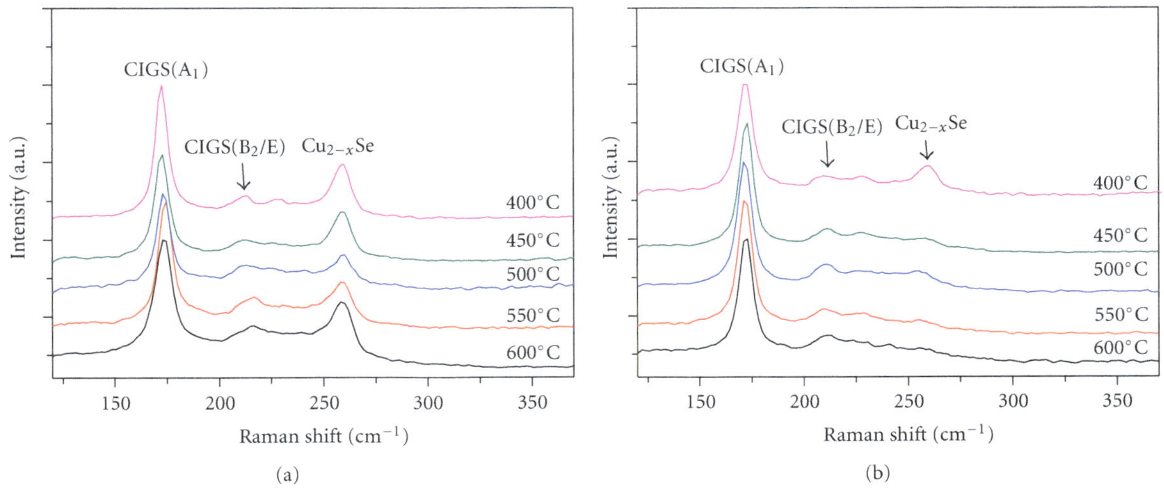

FIGURE 3: Raman spectra of CIGS films selenized at different selenization temperatures for 30 min from (a) Cu-rich precursor films, and (b) Cu-poor precursor films.

Surface SEM images of CIGS films selenized from Cu-rich precursor films and Cu-poor precursor films at different selenization temperatures are presented in Figure 4. The insets show the corresponding cross-section micrographs of the films. From Figure 4(a–e) and Figure 4(f–j), the crystallinity of the films improves noticeably as the selenization temperature increases. The Cu-rich films exhibit a granular surface morphology, whereas a triangular surface morphology is observed obviously for the Cu-poor films selenized at 550°C or 600°C. Low selenization temperature of 400°C results in a significant increase in surface roughness of the Cu-rich and Cu-poor films. In comparison with the Cu-poor films, the Cu-rich films have a larger grain size. When the selenization temperature was 500°C or above, the surface morphology of Cu-rich films exhibit closely packed grains which are approximate 1 μm wide. From the cross-section micrographs, the comparison of film growth between the Cu-rich films and the Cu-poor films is also evident. According to the growth model proposed by Klenk et al. [28], a low-temperature liquid Cu-Se phase is present for the Cu-rich films during selenization and can aid the film growth. Consequently, the difference of Cu content between two films could result in different morphology and grain size. Furthermore, the grain is layered along the cross-section direction, which is more obvious at relatively low selenization temperature. The grain size of the top layer is larger than that of the bottom layer. It could be explained with the degree of selenization. Higher selenization temperature could result in higher degree of selenization [29], which is also consistent with the XRD analysis. Since the reaction proceeds from the surface towards the bulk during selenization, the crystalline quality of the bottom layer increases as the selenization degree increases.

Figure 5 shows the resistivity and carrier concentration of CIGS films selenized from Cu-rich precursor films and Cu-poor precursor films as a function of selenization temperature. All the CIGS films exhibit p-type semiconductor material. In comparison with the Cu-poor films, a significant decrease in resistivity and increase in carrier concentration by several orders of magnitude are observed in the Cu-rich films, which can be explained by the electrical properties of p-type $Cu_{2-x}Se$ phase. The $Cu_{2-x}Se$ is a p-type highly conductive phase, could produce a large number of holes, and thus decreases the resistivity of bulk film. It should be noted that Cu-rich films with such high carrier concentration, which is beyond the optimum range of 10^{16}-10^{17} cm^{-3} [30], cannot be used for devices without further treatment. In both cases of the Cu-rich and Cu-poor films, the resistivity and carrier concentration tend to increase and decrease, respectively, as the selenization temperature increases. Note that the electrical properties of CIGS films are dominated by native defects resulting from deviations from stoichiometric composition, such as vacancies and interstitials [30]. Considering the SEM images presented in Figure 4, increasing selenization temperature results in the improvement of crystallinity and therefore in a decreased total number of defects, which may be related to net carrier concentrations of CIGS films.

4. Conclusions

A two-stage method has been developed to prepare Cu(In, Ga)Se_2 films using Se-containing precursors in Se-vapor selenization process. The effects of precursor composition and selenization temperature on the film properties were investigated. The [Cu]/[In + Ga] and [Ga]/[In + Ga] ratios in CIGS films could be controlled by adjusting the sputtering parameters of the $Cu_{0.8}Ga_{0.2}$ and In_2Se_3 targets. Two sets of films, Cu-poor and Cu-rich films, were intentionally prepared for comparison. It is revealed that the amount of Cu in the precursor films strongly influences the morphology, structural and electrical properties of the subsequent selenized films. XRD analysis shows that Cu(In,Ga)Se_2 solid solution needs higher selenization temperature due to the inclusion of gallium. For Cu-rich films, $Cu_{2-x}Se$ segregation is always

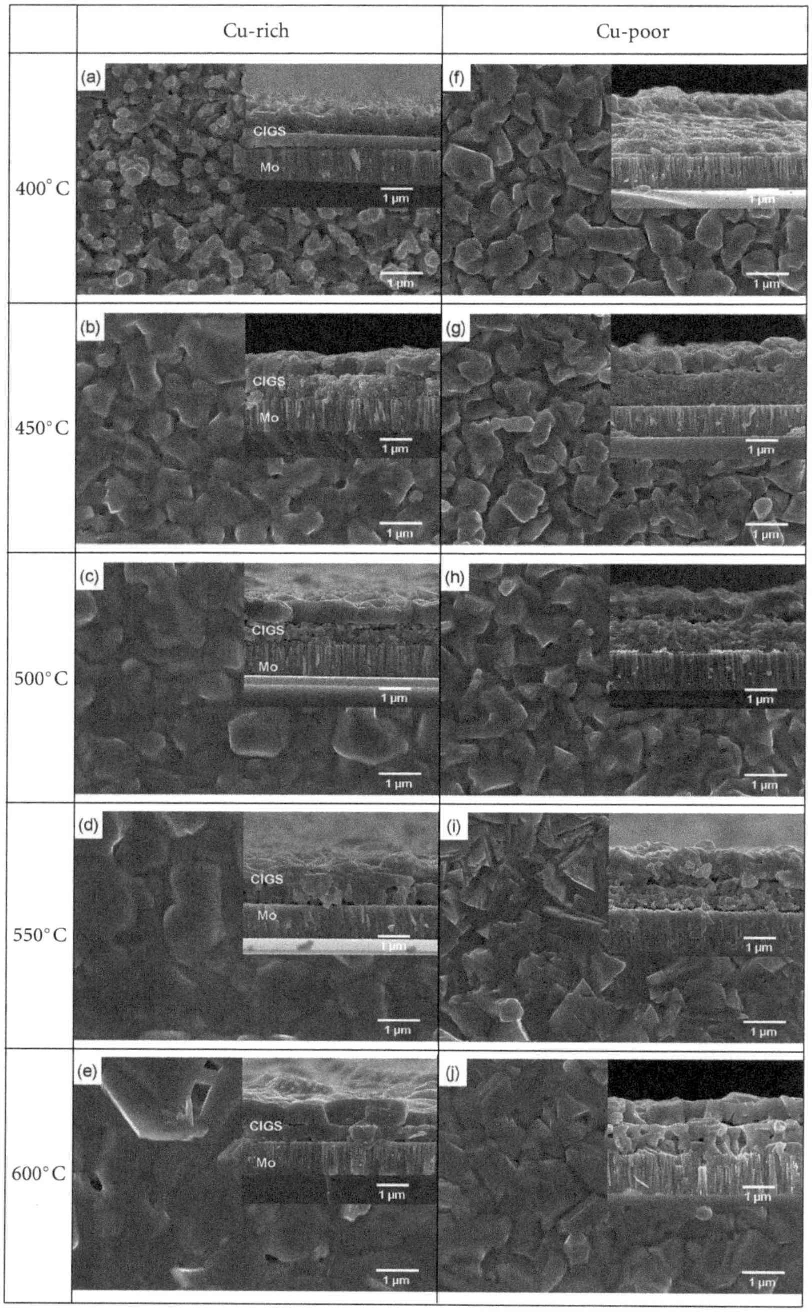

FIGURE 4: Comparison of surface and cross-section (insets) SEM images of CIGS films selenized from Cu-rich precursor films (left column) and Cu-poor precursor films (right column) at different selenization temperatures: (a), (f) 400°C; (b), (g) 450°C; (c); (h) 500°C, (d), (i) 550°C; (e), (j) 600°C.

present, but for Cu-rich films, it exists only when the selenization temperature was 400°C due to inadequate reaction. In comparison with the Cu-poor films, the Cu-rich films have a higher crystallinity due to the presence of low-temperature liquid Cu-Se phase. The crystallinity of the films improves noticeably as the selenization temperature increases. The resistivity of the Cu-rich films is about three orders of magnitude lower than that of the Cu-poor films, which is attributed to the presence of p-type highly conductive $Cu_{2-x}Se$ phase. In both cases of the Cu-rich and Cu-poor films, the dependence of resistivity on selenization temperature is similar and may be due to the change in film crystallinity and native defects.

Acknowledgments

The authors are thankful to Song Jun for equipment maintenance, Li Jun for XRF measurement, and Chen Feng-en

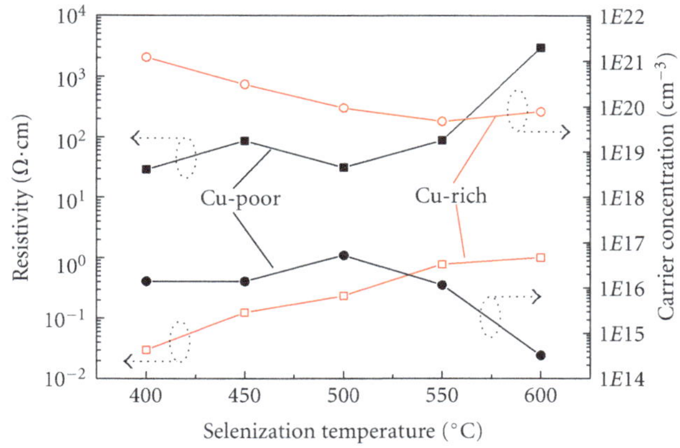

FIGURE 5: Resistivity (squares) and carrier concentration (circles) of CIGS films as a function of selenization temperature. Full and open symbols correspond to CIGS films selenized from Cu-poor precursor films and Cu-rich precursor films, respectively.

for measurement of Raman spectra. This work was financially supported by Initiative Scientific Research Project of Tsinghua University.

References

[1] M. A. Contreras, B. Egaas, K. Ramanathan et al., "Progress toward 20% efficiency in $Cu(In,Ga)Se_2$ polycrystalline thin-film solar cells," *Progress in Photovoltaics*, vol. 7, no. 4, pp. 311–316, 1999.

[2] P. Jackson, D. Hariskos, and E. Lotter, "New world record efficiency for $Cu(In,Ga)Se_2$ thin-film solar cells beyond 20%," *Progress in Photovoltaics*, vol. 19, no. 7, pp. 894–897, 2011.

[3] A. Jager-Waldau, "Progress in chalcopyrite compound semiconductor research for photovoltaic applications and transfer of results into actual solar cell production," *Solar Energy Materials and Solar Cells*, vol. 95, no. 6, pp. 1509–1517, 2011.

[4] M. Kaelin, D. Rudmann, and A. N. Tiwari, "Low cost processing of CIGS thin film solar cells," *Solar Energy*, vol. 77, no. 6, pp. 749–756, 2004.

[5] S. Niki, M. Contreras, I. Repins et al., "CIGS absorbers and processes," *Progress in Photovoltaics*, vol. 18, no. 6, pp. 453–466, 2010.

[6] A. Rao, S. Krishnan, G. Sanjeev, K. Siddappa, H. S. Ullal, and X. Z. Wu, "8 MeV electron irradiation studies on electrical characteristics of $Cu(In,Ga)Se_2$ solar cells," *Solar Energy Materials and Solar Cells*, vol. 93, no. 9, pp. 1618–1623, 2009.

[7] M. A. Green, K. Emery, Y. Hishikawa, and W. Warta, "Solar cell efficiency tables (version 37)," *Progress in Photovoltaics*, vol. 19, no. 1, pp. 84–92, 2011.

[8] O. Lundberg, J. Lu, A. Rockett, M. Edoff, and L. Stolt, "Diffusion of indium and gallium in $Cu(In,Ga)Se_2$ thin film solar cells," *Journal of Physics and Chemistry of Solids*, vol. 64, no. 9-10, pp. 1499–1504, 2003.

[9] C. L. Jensen, D. E. Tarrant, J. H. Ermer, and G. A. Pollock, "Role of gallium in $CuInSe_2$ solar cells fabricated by a two-stage method," in *Proceedings of the 23rd IEEE Photovoltaic Specialists Conference*, pp. 577–580, Louisville, Ky, USA, May 1993.

[10] U. Rau and H. W. Schock, "$Cu(In,Ga)Se_2$ thin-film solar cells," in *Solar Cells: Materials, Manufacture and Operation*, T. Markvart and L. Castaner, Eds., pp. 303–349, Elsevier, Oxford, UK, 2005.

[11] J. R. Tuttle, M. Contreras, A. Tennant, D. Albin, and R. Noufi, "High efficiency thin-film Cu (In, Ga) Se_2-based photovoltaic devices: progress towards a universal approach to absorber fabrication," in *Proceedings of the 23rd IEEE Photovoltaic Specialists Conference*, pp. 415–421, Louisville, Ky, USA, May 1993.

[12] M. Venkatachalam, M. D. Kannan, S. Jayakumar, R. Balasundaraprabhu, and N. Muthukumarasamy, "Effect of annealing on the structural properties of electron beam deposited CIGS thin films," *Thin Solid Films*, vol. 516, no. 20, pp. 6848–6852, 2008.

[13] P. Jackson, R. Würz, U. Rau et al., "High quality baseline for high efficiency, $Cu(In_{1-x},Ga_x)Se_2$ solar cells," *Progress in Photovoltaics*, vol. 15, no. 6, pp. 507–519, 2007.

[14] K. Ramanathan, G. Teeter, J. C. Keane, and R. Noufi, "Properties of high-efficiency $CuInGaSe_2$ thin film solar cells," *Thin Solid Films*, vol. 480-481, pp. 499–502, 2005.

[15] M. A. Contreras, K. Ramanathan, J. AbuShama et al., "Diode characteristics in state-of-the-art ZnO/CdS/ $Cu(In_{1-x}Ga_x)Se_2$ solar cells," *Progress in Photovoltaics*, vol. 13, no. 3, pp. 209–216, 2005.

[16] V. F. Gremenok, E. P. Zaretskaya, V. B. Zalesski et al., "Preparation of $Cu(In,Ga)Se_2$ thin film solar cells by two-stage selenization processes using N_2 gas," *Solar Energy Materials and Solar Cells*, vol. 89, no. 2-3, pp. 129–137, 2005.

[17] J. H. Shi, Z. Q. Li, D. W. Zhang, Q. Q. Liu, Z. Z. Sun, and S. M. Huang, "Fabrication of Cu(In, Ga)Se_2 thin films by sputtering from a single quaternary chalcogenide target," *Progress in Photovoltaics*, vol. 19, no. 2, pp. 160–164, 2011.

[18] V. Alberts, S. Zweigart, and H. W. Schock, "Preparation of device quality $CuInSe_2$ by selenization of Se-containing precursors in H_2Se atmosphere," *Semiconductor Science and Technology*, vol. 12, no. 2, pp. 217–223, 1997.

[19] W. Witte, R. Kniese, and M. Powalla, "Raman investigations of $Cu(In,Ga)Se_2$ thin films with various copper contents," *Thin Solid Films*, vol. 517, no. 2, pp. 867–869, 2008.

[20] N. Barreau, J. Lähnemann, F. Couzinié-Devy, L. Assmann, P. Bertoncini, and J. Kessler, "Impact of Cu-rich growth on the $CuIn_{1-x}Ga_xSe_2$ surface morphology and related solar cells behaviour," *Solar Energy Materials and Solar Cells*, vol. 93, no. 11, pp. 2013–2019, 2009.

[21] W. Li, Y. Sun, Y. Wang, H. Cai, F. Liu, and Q. He, "Effects of substrate temperature on the properties of facing-target sputtered Al-doped ZnO films," *Solar Energy Materials and Solar Cells*, vol. 91, no. 8, pp. 659–663, 2007.

[22] A. A. Kadam and N. G. Dhere, "Highly efficient $CuIn_{1-x}Ga_xSe_{2-y}S_y$/CdS thin-film solar cells by using diethylselenide as selenium precursor," *Solar Energy Materials and Solar Cells*, vol. 94, no. 5, pp. 738–743, 2010.

[23] R. Caballero and C. Guillén, "Comparative studies between Cu-Ga-Se and Cu-In-Se thin film systems," *Thin Solid Films*, vol. 403-404, pp. 107–111, 2002.

[24] W. Witte, R. Kniese, A. Eicke, and M. Powalla, "Influence of the GA content on the Mo/$Cu(In,Ga)Se_2$ interface formation," in *Proceedings of the 4th IEEE World Conference on Photovoltaic Energy Conversion*, vol. 1, pp. 553–556, May 2006.

[25] T. Delsol, A. P. Samantilleke, N. B. Chaure, P. H. Gardiner, M. Simmonds, and I. M. Dharmadasa, "Experimental study of graded bandgap $Cu(InGa)(SeS)_2$ thin films grown on glass/molybdenum substrates by selenization and sulphidation," *Solar Energy Materials and Solar Cells*, vol. 82, no. 4, pp. 587–599, 2004.

[26] E. P. Zaretskaya, V. F. Gremenok, V. Riede et al., "Raman spectroscopy of $CuInSe_2$ thin films prepared by selenization," *Journal of Physics and Chemistry of Solids*, vol. 64, no. 9-10, pp. 1989–1993, 2003.

[27] B. J. Stanbery, "Copper indium selenides and related materials for photovoltaic devices," *Critical Reviews in Solid State and Materials Sciences*, vol. 27, no. 2, pp. 73–117, 2002.

[28] R. Klenk, T. Walter, H. W. Schock, and D. Cahen, "A model for the successful growth of polycrystalline films of $CuInSe_2$ by multisource physical vacuum evaporation," *Advanced Materials*, vol. 5, no. 2, pp. 114–119, 1993.

[29] Y. Goushi, H. Hakuma, K. Tabuchi, S. Kijima, and K. Kushiya, "Fabrication of pentanary $Cu(InGa)(SeS)_2$ absorbers by selenization and sulfurization," *Solar Energy Materials and Solar Cells*, vol. 93, no. 8, pp. 1318–1320, 2009.

[30] R. Noufi, R. Axton, C. Herrington, and S. K. Deb, "Electronic properties versus composition of thin films of $CuInSe_2$," *Applied Physics Letters*, vol. 45, no. 6, pp. 668–670, 1984.

Effect of PdS on Photocatalytic Hydrogen Evolution of Nanostructured CdS under Visible Light Irradiation

Qingyun Chen, Cheng Suo, Shu Zhang, and Yunhai Wang

State Key Laboratory of Multiphase Flow in Power Engineering, School of Energy and Power Engineering, Xi'an Jiaotong University, Xi'an 710049, China

Correspondence should be addressed to Qingyun Chen; qychen@mail.xjtu.edu.cn

Academic Editor: Elias Stathatos

To investigate the effect of PdS as a cocatalyst for photocatalytic hydrogen evolution, nanostructured PdS/CdS were prepared by an in situ coprecipitation and hydrothermal method, respectively. The as-prepared photocatalysts were characterized by transmission electron microscopy (TEM), X-ray diffraction (XRD), UV-visible absorption spectra, and photoluminescence spectra (PL). With PdS highly dispersed in the CdS nanostructures, the photoactivity was evaluated by hydrogen evolution from aqueous solution containing Na_2S/Na_2SO_3 as sacrificial reagents under visible light irradiation. When the concentration of PdS was 1% by weight, PdS/CdS, prepared by the in situ coprecipitation, showed the highest photocatalytic activity, while that prepared by hydrothermal method showed the most stability for hydrogen evolution. The effect of highly dispersed PdS on the photoactivity was discussed.

1. Introduction

Hydrogen, as a kind of renewable energy, has attracted more and more attention due to the depletion of fossil fuels and the pollution caused by continuously increasing energy demands [1–3]. Photocatalytic water-splitting hydrogen evolution using semiconducting photocatalysts is one of the most ideal ways to make full use of solar energy to produce hydrogen [4]. Up to now, a lot of semiconducting photocatalysts have been reported to produce hydrogen, such as TiO_2 [5], CdS [6], and ZnS [7]. Among them, CdS has been widely studied because of its excellent light absorption property in the visible region [8].

CdS alone demonstrates low photoactivity and bad stability for hydrogen evolution under visible light because, in its photocatalytic process, the phenomenon of photocorrosion is prone to occur [9]. Many efforts have been made to overcome the problem of photocorrosion, such as photosensitizing CdS with photosensitizers [4, 10], combining CdS with another semiconductor [11], and exploiting alternative preparation approaches to control the size and shape of CdS [12].

Loading a cocatalyst onto CdS is also a good method to improve its photoactivity and stability. A suitable cocatalyst cannot only promote the separation of photoexcited electrons and holes and suppress their recombination, but also decrease the activation potentials for H_2 evolution [13]. Transition metals and their oxides (sulfides) have been widely used as cocatalysts including Pd [14–16], Pt [17, 18], NiO [19], RuO_2 [20], and MoS_2 [21]. Therefore, it is highly desirable to find suitable cocatalysts to improve the photoactivity and stability of CdS. Recently, Yan et al. reported that CdS loaded with Pt-PdS bicomponent cocatalyst achieved the highest quantum yield of about 93% in the presence of sacrificial reagents under visible light irradiation [22]. Therefore, it would be of great interest to investigate the effect of the cocatalyst on the photocatalytic performance of CdS.

In this paper, nanostructured PdS/CdS photocatalysts were synthesized by in situ coprecipitation and hydrothermal method, respectively. The effect of PdS on the photoactivity and stability of CdS was investigated for hydrogen evolution under visible light from water containing Na_2S and Na_2SO_3 as sacrificial reagents.

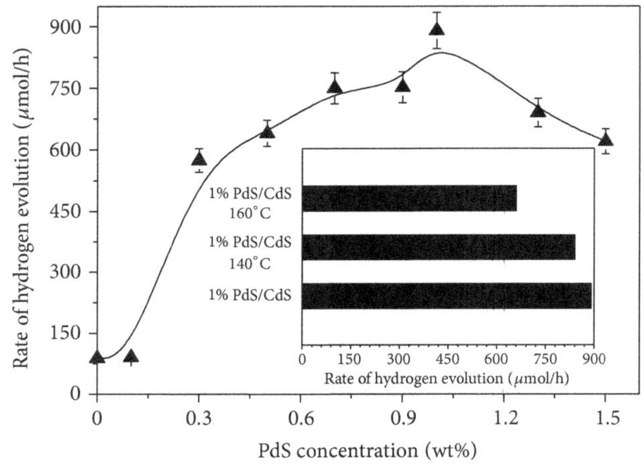

Figure 1: Rate of hydrogen evolution on PdS/CdS samples loaded with different amounts of PdS under visible light; (insert) rate of hydrogen evolution on PdS/CdS samples loaded with 1 wt% amount of PdS prepared by different hydrothermal temperatures and in situ coprecipitation.

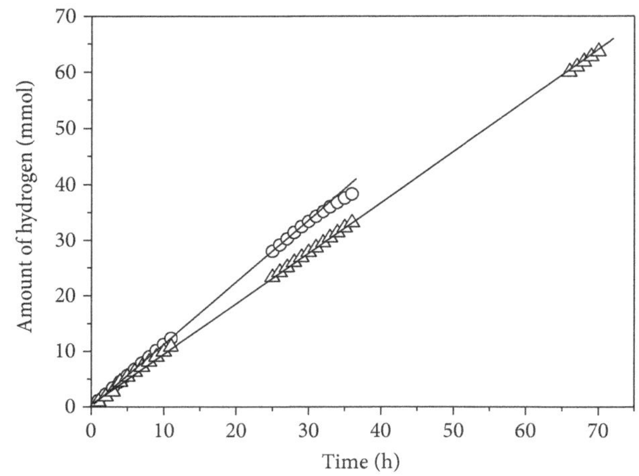

Figure 2: Time courses of hydrogen evolution on PdS/CdS samples loaded with 1 wt% PdS prepared by different methods.

2. Experiments

2.1. Synthesis of PdS/CdS. All of the chemicals were of analytical grade and used without further purification. Deionized water was used in the sample synthesis.

The nanostructured CdS particles were synthesized as follows: an aqueous solution of 0.14 M Na_2S was added slowly to 0.14 M $Cd(OAc)_2$ solution under vigorous stirring at room temperature. The yellow mixture was stirred for 12 h and kept for additional 12 h. The resulting yellow solid was collected by filtration and washed with deionized water several times. The wet solid was suspended in deionized water and transferred into a 100 mL stainless Teflon-lined autoclave. The autoclave was sealed and heated at 200°C for 24 h and then cooled down to room temperature naturally. The yellow solid was filtered and washed with water and ethanol subsequently, followed by drying at 80°C for 24 h.

The loading of PdS on CdS producing PdS/CdS was realized by the following two methods. For the in situ coprecipitation, a $PdCl_2$ aqueous solution was added dropwise under stirring at room temperature to a suspension of CdS nanoparticles (0.2 g, as prepared above) dispersed in Na_2S aqueous solution just before the photocatalytic reaction. For the hydrothermal method, the mixture obtained by the in situ coprecipitation was stirred for 12 h. The resulting precipitates were centrifuged and washed with deionized water several times. The precipitates were suspended in deionized water, transferred into a 100 mL stainless Teflon-lined autoclave, and heated at a certain temperature for 6 h.

2.2. Characterization. The X-ray diffraction (XRD) patterns of catalysts were obtained from a Panalytical X'pert Pro X-ray diffractometer equipped with Cu Kα irradiation with the scanning step of 0.05°/s; the operation voltage and current were 45 kV and 40 mA, respectively. UV-visible absorption spectra were measured by a HITACHI UV4100 instrument, with the scanning range from 300 to 800 nm. The photoluminescence spectra (PL) were measured at room temperature using a PTI QM-4 fluorescence spectrophotometer. The crystallite morphologic micrograph was observed on a high resolution transmission electron microscopy (HRTEM) JEOL JEM-3010 instrument and a field emission scanning electron microscopy (SEM) JSM-6700F (Japan).

2.3. Evaluation of Photocatalytic Activity. Photocatalytic reaction was carried out in a side-irradiation Pyrex cell. The effective irradiation area for the cell is 12.56 cm^2. The powder of photocatalyst (0.2 g) was dispersed by a magnetic stirrer in an aqueous solution (200 mL) consisting of Na_2S (0.5 M) and Na_2SO_3 (0.5 M) as electron donors in the cell. The photocatalysts were irradiated with visible light through a cutoff filter ($\lambda > 430$ nm, $T = 65\%$) from a 300 W Xe lamp. The amount of H_2 gas was determined by an online thermal conductivity detector (TCD) gas chromatograph (NaX zeolite column, nitrogen as a carrier gas).

3. Results and Discussion

Figure 1 shows the rate of hydrogen evolution on PdS/CdS photocatalysts loaded with different amounts of PdS prepared by the in situ coprecipitation under visible light, together with those prepared by the hydrothermal method (insert in Figure 1). As the amount of PdS increases from 0 to 1%, the hydrogen evolution rate on PdS/CdS increases markedly from ~88.1 μmol/h to ~890.3 μmol/h, about 10 times higher. However, when the amount of PdS increases to 1.3%, the photocatalytic activity decreases slightly. The surplus PdS can work as an optical filter to shield incident light and hence suppress further enhancement of photocatalytic activity for hydrogen evolution. For comparing, the insert in Figure 1 shows the rate of hydrogen evolution on the PdS/CdS photocatalyst loaded with 1 wt% PdS prepared by hydrothermal

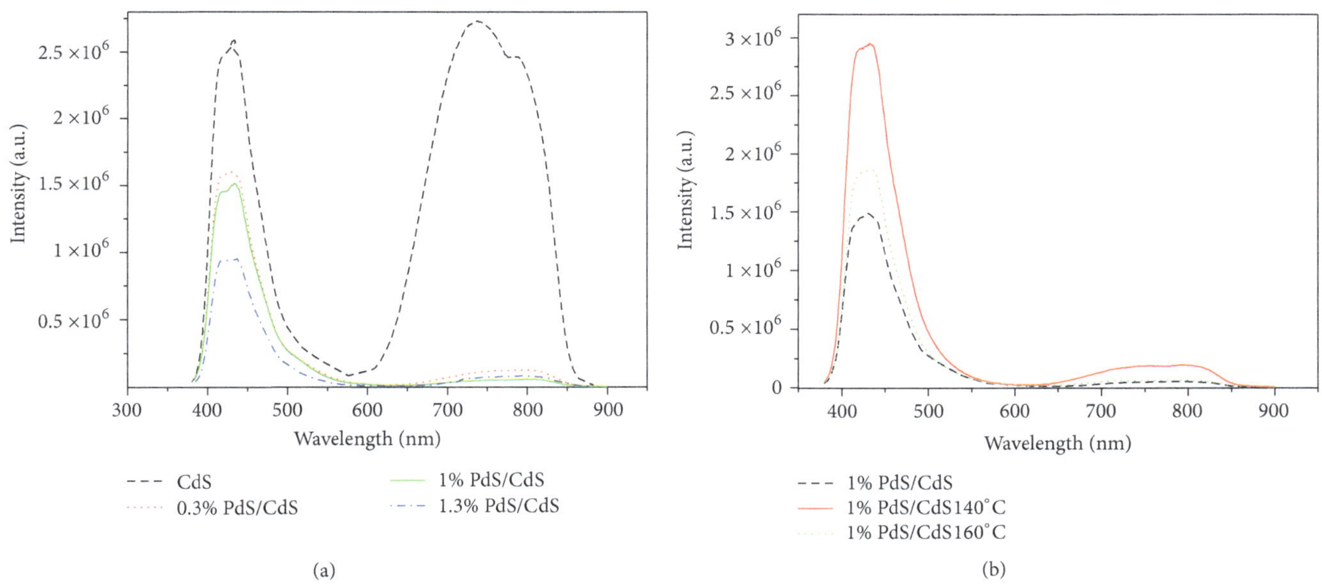

FIGURE 3: (a) Photoluminescence spectra of PdS/CdS samples loaded with different amounts of PdS. (b) Photoluminescence spectra of PdS/CdS samples loaded with 1 wt% PdS prepared by different methods.

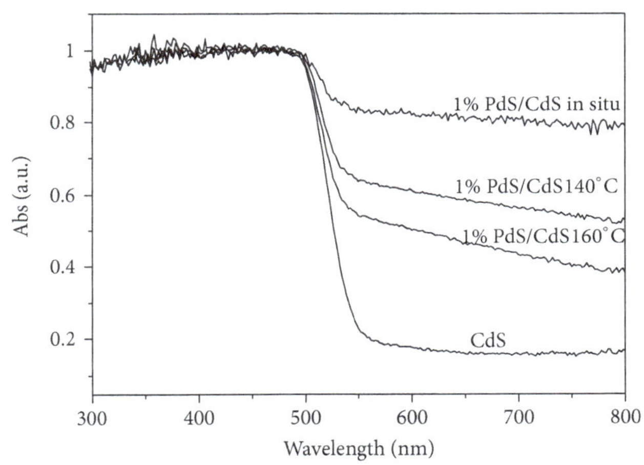

FIGURE 4: UV-visible absorption spectra of CdS and PdS/CdS samples.

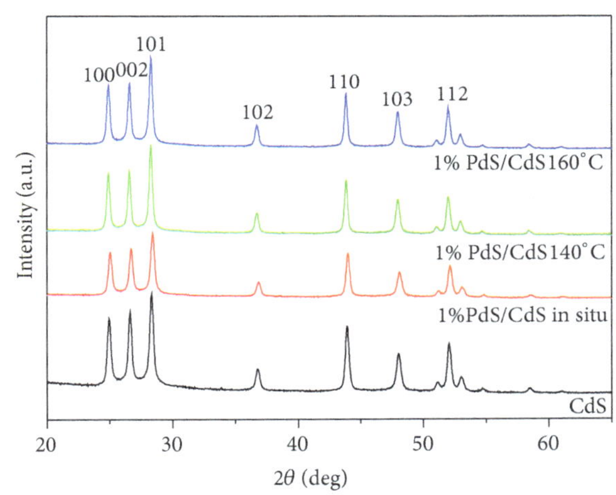

FIGURE 5: XRD patterns of PdS/CdS samples.

method. It appears that the photocatalytic activity also decreases slightly. But the photocatalytic stability improves, as shown in Figure 2 which shows the hydrogen evolution on PdS/CdS photocatalysts loaded with 1% PdS prepared by different methods as long as Na_2S/Na_2SO_3 is present. PdS/CdS photocatalysts prepared by the in situ coprecipitation can act stably for less than 40 hours. It is noted that PdS/CdS photocatalysts prepared by hydrothermal method can exhibit significantly improved stability, showing steady hydrogen production more than 70 hours. The observation suggests that the dispersion of PdS in nanostructured CdS can improve the photoactivity, but the stable photocatalytic activity for hydrogen evolution can be influenced by the synthesis method.

To analyze the reason, Figure 3 shows the photoluminescence spectra of PdS/CdS photocatalysts. The photoluminescence spectroscopy can give information about the photoexcited energy/electron transfer and recombination process, and it has been widely used to investigate the photophysical and photochemical properties of photocatalysts [23]. As shown in Figure 3(a), the pure CdS exhibits broad fluorescence peak related to Stoke's shifts ~435 nm and ~750 nm [4]. The band at ~432 nm is due to intrinsic emission, whereas that at 750 nm originates from transition of electrons rapped at surface state to the valance band of CdS. This fact suggests that nanostructured CdS has defects. By loading PdS, the fluorescence intensities are quenched, which reveals that the recombination of photoelectrons and holes is efficiently suppressed resulting in the improved photoactivity.

FIGURE 6: TEM image of (a and b) CdS, (c) PdS/CdS prepared by in situ coprecipitation, and (d) PdS/CdS prepared by hydrothermal method at 160°C.

However, the fluorescence intensities of PdS/CdS prepared by hydrothermal method are slightly enhanced as shown in Figure 3(b). This fact might be beneficial for the stability of photocatalysts. Then it can be concluded that the weaker the fluorescence intensity of the photocatalyst is, the higher the photoactivity is.

The UV-visible absorption spectra for PdS/CdS are shown in Figure 4. The pure CdS has an absorption edged at about 550 nm, with energy gap estimated to be 2.3 eV. The spectra of PdS/CdS show an enhanced absorption in visible light region. Such an enhancement can be assigned to the existence of PdS phase. It was reported that PdS could not absorb visible light [4]. Therefore, the fact suggests that the PdS phase can enhance the absorption of CdS phase. From Figure 4, it can be seen that the enhancement of the absorption in visible light region is different which is dependent on the preparation condition. It suggests that the higher the visible adsorption is, the higher the photoactivity is.

The XRD patterns of various samples displayed in Figure 5 can be readily indexed as the (100), (002), (101), (110), (103), and (112) planes of the hexagonal phase of the wurtzite CdS structure, which are well matched with the standard values (JCPDS Card No. 02-0549) according to the main diffraction peaks at 24.9°, 26.6°, 28.4°, 43.7°, 48.1°, and 52.0°. It is worth noting that there are no characteristic peaks associated with PdS in these XRD patterns, which may be due to the low concentration of PdS and its relatively low crystallinity for the PdS/CdS samples prepared by the in situ coprecipitation and due to the high dispersion of PdS particles in the nanostructured CdS and the low concentration of PdS for the PdS/CdS samples prepared by hydrothermal method. By the way the diffraction peak would not shift when Pd^{2+} was doped into the lattice of CdS, considering that an ionic radius of Pd^{2+} (0.88 Å) is lower than that of Cd^{2+} (0.97 Å). The fact indicates that no significant lattice deformation occurred with different treatments of PdS/CdS samples.

Morphology of the as-prepared photocatalysts is shown in Figure 6. From Figure 6(a), it can be seen that the nanostructured CdS is mainly composed of nanoshperes. As shown in Figure 6(b), for the higher magnification of CdS, these nanospheres are well crystallized. When PdS is loaded, there appears spherical nanoparticles on the surface of CdS which disperse more uniformly by comparing Figures 6(b) and 6(c). PdS by itself displays no photoactivity for H_2 production, but PdS can act as a cocatalyst together with CdS. Then it suggests that PdS exists on the surface of CdS which can enhance the photoactivity of CdS. However, there appears a nanoparticle layer for PdS/CdS prepared by hydrothermal method as shown in Figure 6(d). This indictes that the loading of PdS can improve the photoactivity of CdS, but the nanoparticle layer may decrease the photoactivity in spite of the increase in the stability of CdS.

4. Conclusion

Nanostructured PdS/CdS photocatalysts were synthesized by an in situ coprecipitation and a hydrothermal method, respectively. With a concentration of PdS approximately 1.0%, PdS/CdS photocatalysts synthesized by the in situ coprecipitation displayed the highest photocatalytic activity for hydrogen evolution under visible light irradiation from an aqueous solution containing sulfide and sulfite. However, PdS/CdS photocatalysts synthesized by the hydrothermal method showed the highest stability for hydrogen evolution. This study indicates that it is an effective method for preparing a stable, efficient photocatalyst by using a suitable cocatalyst.

Acknowledgments

This work was financially supported by the National Basic Research Program of China (2009CB220000), Research Fund for the Doctoral Program of Higher Education of China (20110201120042), and Natural Science Foundation of China (21206133).

References

[1] X. B. Chen, S. H. Shen, L. J. Guo, and S. S. Mao, "Semiconductor-based photocatalytic hydrogen generation," *Chemical Reviews*, vol. 11, pp. 6503–6570, 2010.

[2] C. L. Li, J. Yuan, B. Y. Han, and W. F. Shanguan, "Synthesis and photochemical performance of morphology-controlled CdS photocatalysts for hydrogen evolution under visible light," *International Journal of Hydrogen Energy*, vol. 36, no. 7, pp. 4271–4279, 2011.

[3] S. H. Shen, P. H. Guo, J. W. Shi, and L. J. Guo, "Visible-light-driven photocatalytic water splitting on nanostructured semiconducting materials," *International Journal of Nanotechnology*, vol. 8, no. 6-7, pp. 523–591, 2011.

[4] S. Zhang, Q. Y. Chen, D. W. Jing, Y. H. Wang, and L. J. Guo, "Visible photoactivity and antiphotocorrosion performance of PdS-CdS photocatalysts modified by polyaniline," *International Journal of Hydrogen Energy*, vol. 37, pp. 791–796, 2012.

[5] A. Fujishima and K. Honda, "Electrochemical photolysis of water at a semiconductor electrode," *Nature*, vol. 238, no. 5358, pp. 37–38, 1972.

[6] E. Borgarello, N. Serpone, E. Pelizzetti, and M. Barbeni, "Efficient photochemical conversion of aqueous sulphides and sulphites to hydrogen using a rhodium-loaded CdS photocatalyst," *Journal of Photochemistry*, vol. 33, no. 1, pp. 35–48, 1986.

[7] I. Tsuji and A. Kudo, "H_2 evolution from aqueous sulfite under visible-light irradiation over Pb and halogen-codoped ZnS photocatalysts," *Journal of Photochemistry and Photobiology A*, vol. 156, no. 1–3, pp. 249–252, 2003.

[8] M. Matsumura, S. Furukawa, Y. Saho, and H. Tsubomura, "Cadmium sulfide photocatalyzed hydrogen production from aqueous solutions of sulfite: effect of crystal structure and preparation method of the catalyst," *Journal of Physical Chemistry*, vol. 89, no. 8, pp. 1327–1329, 1985.

[9] K. G. Kanade, B. Jin-OoK, D. P. Amalnerkar, and B. B. Kale, "Nano-CdS by polymer-inorganic solid-state reaction: visible light pristine photocatalyst for hydrogen generation," *Materials Research Bulletin*, vol. 41, no. 12, pp. 2219–2225, 2006.

[10] H. Zhang and Y. F. Zhu, "Significant visible photoactivity and antiphotocorrosion performance of CdS photocatalysts after monolayer polyaniline hybridization," *Journal of Physical Chemistry C*, vol. 114, no. 13, pp. 5822–5826, 2010.

[11] J. S. Jang, S. M. Ji, S. W. Bae, H. C. Son, and J. S. Lee, "Optimization of CdS/TiO_2 nano-bulk composite photocatalysts for hydrogen production from Na_2S/Na_2SO_3 aqueous electrolyte solution under visible light ($\lambda \geq 420$ nm)," *Journal of Photochemistry and Photobiology A*, vol. 188, no. 1, pp. 112–119, 2007.

[12] A. Eychmuller, "Structure and photophysics of semiconductor nanocrystals," *The Journal of Physical Chemistry B*, vol. 104, pp. 6514–6528, 2000.

[13] J. S. Jang, D. J. Ham, N. Lakshminarasimhan, W. Y. Choi, and J. S. Lee, "Role of platinum-like tungsten carbide as cocatalyst of CdS photocatalyst for hydrogen production under visible light irradiation," *Applied Catalysis A*, vol. 346, no. 1-2, pp. 149–154, 2008.

[14] I. B. Rufus, V. Ramakrishnan, B. Viswanathan, and J. C. Kuriacose, "Photoelctrochemical and photocatalytic studies on Pd/CdS," *Indian Journal of Technology*, vol. 27, pp. 171–173, 1989.

[15] J. B. Zhong, Y. Lu, W. D. Jiang et al., "Characterization and photocatalytic property of Pd/TiO_2 with the oxidation of gaseous benzene," *Journal of Hazardous Materials*, vol. 168, no. 2-3, pp. 1632–1635, 2009.

[16] A. Khalil, M. A. Gondal, and M. A. Dastageer, "Augmented photocatalytic activity of palladium incorporated ZnO nanoparticles in the disinfection of Escherichia coli microorganism from water," *Applied Catalysis A*, vol. 402, no. 1-2, pp. 162–167, 2011.

[17] B. Neppolian, A. Bruno, C. L. Bianchi, and M. Ashokkumar, "Graphene oxide based $Pt-TiO_2$ photocatalyst: ultrasound assisted synthesis, characterization and catalytic efficiency," *Ultrasonics Sonochemistry*, vol. 19, pp. 9–15, 2011.

[18] N. Sahu, S. N. Upadhyay, and A. S. K. Sinha, "Kinetics of reduction of water to hydrogen by visible light on alumina supported Pt-CdS photocatalysts," *International Journal of Hydrogen Energy*, vol. 34, no. 1, pp. 130–137, 2009.

[19] J. G. Yu, W. G. Wang, and B. Cheng, "Synthesis and enhanced photocatalytic activity of a hierarchical porous flowerlike p-n junction NiO/TiO_2 photocatalyst," *Chemistry*, vol. 5, no. 12, pp. 2499–2506, 2010.

[20] A. A. Ismail, L. Robben, and D. W. Bahnemann, "Study of the efficiency of UV and visible-light photocatalytic oxidation of methanol on mesoporous RuO_2-TiO_2 nanocomposites," *ChemPhysChem*, vol. 12, no. 5, pp. 982–991, 2011.

[21] X. Zong, H. J. Yan, G. P. Wu et al., "Enhancement of photocatalytic H_2 evolution on CdS by loading MoS_2 as cocatalyst under visible light irradiation," *Journal of the American Chemical Society*, vol. 130, no. 23, pp. 7176–7177, 2008.

[22] H. J. Yan, J. H. Yang, G. J. Ma et al., "Visible-light-driven hydrogen production with extremely high quantum efficiency on Pt-PdS/CdS photocatalyst," *Journal of Catalysis*, vol. 266, no. 2, pp. 165–168, 2009.

[23] H. P. Liu, K. Zhang, D. W. Jing, G. J. Liu, and L. J. Guo, "SrS/CdS composite powder as a novel photocatalyst for hydrogen production under visible light irradiation," *International Journal of Hydrogen Energy*, vol. 35, no. 13, pp. 7080–7086, 2010.

23

Optimal *I-V* Curve Scan Time of Solar Cells and Modules in Light of Irradiance Level

Matic Herman, Marko Jankovec, and Marko Topič

Laboratory of Photovoltaics and Optoelectronics, Faculty of Electrical Engineering, University of Ljubljana, Tržaška Cesta 25, 1000 Ljubljana, Slovenia

Correspondence should be addressed to Matic Herman, matic.herman@fe.uni-lj.si

Academic Editor: Wayne A. Anderson

High-efficiency solar cells and modules exhibit strong capacitive character resulting in limited speed of transient responses. A too fast *I-V* curve measurement can thus introduce a significant error due to its internal capacitances. This paper analyses the *I-V* curve error of a measured solar cell or module in light of scan time and irradiance level. It rests on a two-diode solar cell model extended by two bias-dependent capacitances, modelling the junction, and the diffusion capacitance. A method for determination of all extended model parameters from a quasistatic *I-V* curve and open-circuit voltage decay measurement is presented and validated. Applicability of the extended model and the developed parameter extraction method to PV modules is demonstrated and confirmed. SPICE simulations of the extended model are used to obtain the *I-V* curve error versus scan time dependence and the *I-V* curve hysteresis. Determination of the optimal scan time is addressed, and finally the influence of the irradiance level on the *I-V* curve scan time and error is revealed. The method is applied but is not limited to three different wafer-based silicon solar cell types.

1. Introduction

The *I-V* curve measurement is an essential performance characterization technique for solar cells and modules as two-terminal DC generators. Either in dark or under illumination, the measurement concept is simple: during the sweep of the terminals of the device under test (DUT) from open circuit to short circuit or vice versa, voltage and current should be measured in quasi static conditions. It can be mistakenly assumed that PV generators exhibit negligible internal capacitance. In particular, the last generation thin film solar cells (SCs) and high-efficiency crystalline silicon SCs exhibit extremely high internal capacitances [1]. The internal capacitance together with the series, parallel, and diode differential resistance form an RC circuit that introduces a transient time constant into the measurement process. The time constant determines the quasi static condition. Its magnitude depends on various parameters such as the operating point (voltage and current), temperature, irradiance level, minority carrier lifetime, and other semiconductor parameters only to mention the most important ones.

While sweeping the *I-V* curve from point to point, the transient behaviour needs to fade out prior to the measurement being error-free. The charge of the capacitor in the DUT needs to equalize at every point of the measurement. An inappropriate *I-V* scan time can introduce a significant error into the *I-V* measurement. When the targeted error is to be kept below a predefined threshold, the *I-V* curve scan time is downward limited and determined by the SC's dynamic properties. On the other hand, there are several factors determining the scan time upper limit. Foremost, the scan time is dictated by the *I-V* curve measurement device. For resistive/semiconductor-based curve tracers, the upper limit is defined by the safe operating area (SOA) of the load (usually an MOSFET). For capacitive curve tracers, the capacitor charging time is a fixed quotient of the measurement load capacitance and the current capabilities of the DUT. At indoor measurements, the *I-V* curve scan time is upward limited by the duration of homogenous light pulse guaranteed by a flash solar simulator. Measurements under a continuous solar simulator or under the sun can further suffer from errors induced by radiation heating of the specimen [2]; thus, the scan time should not be too long.

To sum up, the optimal scan time should be as short as possible but long enough for the dynamic error to be kept below the predefined error threshold. A too short scan time can be easily detected by scanning the curve in both directions [3]: forward—a scan from short circuit current I_{SC} to open circuit voltage V_{OC}; reverse—from V_{OC} to I_{SC}. A hysteresis can be observed demonstrating underestimation of the real *I-V* curve in the forward direction and overestimation in the reverse direction (Figure 1(a)).

The resulting error cannot be eliminated neither by averaging both *I-V* curves by voltage nor summing the errors (determined by the error function given in Section 2) of both scan directions (Figure 1(b)). In particular, at too short scan times, the mismatch is high due to $R_s dI/dt$ term, where R_s is the series resistance of the DUT. The term that determines the error at a fixed time constant value varies at different sweep directions [3]. Furthermore, in several cases (capacitive or inductive load), the detection of the error by sweeping in both directions is not applicable; thus, the determination of required scan time or characterization of dynamic error is even more important.

Measurements under solar simulators are based on a calibrated light source that emits as close as possible to $1000\,W/m^2$ as specified in the standard test conditions (STCs) [4]. However, in measurements under the sun, the irradiance levels can vary over a broad range. The IEC 60409-10 standard defines the procedures for scaling the *I-V* curves to STC [5], but does the irradiance level impact the *I-V* measurement error or is there a need for a different scan time? We will address these aspects in our quest for the optimal scan time.

The hysteresis and the effect of the capacitive character of solar cells are well known and reported [3, 6]. Several variations of the dynamic model exist, where most authors use a one-diode model including one or two capacitances [1, 6, 7], while one author uses a two-diode model and two capacitances [8]. None of the presented models were applied and verified on different solar cell technologies. Our previous research has shown that a two-diode model should be used for the optimal extraction of the junction capacitance [9]. The aim of this paper is to analyse the *I-V* curve error of a solar cell or module under test in light of scan time and irradiance level. The analysis is accomplished through SPICE simulation utilizing a two-diode model extended by two nonlinearly bias-dependent capacitors modelling the dynamic character of PV generators. A method for parameter extraction of the presented extended model is developed and evaluated. Whether the model suitably characterizes not only solar cells but also PV modules is discussed in theory and practice. An algorithm for scan time determination is developed, and finally the influence of the irradiance level on the *I-V* curve scan time and error is revealed. Results are demonstrated on three different solar cells of three different technologies: polycrystalline silicon (pc-Si), monocrystalline silicon (mc-Si) and back-contacted monocrystalline silicon (BC mc-Si).

2. Measurement and Modelling Methodology

2.1. Error Determination. The optimal scan time is the minimal time at which the error between a reference curve measured in true quasi static conditions and a curve under the influence of a transient error does not exceed a predefined error threshold. Determination of the error between the reference and the compared *I-V* curve is an essential yet delicate procedure that needs to be established for further analysis. The comparison of *I-V* curves is difficult due to the wide curve slope span (dI/dV) from I_{SC} to V_{OC}. Neither the current difference at equidistant voltage intervals nor the voltage difference at equidistant current intervals is appropriate. A possible comparison quantity is the maximum peak power P_{MPP}, although it only relates to single points on both curves and does not provide alignment information of the entire *I-V* curve. We suggest a method that delivers overall *I-V* curve comparison, with the error equally weighted throughout the curve form I_{SC} to V_{OC}.

Let the reference curve be stored in the $\mathbf{M}_{IV}$ matrix, and let the compared curve be stored in the $\mathbf{N}_{IV}$ matrix, where I denotes the current, and V denotes the voltage column of the two-column matrices. The error is derived from a perspective of the available power of the PV generator at a specified load. The points for error calculation are selected as two intersections of a variable linear resistive load characteristics and both curves being compared (ith point $\mathbf{O}_{IV}(i)$ on the reference and ith point $\mathbf{N}_{IV}(i)$ on the compared curve, Figure 2).

The characteristics of the variable resistive load is defined as $I = \mathbf{g}(i) \cdot V$. Foremost, the vector $\mathbf{g}$ is calculated as $\mathbf{g}(i) = \mathbf{N}_I(i)/\mathbf{N}_V(i)$ for each of N points of the compared curve. The reference curve is defined discretely; thus, interpolation is required. For every line $I = \mathbf{g}(i) \cdot V$ going through the $\mathbf{N}_{IV}(i)$, the intersection point on the reference curve is interpolated (1st order) from the nearest points ($\mathbf{M}_{IV}(j)$ and $\mathbf{M}_{IV}(j+1)$) and stored into the $\mathbf{O}_{IV}(i)$ matrix. Finally, the components of the error vector can be calculated from $\mathbf{N}_{IV}(i)$ and $\mathbf{O}_{IV}(i)$ point pairs:

$$\mathbf{E}(i) = \sqrt{\left(\frac{\mathbf{N}_V(i) - \mathbf{O}_V(i)}{V_{OCref}}\right)^2 + \left(\frac{\mathbf{N}_I(i) - \mathbf{O}_I(i)}{I_{SCref}}\right)^2}, \quad i = 1, \ldots, N, \qquad (1)$$

where V_{OCref} and I_{SCref} denote the open circuit voltage and the short circuit current of the reference curve.

The maximum value of the error vector $E = \max(\mathbf{E}(i))$ is chosen as a qualitative comparison factor to obtain information on the maximum error between the two curves with i typically appearing close to the maximum power point (also seen in Figure 1).

2.2. The Solar Cell Model. The concept of scan time determination is based on a two-diode solar cell model extended by two parallel capacitances modelling the dynamic processes in

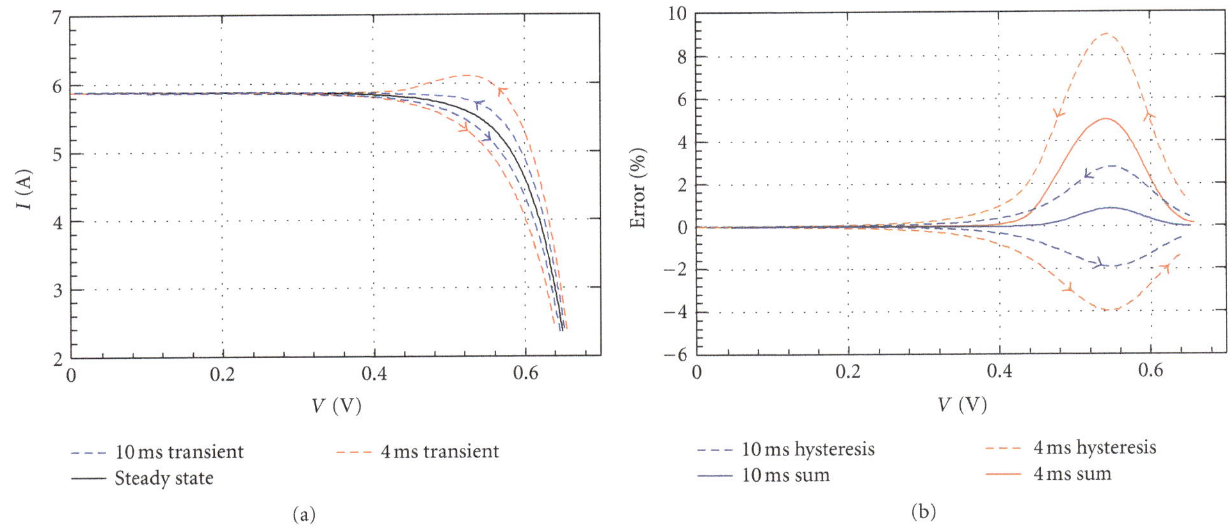

FIGURE 1: The I-V curve hystereses due to the forward and reverse sweep (a) and the corresponding errors caused by the internal capacitance (b) of a back-contacted monocrystalline silicon solar cell (arrows indicate the direction of the sweeps).

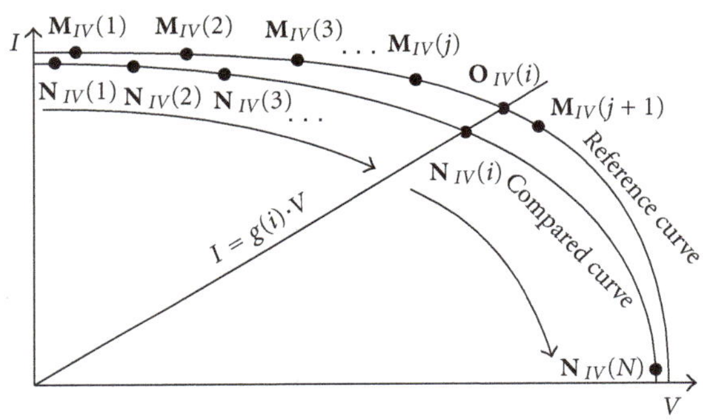

FIGURE 2: Error calculation geometry for a forward sweep.

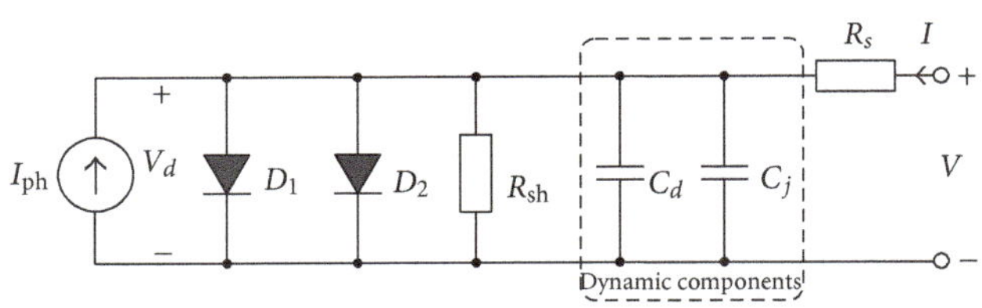

FIGURE 3: Two-diode solar cell model extended by two capacitances.

a solar cell (Figure 3). The two-diode model instead of a more common one-diode model is required for better accuracy of dynamic parameter extraction [9].

The model in Figure 3 consists of diodes D_1 and D_2 representing the Shockley diffusion direct current and the space charge recombination current, respectively, lumped shunt resistance R_{sh}, lumped series resistance R_s, junction capacitance C_j, diffusion capacitance C_d, and the photo-generated current under illumination I_{ph}. The diodes D_1/D_2 are modelled with the saturation currents I_{s1}/I_{s2} and the diode ideality factors n_1/n_2.

The capacitances C_d and C_j determine the dynamic properties of the SC under investigation. Assuming an abrupt p^+-n or n^+-p junction, an analytical approximation of the C_j is derived from the $C_j(V_d) = -dQ/dV_d$ equation [10]:

$$C_j(V_d) = \frac{C_{j0}}{\sqrt{1 - V_d/V_{bi}}}; \quad V_d < V_{bi}, \tag{2}$$

where C_{j0} is the junction capacitance at zero applied voltage, V_d is the internal solar cell voltage, and V_{bi} is the built-in potential. Such voltage-dependent capacitance, although limited as V approaches V_{bi}, is implemented into the SPICE diode model.

The C_d, as the key component of the SC's forward-biased dynamic character, is an outcome of minority charge carrier storage in the neutral region and its diffusion ability. C_d is proportional to the base minority carrier lifetime τ_d. For $V_d \gg kT/q$ (forward-biased solar cell), the C_d can be written as a function of temperature, applied voltage, and carrier lifetime using the following simplified equation [10]:

$$C_d(V_d) = \frac{q}{2n_1kT} I(V_d)\tau_d = \frac{q}{2n_1kT} I_S e^{qV_d/n_1kT} \tau_d, \tag{3}$$

where q is the electron charge, τ_d is the effective diode base minority carrier lifetime, and $I_S e^{(qV_d/n_1kT)}$ is the diode forward DC current. In the voltage range where $C_d \geq C_j$, the forward current of the diode D_1 is a few orders of magnitude higher than that of D_2; thus, only the current of the diode D_1 can be used in (3).

Measurement of C_d using direct C-V measurement is aggravated due to high solar cell currents in forward bias. More often, C_d is calculated from (3), while the lifetime τ_d is measured using one of the following methods: reverse recovery [11], short circuit current decay [12], or open circuit voltage decay (OCVD) [8]. During the OCVD measurement, the charge stored in the SC structure diminishes through different loss mechanisms. The OCVD measurement starts with application of a predefined external current or voltage (forward bias) to the device under test (DUT). At $t = 0$, the forward bias is abruptly switched off. The voltage decay on the DUT's terminals is measured as the charge stored within the DUT diminishes with time. Voltage should be monitored as close to the DUT as possible to avoid the error caused by the series resistance of the measurement setup.

Three characteristic regions can be defined when analysing the OCVD signal of crystalline silicon SCs (see Figure 5 for OCVD shapes of different technologies) [13]:

(i) the voltage drop due to the lumped series resistance,

(ii) the quasilinear region due to the charge storage in neutral region (diffusion component),

(iii) the exponential decay that starts with the diffusion capacitance dropping below the junction capacitance. The charge stored in the space charge region is released through the recombination process and the lumped shunt resistance R_{sh}.

With both capacitances implemented and the static two-diode model parameters determined, we will present a method to extract both the τ_d and the C_{j0} parameters from the OCVD signal.

2.3. Extended Model Parameter Extraction. Two basic measurements—the *I-V* characteristic and the OCVD signal—are required for determination of the parameters of the extended model shown in Figure 3. As the OCVD signal is measured without the illumination, the method gives best results with the dark quasi static *I-V* measurement. Only the extraction of the photo-generated current I_{ph} is required under illumination.

Our measurement setup consists of an *I-V* curve tracer and an OCVD measurement setup. The curve tracer is constructed from parallel connection of three independent channels of two Keithley SMU 2602A units. One channel is configured as a voltage source determining the measurement voltage, whereas the other two channels are configured as current sources connected in parallel, extending the current range of the setup. All channels are interconnected with TSP link to provide single script output control. We programmed a measurement script that enabled automated measurements up to 1 V and 9 A with the worst-case voltage measurement accuracy of ±0.035% and the worst-case current measurement accuracy of ±0.18% [14]. The OCVD measurement setup is based on a 16-bit PCI6014 DAQ card controlling the switch and measuring the voltage decay. Sampling frequency is set to 200 kHz with the worst-case voltage measurement accuracy of ±0.031% [15].

Measurements under illumination were carried out on a temperature-controlled platform under a Newport class A continuous solar simulator. The STC conditions were met; the cell temperature regulation was maintained at 25°C by a temperature-stabilized measuring table [16], and the solar irradiance of 1000 W/m^2 was set using a calibrated reference solar cell with ±2% accuracy [17]. Following the acquisition of the required measurements, the algorithm starts with extraction of the two-diode static model parameters (I_{s1}, n_1, I_{s2}, n_2, R_s, and R_{sh}) from the measured *I-V* curve. Extraction of R_s is based on a method given in [18], where further improvements are implemented to optimize the accuracy of the dynamic model. A fully automated algorithm in MATLAB is developed that comprehends the following steps.

I. The influential range for the series resistance R_s, diode D_1, and diode D_2 on the input I-V curve is determined.

II. The series resistance R_s and the diode ideality factor n_1 are extracted using a bijective transformation and linear regression [18]. The inner solar cell potential $V_d = V - R_s I$ is calculated for further parameter extraction.

III. The D_1 diode's saturation current I_{s1} is calculated from the upper exponential region of the I-V curve. A derivative-free unconstrained simplex multivariable minimum search method is used to obtain the best fit [19] in the selected voltage range determined by the curve alignment function

$$\theta_{RMS} = \sqrt{\frac{\sum_{i=1}^{n} 2\log|I_{meas}(U_i)/I_{model}(U_i)|}{n}}. \quad (4)$$

The modelled curve is calculated at voltage points U_i of the measured curve. Index i runs through n measured points of the selected voltage range.

IV. The current of D_1 is subtracted from the total current. Both diode D_2 parameters are calculated from the lower exponential region of the remaining current using the same derivative-free unconstrained simplex multivariable minimum search method.

V. The shunt resistance R_{sh} is calculated as the derivative of the I-V curve at $V = 0$ and subtracted by the derivative of D_1 and D_2 characteristics [18].

VI. D_2 parameters are recalculated with a new target current. Both the D_1 and R_{sh} currents are subtracted from the total current.

VII. An additional step is added to fine-tune the diode D_1 parameters. I_{s1} and n_1 are recalculated to minimize the error in the range where the currents of both diodes are of comparable magnitudes.

Once the static parameters are known, we proceed with the extraction of the dynamic parameters (τ_d for C_d, C_{j0} for C_j). The effective lifetime τ_d can be extracted from the OCVD signal using the equation [20]

$$\tau_d = \frac{nkT/q}{dV/dt}, \quad (5)$$

where dV/dt is the slope of the linear part (region ii) of the OCVD signal. In the case of solar cells, where diffusion lengths approach the length of the base, the voltage decay is influenced by the base-emitter coupling effect causing the region ii to deviate from a straight line [20]. Extraction of τ_d from (5) is unreliable in such case. Our method uses (5) only for initial approximation of the simplex algorithm.

The initial approximation of the C_{j0} parameter can be calculated form (2) considering typical semiconductor parameters of the solar cell under investigation. Higher precision of C_{j0} is obtained from the low-bias region of the OCVD signal (region iii), where the exponential decay is driven only by C_j.

An unconstrained simplex multivariable minimum search optimization algorithm is used to minimize the error between the OCVD simulation and measurement. Best fit is obtained for both the τ_d and the C_{j0} parameters at the same time. The algorithm is implemented in MATLAB, where SPICE netlist is generated, and transient analysis is executed for each optimization step. Error between the measurement and simulation is calculated and passed back to the optimization algorithm. The extraction algorithm applied to the three test cells with the basic parameters shown in Table 1 yields the extended two-diode model parameters shown in Table 2 together with the static I-V RMS curve alignment factor.

2.4. Evaluation of the Extraction Method and the Extended Model. An accurate static model is required for accurate dynamic parameter extraction. However, there is no direct relation between the model and the parameter accuracies. The accuracy of the presented parameters in Table 2 is not relevant as the dynamic parameters are not calculated straightforward from the static parameters. It is the static I-V curve matching that is essential for good OCVD simulation agreement. We can show good static I-V curve alignment between model simulations and measurements, thus proving that such model is suitable for use over wide operating range of solar cells as seen in Figure 4. Accompanying curve alignment RMS errors defined by (4) are listed in Table 2 and range in the three cases from 0.0279 to 0.0491.

The quality of the dynamic model parameters can first be estimated by observing the degree of alignment between the simulated OCVD signal and the measured one. In the case of inaccurate dynamic (or static) model parameters, the simulated OCVD signal does not match the measured OCVD signal. Figure 5 shows a good alignment with slight deviations in the higher voltage range of mc-Si and BC mc-Si cells. Analysis showed that the deviations are a result of the base-emitter coupling effect that is not included in the dynamic model.

The minority carrier lifetimes are injection level dependent, especially when high injection levels are reached. However, in the range from 0.2 to 1, sun irradiance silicon-based solar cells are found in the low injection regime, where the lifetime does not vary significantly [21]. To verify the assumption, OCVDs of all test cells were measured at initial currents of 6 A, 4 A, and 2 A and compared to simulations using parameters extracted at short-circuit currents under STC (Table 2). The simulations were performed under equal conditions as the measurements, and the alignment is comparable to the alignment seen in Figure 5. From that, we can assume that a constant minority carrier lifetime can be used to analyse the effect of irradiance in the range from $1100\,\mathrm{W/m^2}$ down to $200\,\mathrm{W/m^2}$.

For the purpose of further evaluation, errors between several I-V scans with varied scan speeds versus a slower transient reference scan have been compared between measurement and simulation. A simple approach to vary the scan time of the I-V curve sweep from I_{SC} to V_{OC} is to vary the capacitance of a capacitive load. At $t = 0$, a power MOSFET connects the load capacitor to the DUT. The load capacitor starts charging with the initial current limited by the series resistance of the measurement circuitry including the series resistance of the load capacitor. To compensate the resistance and attain the short circuit current, a small negative initial charge is applied to the load capacitor prior to the measurement. During the I-V curve sweep, the current and voltage of the DUT are measured using two channels of the DAQ interface.

A series of measurements and simulations were performed on the three cells. Figure 6 presents some of them for the mc-Si cell (measurement B is omitted for better readability of the chart). The load capacitor values, the scan times, and the errors of both simulations and measurements are shown in Table 3.

The agreement between measurement and simulation is within the acceptable tolerance of $\pm 0.1\%$ with an exception of measurements C and D of the BC mc-Si cell. Solar cells with diffusion lengths approaching the physical length of the base encounter the base-emitter coupling effect that decreases the accuracy of the C_d model. The selected C_d model causes the error of such solar cells to be overestimated. The misalignment arises at higher error values and shorter scan times, whereas at measurement B, at error value of 1.4%, the mismatch error between measurement and simulation is still within $\pm 0.1\%$.

In capacitive I-V curve measurement, as well as in semiconductor/resistive measurement, the scan speed has almost no effect on I_{SC}. Solar cell's dynamic error manifests in somewhat lower voltage reading on the DUT terminals. The voltage mismatch in the area near I_{SC} is detectable only in low-quality cells, where the shunt resistance creates a significant current slope. In regular quality cells, even though the voltage mismatch does exist, it is undetectable due to the almost constant I-V curve slope near I_{SC}. The V_{OC} on the other hand is strongly affected in semiconductor/resistive curve tracer systems but is not affected in capacitive-based tracer systems. As the capacitor is reaching its top charge, the current rate decreases, and the transient error fades away. Therefore, we can only observe the effect of scan speed on P_{MPP} shown in Table 3. As expected, the P_{MPP} is hardly affected in the pc-Si cell with mild dynamic character, whereas in the BC mc-Si cell due to the expressed diffusion capacitance, the measured maximal power at the fastest scan speed is significantly reduced.

2.5. The Generalized PV Module Model. PV modules are typically constructed from single cells connected in series. An equivalent generalized model can be introduced to model the PV module I-V characteristics using a small amount of circuit parameters [22]. In general, the characteristics of solar cells within a module should be as similar as possible

TABLE 1: Basic parameters of solar cells under investigation under STC.

Name	A, cm^2	I_{SC}, A	V_{OC}, V	FF, %	P_{mpp}, W	η, %
pc-Si	240	7.46	0.586	75.3	3.29	12.9
mc-Si	240	8.59	0.631	79.5	4.31	17.9
BC mc-Si	153	5.88	0.660	76.8	2.98	19.5

TABLE 2: Extracted model parameters of solar cells under investigation under STC.

Name	R_s, mΩ	R_{sh}, kΩ	I_{s1}, μA	n_1	I_{s2}, μA	n_2	τ_d, μs	C_{j0}, μF	θ_{RMS}
pc-Si	3.41*	2.85	0.104	1.293	1.68	4.240	5.0	12.5	0.0279
mc-Si	0.371	10.4	0.0288	1.260	0.150	3.588	35.0	5.0	0.0287
BC mc-Si	2.01*	0.115	0.204	1.543	0.420	4.550	240	1.0	0.0491

*Measured with soldered ribbons.

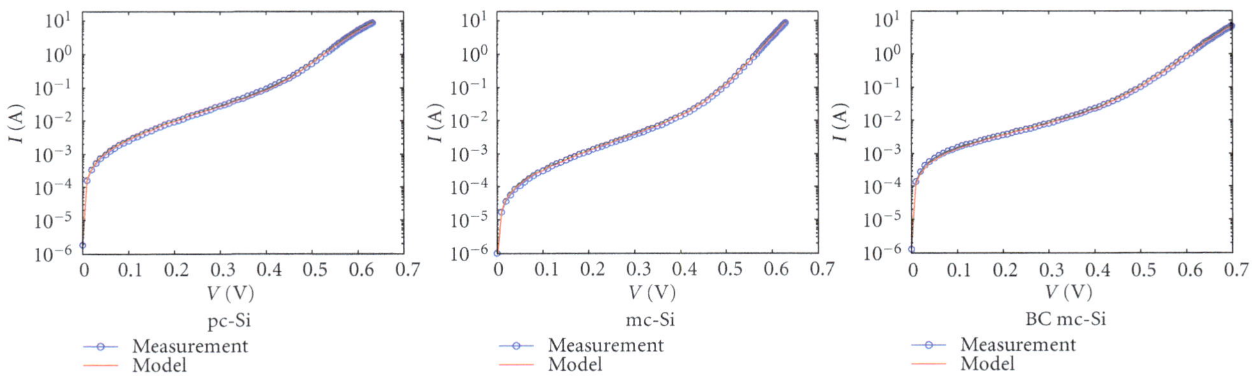

FIGURE 4: Alignment between modelled and measured I-V curves of selected solar cells shown in semilogarithmic scale.

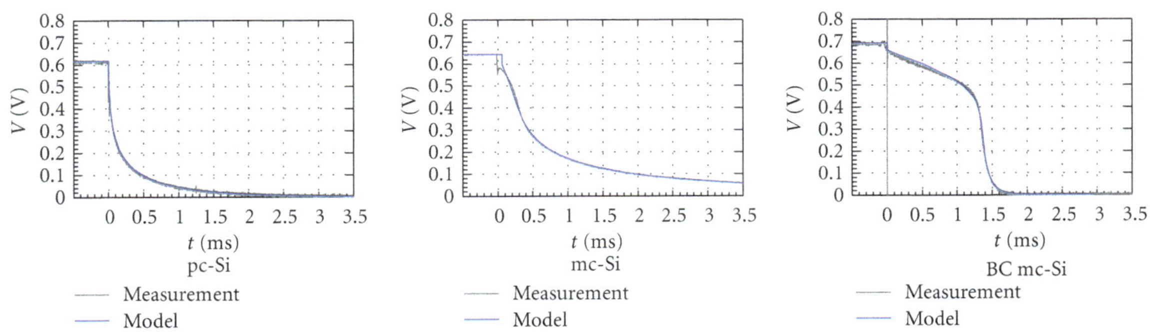

FIGURE 5: Alignment between modelled and measured OCVD signals of selected solar cells.

TABLE 3: Capacitive curve tracer I-V results: measurement versus simulation.

		Scan time/maximum power/measurement error/simulation error t, ms/P_{mpp}, W/E_{meas}, %/E_{sim}, %											
C_m [mF]	Meas.	pc-Si				mc-Si				BC mc-Si			
39.1 ± 2%	Ref.	22.0	3.30	—	—	20.5	4.29	—	—	27.1	2.90	—	—
28.5 ± 2%	A	17.1	3.29	0.2*	0.01	14.8	4.28	0.14	0.17	19.0	2.87	0.76	0.8
17.7 ± 2%	B	10.4	3.29	0.2*	0.05	9.1	4.26	0.21	0.26	10.8	2.82	1.34	1.41
11.6 ± 2%	C	7.2	3.28	0.2*	0.9	6.1	4.23	0.61	0.57	7.4	2.77	2.1	2.4
5.0 ± 2%	D	3.1	3.26	0.38	0.31	2.8	4.16	1.2	1.17	3.2	2.62	4.15	4.91

* s/n ratio too low for accurate error determination.

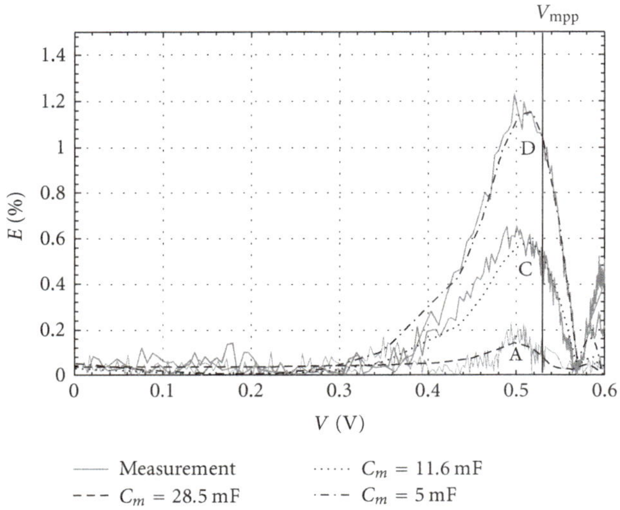

FIGURE 6: Voltage distribution of transient error for mc-Si cell.

to obtain the best module efficiency; thus, the theory of the generalized model assumes a PV module is constructed of N_s equal cells connected in series. The generalized one-diode static model characteristics are written as follows [22]:

$$I_M = I_{phM} - I_S\left(e^{((qV_M - I_M R_{sM})/N_s nkT)} - 1\right) - \frac{1}{N_s} R_{shM} I_M, \quad (6)$$

where indices M denote quantities referred to the module. The generalized model equals the diode model up to the factor of N_s. The same applies if a two-diode model is used instead.

In theory, the total module capacitance is $C_{dM}(V_M) + C_{jM}(V_M) = \left(\sum_{i=1}^{N_s} (C_{di}(V) + C_{ji}(V))^{-1}\right)^{-1}$. The junction capacitance at zero voltage equals $C_{j0M} = C_{j0}/N_S$ in the case of identical cells. In the case of nonidentical cells' C_{j0}, the total C_{j0M} equals parallel connection of individual C_{j0} values.

From (3) and (6), presuming equal τ_d of all cells, the C_d can be expressed as

$$C_{dM}(V_M) = \frac{q}{2N_S nkT} I_S \, e^{((qV_M - I_M R_{sM})/N_S nkT)} \tau_d, \quad (7)$$

in the case of a PV module. As expected, the value of the minority carrier lifetime withholds the value of a single cell, whereas the total capacitance reduces by the factor N_S.

A test module with access to all interconnections between the individual cells (built of 4 monocrystalline silicon SCs shown in Figure 7) was laminated in our laboratory. Validation of the method for the dynamic parameter extraction of a PV module is carried out through the distribution of the dynamic components along the module. Figure 8 displays the OCVD measurement of the individual cells, the sum of the individual measurements, and the OCVD measurement of the entire module.

Very good agreement between the sum and the module OCVD response with deviation only at times $t < 0$ s can be observed. This deviation arises from the series resistance of bus bars used to contact the cell interconnections. The

FIGURE 7: An experimental 4-cell mc-Si module (dashed circles mark access to external and individual cell contacts).

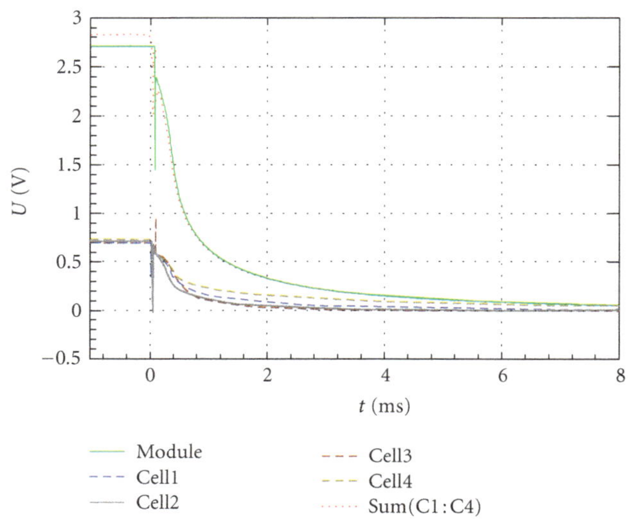

FIGURE 8: OCVD signal of the 4-cell mc-Si module and its individual cells.

actual series resistance of the module is determined from the I-V measurement and is properly considered in the dynamic parameter extraction method. The OCVD signal of the module equals the sum of the individual responses of SCs involved; therefore, in the OCVD signal of the PV module each solar cell contributes its share of charge proportional to its total internal capacitance. The OCVD signal of the PV module can thus be used to extract C_{j0M} and τ_{dM} parameters, where C_{j0M} should equal the parallel connection of the individual C_{j0} values, and τ_{dM} should equal the average value of the individual τ_d values. The individual values of C_{j0M} and τ_{dM}, however, should not diverge considerably from each other.

Table 4 displays results of the parameter extraction method applied to the individual cells and to the module. Most results calculated from individual cells are in good agreement with the theory. The discrepancy in the R_s can be explained through the resistance of additional bus bar length required to contact individual cells. The second discrepancy, found in I_{S2}, is caused by cell 4 with parameter I_{S2} unfortunately significantly smaller compared to the other

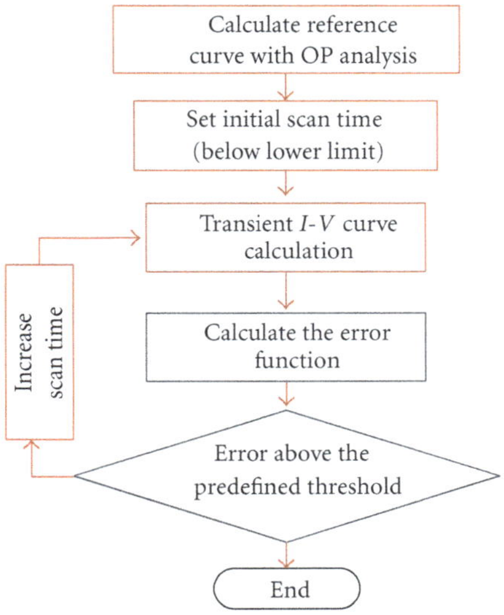

FIGURE 9: Scan time extraction algorithm.

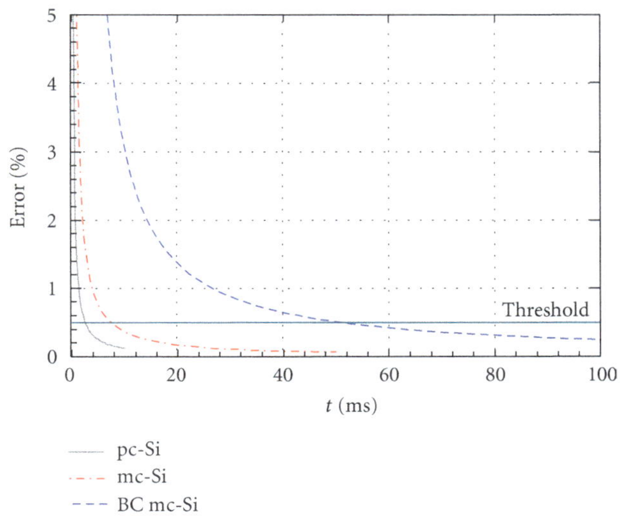

FIGURE 10: Error versus scan time for the three cells under investigation.

cells. In such case, the effective I_S is not determined by the smallest I_S value.

However, the extracted dynamic parameters are determined within the expected range of tolerance, and extraction process can be used to determine the generalized two-diode model for PV modules when cells with similar properties are involved, which is the case in all high-efficiency PV modules.

3. Scan Time Determination

The minimum scan time of a single cell or a PV module at a predefined error threshold can be determined by an iterative algorithm from a series of transient I-V curve simulations. The dynamic model of the DUT is used for both the reference and the compared curve simulations. The reference curve is simulated by sweeping the operating point (OP) analysis to eliminate the capacitance effects, whereas the compared curves are simulated using transient analysis. The numerical computation and netlist generations are handled in MATLAB, whereas SPICE is executed within MATLAB for circuit simulations (Figure 9).

In the process of determination of the maximal tolerable predefined error threshold, other sources of I-V curve errors as well as the ratio of the required scan time to the I-V curve error should be considered. It is reported that specialized laboratories are able to achieve I-V curve measurement uncertainties from 1.6% to 3% [3]. Also in Figure 10, we can observe a quasiexponential decrease of error versus scan time leading to very long scan times at low error thresholds. Determination of the predefined error depends strongly on the purpose of the measurement. To stay at least a factor of 3 below other uncertainties and to avoid too long scan times (Figure 10), the predefined error threshold was set at 0.5%. Rendered optimum scan times are 2.2 ms, 7.3 ms, and 43.8 ms for the pc-Si, mc-Si, and BC mc-Si solar cells, respectively.

4. Effect of Irradiance Level

4.1. Experiment Background and Expectations. In the study of the influence of irradiance (G) on the scan time, the RC time constant τ_C, causing the delay when sweeping the I-V curve from point to point, should be calculated. In forward bias, the C_d dominates the total cell capacitance, and the current through the D_2 is negligible. When sweeping the I-V curve using a voltage source, the time constant is defined as

$$\tau_C = RC = \frac{C_d}{g_s + g_{sh} + g_d}, \quad (8)$$

where $gs = 1/R_s$, $g_{sh} = 1/R_{sh}$, and g_d is differential conductance of the diode D_1 between the neighbouring points. The following equation can be derived using simplified equations for C_d and g_d [10] and assuming that $g_{sh} \ll g_s$:

$$\tau_c = \frac{\tau_d}{2} \frac{I_{d1}}{(I_{d1} + g_s V_T)}. \quad (9)$$

I_{d1} is the current flowing through the diode D_1 (Figure 3), and V_T is the thermal voltage. Equation (9) implies that the time constant is slightly increasing with diode current. The current I_{d1} is the difference between the photogenerated current and the current flow out of the device terminals ($I_{d1} = I_{ph} - I$). At V_{OC}, where $I = 0$ A, the decrease in irradiance yields lower time constants and thus smaller transient errors.

The trend of τ_C can be observed in a step response of the measured device. A simulation in V_{OC}, where the time constant is maximal, with a step amplitude of 0.01 V, has been performed at different currents I_{d1}. A sweep of I_{d1} is used to model the change in irradiance. The stationary V_{OC} conditions have been established by a large capacitor in series with a step voltage source.

Simulation in Figure 11 confirms the (9). A decrease in irradiance yields shorter step responses and shorter time

Table 4: Extracted parameters of individual cells and module (monocrystalline silicon solar cells).

Cell	R_s, mΩ	R_p, kΩ	I_{s1}, μA	n_1	I_{s2}, μA	n_2	τ_d, μs	C_{j0}, μF
Cell 1	6.251	5.0	5.65	1.15879	141.0	3.64096	35.0	4.50
Cell 2	9.693	1.2	5.77	1.16285	446.0	4.38751	35.0	5.70
Cell 3	7.086	5.0	3.80	1.13891	696.0	5.10632	45.0	5.30
Cell 4	10.216	6.3	5.15	1.15259	33.9	2.98760	45.0	4.70
Module calculated	33.246	17.5	3.80	4.61314	33.9	16.12239	42.5	1.25
Module measured	23.039	17.0	3.49	4.53281	183.0	14.32319	41.3	1.43

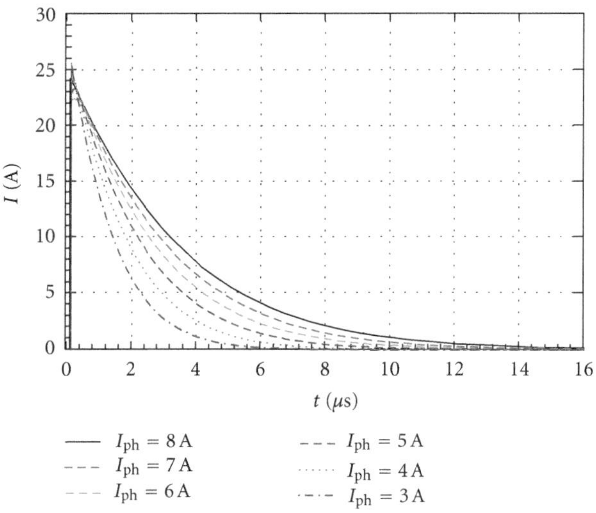

Figure 11: Simulated step response of mc-Si solar cell.

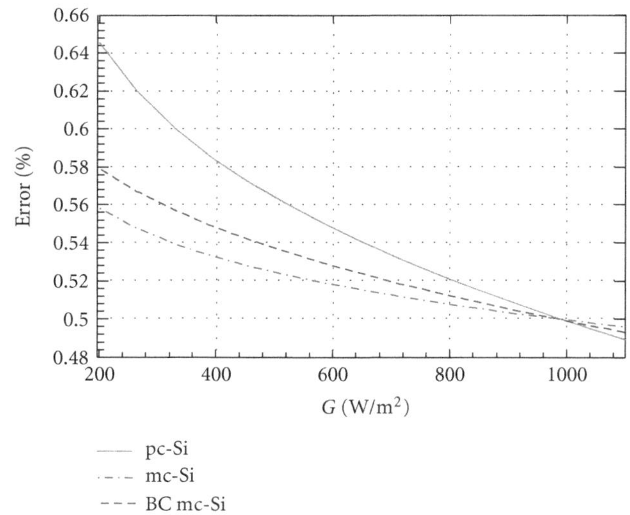

Figure 12: Effect of irradiance variation on transient error.

constants. At first sight, the decrease of the time constant should also decrease the *I-V* curve error. However, as follows, there is more to the transient error than just the time constant.

4.2. Effect of Irradiance on I-V Curve Error. The source of the transient error is the capacitance that introduces the time constant τ_C. As soon as the capacitance is present in the circuit, there are several factors aside the time constant affecting the magnitude of the error.

(i) The duration of *I-V* scan severely impacts the error: each time the load changes, the capacitor charge needs to be equalized. The transient error directly depends on dV/dt between consequent scan points. When the same sweep algorithm is involved, what matters is the total time to sweep from I_{SC} to V_{OC}, regardless of the number of points.

(ii) The *I-V* curve sweep algorithm: achieving a homogenous density of *I-V* curve points is a common issue. Constant current (or voltage) step yields approximately equidistant point distribution from the maximum power point to V_{OC} (or to I_{SC}) and only a few points in the other direction. Different approaches are used to acquire equidistant point distribution along the curve such as logarithmic voltage step, combined voltage and current stepping, adaptable on the fly step calculation. Whichever method is used, they all vary dV/dt throughout the *I-V* curve.

(iii) Lastly, the method of error determination also affects the *I-V* curve error: strictly considering the time constant, both points being compared should be selected at the same time of the measurement. However, the *I-V* curve does not contain the time information and has to be compared according to the equal power availability. Furthermore the direct comparison of two curves at different irradiance is impossible due the modified I_{SC}, the V_{OC}, and the entire curve path.

The error is thus determined using the method from Section 2 between a transient simulation and a quasi static operating point analysis (the ideal reference *I-V* curve). The scan time for each DUT is determined from the predefined error threshold of 0.5% at STC. The variation of irradiance in simulations is modelled through variation of the photogenerated current I_{ph}.

The *I-V* curve error versus irradiance for all three cells under test (Figure 12) surprisingly yields a result of reversed error/irradiance dependency. The sweep is performed at a high-density constant voltage step rate to exclude the error arising from this source. Although the time constant decreases with decreasing irradiance, the error and consequently the optimal scan time moderately increase. The reverse trend arises from calculation of the relative error value. The effects of V and I can be evaluated separately. First, we focus on the relative current error calculation. At decreasing irradiance, the absolute error of current (δI) slightly decreases due to the fall of τ_C (9). For calculation of the relative error, the δI is divided by the short circuit current I_{SC}, which also decreases with the irradiance. Assuming, $I_{ph} \approx I_{SC}$, the reduction of I_{SC} is more rapid than the reduction of δI, which is proportional to the τ_C. Consequently, the overall error increases although the time constant decreases with irradiance. The actual error defined in (1) is an aggregate of the relative current error and the relative voltage error. The open circuit voltage varies only minor in observed irradiance range; thus, the trend of error versus irradiance mostly reflects the trend of relative current versus irradiance.

As the desired output of an error function is certainly a relative value of the I-V curve error, a moderate increase of error at lower irradiance levels should be considered at scan time determination procedure, in particular at outdoor monitoring sites. In scope of the three tested cells of different silicon technologies, the maximum relative deviation of error was 32% for the polycrystalline silicon solar cell at irradiance level of 200 W/m^2 compared to 1100 W/m^2.

5. Conclusions

A method of optimal I-V curve scan time determination of solar cells and modules based on the extended two-diode model was presented. An algorithm acquiring the parameters of the presented model from two simple measurements, an I-V scan and an OCVD signal, has been introduced. A newly introduced error determination function is used to estimate the mismatch between two I-V curves. Different aspects on determination of the tolerable predefined I-V curve error threshold caused by the PV generator's dynamic character were analysed. The predefined error threshold served as a base for the optimal scan time determination. The proposed extraction method has been applied but is not limited to different wafer-based solar cells and modules. Validation was achieved using capacitor-based I-V curve measurement, where different scan times were accomplished by variation of the load capacitance. The extraction method gives good results, whereas there is still some space for dynamic model improvement (PV generators with high minority carrier lifetimes). I-V curve error versus scan time dependency has been analysed where at error threshold of 0.5% scan times diverge from 2.2 ms up to 43.8 ms for DUTs of different silicon solar cell technologies. Therefore, the capacitive character of a DUT should certainly be considered prior to I-V curve measurement.

Even though the capacitance and the time constant of PV generators decrease with decreased irradiance, an increase of the error was ascertained. Mostly due to the relative current error, a moderate relative increase of 32% above the predefined 0.5% error threshold for the poly-crystalline solar cell under test was determined. When very high-precision measurements are required, such increase of error can affect the final measurement. The scan time should either be determined at the lowest expected irradiance level or a sufficient threshold error tolerance should be considered at the highest level of irradiance.

Acknowledgment

The authors would like to thank for the financial assistance given by the Slovenian Research Agency (P20197). M. Herman would personally like to acknowledge the Slovenian Research Agency for providing him with Ph.D. funding.

References

[1] C. Monokroussos, R. Gottschalg, A. N. Tiwari, G. Friesen, D. Chianese, and S. Mau, "The effects of solar cell capacitance on calibration accuracy when using a flash simulator," in *Proceedings of the IEEE 4th World Conference on Photovoltaic Energy Conversion*, pp. 2231–2234, May 2006.

[2] M. Prorok, W. Kołodenny, T. Zdanowicz, R. Gottschalg, and D. Stellbogen, "Reducing uncertainty of PV module temperature determination based on analysis using data gained during outdoor monitoring," in *Proceedings of the 23rd European Photovoltaic Solar Energy Conference*, D. Lincot, H. Ossenbrink, and P. Helm, Eds., 2008.

[3] N. Taylor, *Guidelines for PV Power Measurement in Industry*, Publications Office of the EU, 2010.

[4] International standard: IEC 60904-1 Ed.2: Photovoltaic devices—part 1: measurement of photovoltaic current-voltage characteristics. 2006.

[5] International standard: IEC 60904-10 Ed. 1.0 b:1998, Photovoltaic devices—part 10: methods of linearity measurement. Multiple. Distributed through American National Standards Institute (ANSI), 2007.

[6] G. Friesen and H. A. Ossenbrink, "Capacitance effects in high-efficiency cells," *Solar Energy Materials and Solar Cells*, vol. 48, no. 1–4, pp. 77–83, 1997.

[7] F. Recart and A. Cuevas, "Application of junction capacitance measurements to the characterization of solar cells," *IEEE Transactions on Electron Devices*, vol. 53, no. 3, pp. 442–448, 2006.

[8] R. Salach-Bielecki, T. Pisarkiewicz, T. Stapiński, and P. Wójcik, "Influence of junction parameters on the open circuit voltage decay in solar cells," *Opto-Electronics Review*, vol. 12, no. 1, pp. 79–83, 2004.

[9] M. Herman, M. Jankovec, and M. Topič, "A method for determination of I-V curve scan time of solar cells," in *Proceedings of the 47th International Conference on Microelectronics, Devices and Materials and the Workshop on Organic Semiconductors, Technologies and Devices*, pp. 207–211, 2011.

[10] S. Sze, *Physics of Semiconductor Devices*, Wiley-Interscience, Hoboken NJ, USA, 3rd edition, 2007.

[11] S. R. Dhariwal, "Measurement of lifetime of photoinjected carriers in solar cells by reverse voltage pulse response," *Solid-State and Electron Devices, IEE Proceedings I*, vol. 127, no. 1, pp. 20–24, 1980.

[12] K. Burgard, W. Schmidt, and W. Warta, "Applicability of electrical short circuit current decay for solar cell characterization: limits and comparison to other methods," *Solar Energy Materials and Solar Cells*, vol. 36, no. 3, pp. 241–259, 1995.

[13] M. P. Deshmukh and J. Nagaraju, "Measurement of silicon and GaAs/Ge solar cells ac parameters," *Solar Energy*, vol. 78, no. 1, pp. 1–4, 2005.
[14] "Keithley Model 2601A/2602A SourceMeter Specifications," 2012, http://www.keithley.com/data?asset=52059.
[15] "NI PCI-6013, PCI-6014 Multifunction DAQ specifications," 2012, http://digital.ni.com/worldwide/china.nsf/c362761bd12fc417862564bc006784d3/06897f85067a745686256c1b006cb50d/$FILE/PCI-6013-6014.pdf.
[16] F. Granek and T. Zdanowicz, "Advanced system for calibration and characterization of solar cells," *Opto-Electronics Review*, vol. 12, no. 1, pp. 57–67, 2004.
[17] WPVS Reference Cells—Fraunhofer ISE, 2012, http://www.ise.fraunhofer.de/en/service-units/callab-pv-cells-callab-pv-modules/wpvs-reference-cells.
[18] A. Kaminski, J. J. Marchand, A. Fave, and A. Laugier, "New method of parameters extraction from dark I-V curve," in *Proceedings of the IEEE 26th Photovoltaic Specialists Conference*, pp. 203–206, Calif, USA, October 1997.
[19] W. H. Press, S. A. Teukolsky, W. T. Vetterling, and B. P. Flannery, *Numerical Recipes: The Art of Scientific Computing*, Cambridge University Press, 3rd edition, 2007.
[20] P. W. Hawkes, *Advances in Electronics and Electron Physics*, Academic Press, 1986.
[21] J. O. Schumacher and W. Wettling, "Device physics of silicon solar cells," in *Clean Electricity from Photovoltaics*, Imperial College Press, 2001.
[22] L. Castañer and S. Silvestre, *Modelling Photovoltaic Systems Using PSpice*, J. Wiley, Hoboken, NJ, USA, 2002.

An Analysis and Research on the Transmission Ratio of Dye Sensitized Solar Cell Photoelectrodes by Using Different Etching Process

Chin-Guo Kuo,[1] Cheng-Fu Yang,[2] Mu-Jung Kao,[3] Wen-Pin Weng,[4] Chi-Cheng Chang,[5] Lih-Ren Hwang,[6] and Jian-Lan Nil[7]

[1] *Department of Industrial Education, National Taiwan Normal University, Taipei 10610, Taiwan*
[2] *Department of Chemical and Materials Engineering, National University of Kaohsiung, Kaohsiung 81148, Taiwan*
[3] *Department of Vehicle Engineering, National Taipei University of Technology, Taipei 10608, Taiwan*
[4] *Department of Chemical and Materials Engineering, Lunghwa University of Science and Technology, Taoyuan 33306, Taiwan*
[5] *Department of Information Management, Lunghwa University of Science and Technology, Taoyuan 33306, Taiwan*
[6] *Department of Automation and Control Engineering, Chung Chou University of Science and Technology, Yuanlin 51003, Taiwan*
[7] *Department of Mechatronic Technology, National Taiwan Normal University, Taipei 10610, Taiwan*

Correspondence should be addressed to Mu-Jung Kao; mjkao@ntut.edu.tw

Academic Editor: Ho Chang

Classical photoelectrodes for Dye Sensitized Solar Cells (DSSCs) were fabricated by using the electrochemical method on the titanium (Ti) template, for that the fabrication process would influence the characteristics of the DSSCs. In this study, at first three different methods were used to etch Ti templates from 10 to 17 min, (1) polishing-chemical etching: Ti template was annealed at 450°C for 1 h, abraded using number 80 to 1500 SiC sheet, and then etched in a solution of 5% HF + 95% H_2O; (2) electrochemical polishing-chemical etching: Ti template was annealed at 450°C for 1 h, electrolytic polishing with 42% CH_3OH + 5% $HClO_4$ + 53% $HOCH_2CH_2OC_4H_9$ solution, and the chemical-etching in a solution of 5% HF + 95% H_2O; (3) chemical etching: Ti template was etched in a solution of 5% HF + 95% H_2O and annealed at 450°C for 1 h. When the etching time was changed from 10 to 17 min, the thicknesses of Ti templates decreased from 75.3 μm to 14.8 μm, depending on the etching method. After etching process, the TiO_2 nanotube arrays were fabricated as the photoelectrode of DSSCs by electrochemical process, in which the Ti as anode and platinum (Pt) as cathode. The electrolyte solution included $C_2H_4(OH)_2$, NH_4F, and deionized water. After annealing the grown TiO_2 nanotube arrays at 450°C for 3 h, we would show that the etching process had large effect on the structure and transmittance ratio of the TiO_2 nanotube arrays.

1. Introduction

Tubular inorganic nanostructures offer great potential for use in heterojunction solar cells, fuel cells, molecular filtration, tissue engineering, and Dye Sensitized Solar Cells (DSSCs). In contrast to random nanoparticle systems where slow electron diffusion typically limits device performance [1, 2], the precisely oriented nature of the crystalline (after annealing) nanotube arrays makes them excellent electron percolation pathways for vectorial charge transfer between interfaces [2, 3]. Highly ordered vertically oriented nanotube arrays fabricated by different metal-oxide can be used as photoelectrode materials of the DSSCs, which include titanium dioxide (TiO_2) [4], ferric oxide (Fe_2O_3) [5], zinc oxide (ZnO) [6], and so forth. Among those proposed photoelectrode materials, TiO_2 is best because it has the merits of low cost, chemical stability, and good charge transport [7].

As numerous major advances in research and technology over the past decade have been made possible by the successful development of nanostructures, various avenues have been used to fabricate a diversity of TiO_2 nanostructure photoelectrodes, including the sol-gel method [8], metalorganic

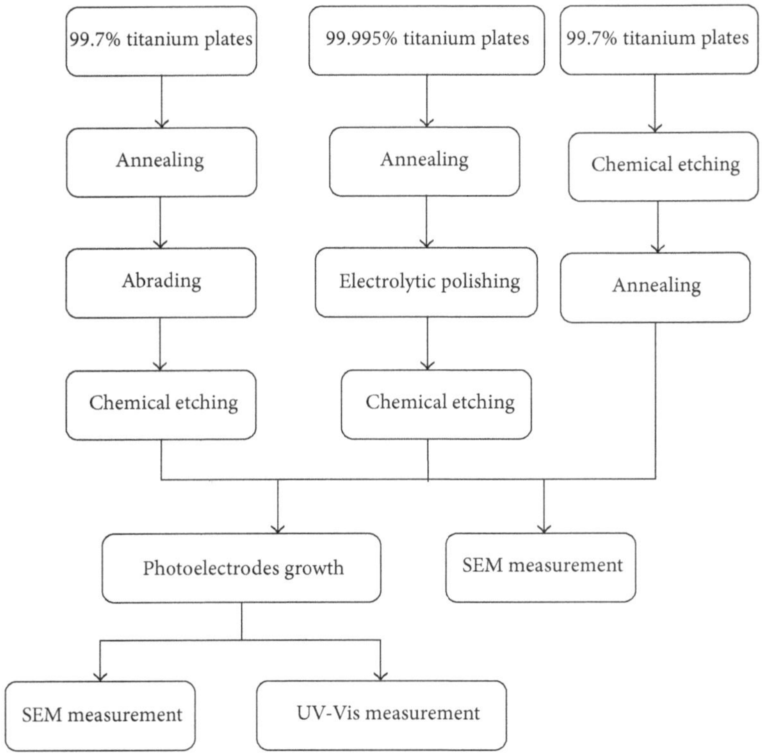

FIGURE 1: Experimental flow chart.

chemical vapor deposition (MOCVD) [9], templating [10], and electrochemical method [11]. Although many of those fabrication routes are complicated due to the use of templates or the nature of the involved chemical processes, it has been demonstrated that self-organized vertically oriented titanium dioxide (TiO_2) nanotube arrays can be fabricated using a simple anodization technique. Using electrochemical method to fabricate the TiO_2 nanotube arrays as photoelectrodes is first one using hydrofluoric acid (HF) electrolyte by Zwilling et al. [12]. After that, many electrolytes are also developed to fabricate the TiO_2 nanotube arrays as photoelectrodes, such as HF/H_2O, $HF/H_2SO_4/H_2O$ [13], and EG (ethylene glycol)/NH_4F/H_2O [14]. When the EG/NH_4F/H_2O electrolyte is used to fabricate TiO_2 nanotube arrays, the length of TiO_2 photoelectrodes has the value between 10 μm and 100 μm [14].

When using as the photoelectrodes, the property of the anatase phase TiO_2 is believed to be superior than that of the rutile phase TiO_2. Because the band gap of anatase phase TiO_2 is 3.2 eV and rutile phase TiO_2 is 3.0 eV, the light absorption edge of anatase phase TiO_2 is located at lower wavelength. When the ultraviolet light is irradiated, the photoelectrodes of TiO_2 nanotube arrays are proceeded photocatalytic activities to make light through photoelectrodes to dye layer [15]. Except the crystalline phase, when the TiO_2 nanotube arrays are used as the photoelectrodes, the transmission ratio of TiO_2 nanotube arrays has large effect on the efficiency of the fabricated DSSCs. In this study, Dye Sensitized Solar Cells (DSSCs) photoelectrodes, TiO_2 nanotube arrays, were fabricated by using electrochemical method and using UV-Vis examination transmission ratio in the DSSCs structure. Therefore, the purpose of this study was to investigate the effect of etching process on transmission ratio of the TiO_2 nanotube arrays after treating by using different etching methods.

FIGURE 2: Cross-section observation of annealed titanium plate.

2. Experimental

The Ti metals were cut into Ti-plates with a size of 2.5 cm × 2.5 cm and thickness of ~142 μm. The flow chart for the different etching process, the process for formation of TiO_2 nanotube arrays, and the characteristic measurements of the TiO_2 nanotube arrays was shown in Figure 1 and described below.

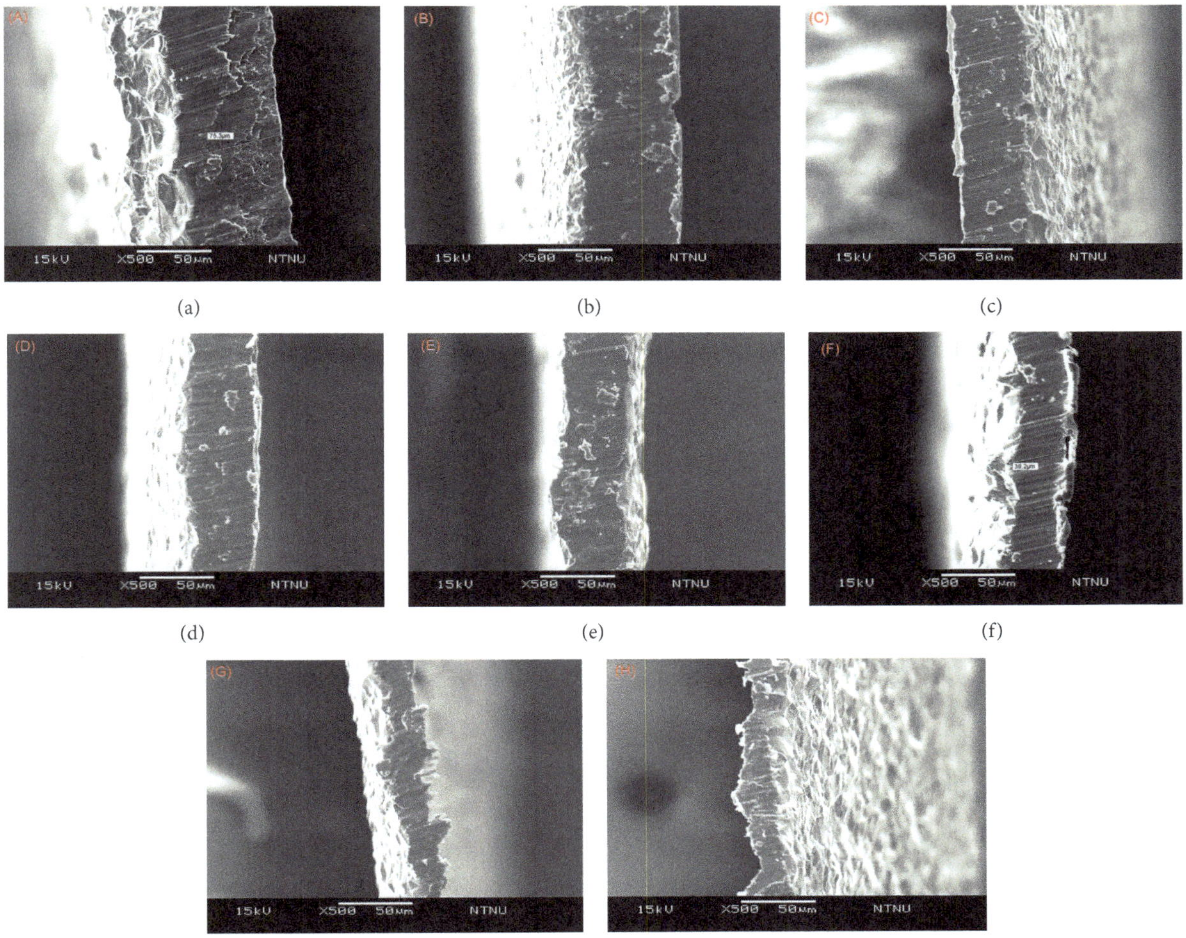

(a) (b) (c) (d) (e) (f) (g) (h)

FIGURE 3: Cross-section morphologies of polishing-chemical etching under different times (a) 10 min, (b) 11 min, (c) 12 min, (d) 13 min, (e) 14 min, (f) 15 min, (g) 16 min, and (h) 17 min.

2.1. Preprocess

2.1.1. Polishing-Chemical Etching. A pure Ti plate with 99.7% purity was used as the template for the polishing-chemical etching process. The template was first annealed 1 h at 450°C and cooled in air, then the template was abraded using SiC sheets (Nos 80 to 1500), and then it was ultrasonicated for 30 min in distilled water. After that, the mixture of 5% HF + 95% H_2O was used as the etching solution and the etching time was changed from 10 to 17 min.

2.1.2. Electrochemical Polishing-Chemical Etching. Another titanium plate with higher purity 99.995% was also used as the template, too. The sample was first annealed 1 h at 450°C and then it was ultrasonicated for 30 min in distilled water. The polishing electrolytic solution was mixed with 42% methanol (CH_3OH) + 5% perchloric acid ($HClO_4$) + 53% ethylene glycol monobutylether ($HOCH_2CH_2OC_4H_9$), and the etching solution was mixed by 5% HF + 95% H_2O; the etching time was changed from 10 to 17 min.

2.1.3. Chemical Etching. A pure titanium plate with 99.7% purity was also used as the template for the chemical etching process. The mixture of 5% HF + 95% H_2O was used as the etching solution and the etching time was changed from 10 to 16 minutes. After the etching process the titanium plate was annealed at 450°C for 1 h.

2.2. Preparation of Photoelectrodes. We fabricated ordered nanochannel TNT arrays at 25°C on the prepared titanium (Ti) square foils (2 cm × 2 cm) at a constant voltage of 50 V and Ti foils as an anode. The electrolyte solutions contained 0.2 M ammonium fluoride (NH_4F) + 2% H_2O + ethylene glycol ($C_2H_4(OH)_2$) with anodization. Ti foils were degreased by ultrasonication in acetone and then isopropanol, respectively, for about 30 min, followed by rinsing with deionized (DI) water, and finally dried in the air before used. Highly ordered TiO_2 nanotube arrays over large area were prepared by a potentiostatic anodization in a two-electrode electrochemical cell with a platinum (Pt) sheet as counter electrode. All anodization experiments were carried

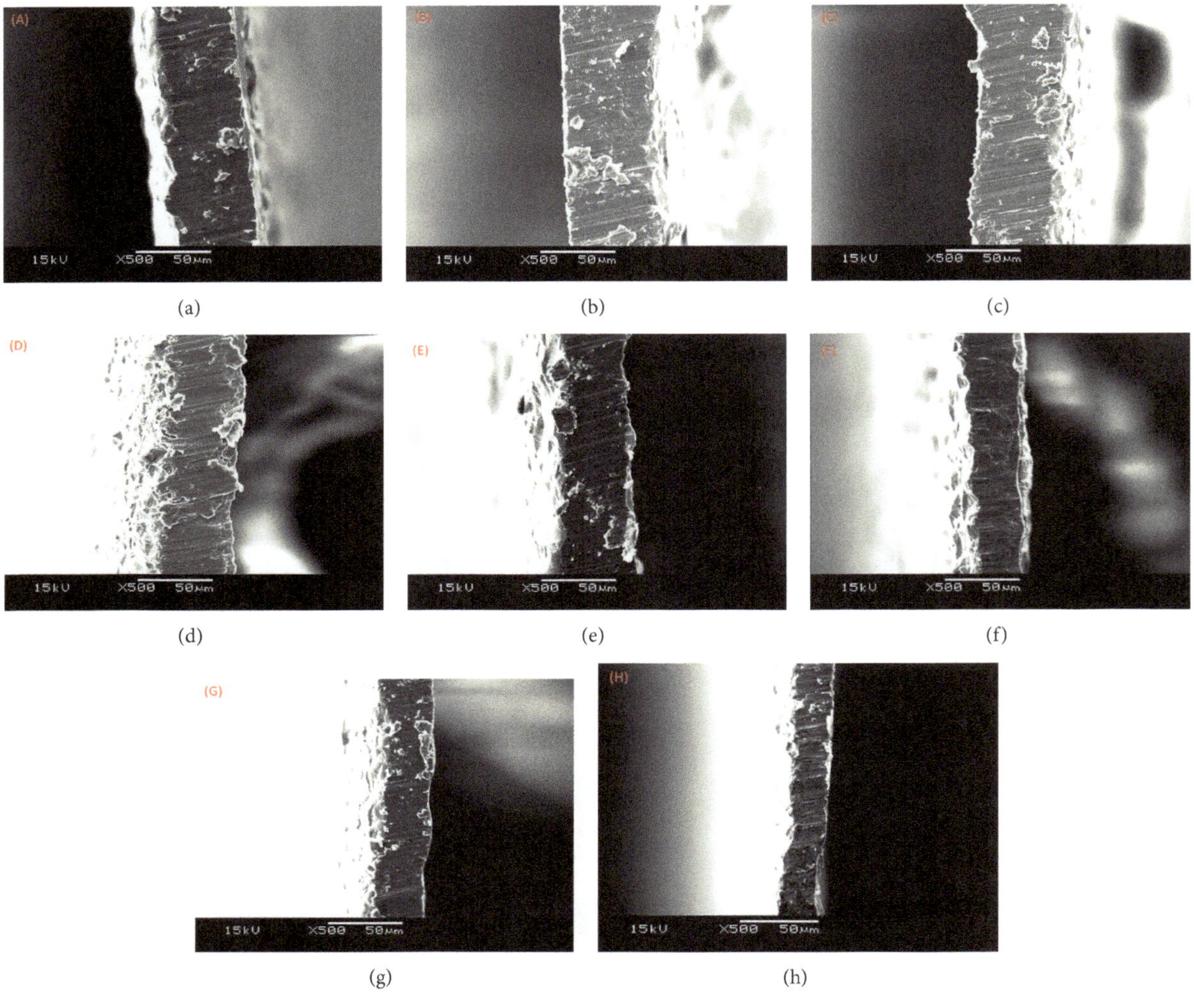

Figure 4: Cross-section morphologies of electrochemical polishing-chemical etching under different times (a) 10 min, (b) 11 min, (c) 12 min, (d) 13 min, (e) 14 min, (f) 15 min, (g) 16 min, and (h) 17 min.

out at room temperature. After the electrochemical process, the foils were annealed at 450°C for 3 h.

3. Results and Discussion

3.1. Morphologies for the Cross-Section of Nonetched and Etched Ti Specimens. Figure 2 shows the cross-section of the nonetched and annealed Ti specimen, the nonetched Ti specimen showed a densified structure and no defect and porous were observed. After different etching processes were used, the etch-treated Ti plates were thoroughly rinsed with distilled water for clean at room temperature and morphologies for the cross-section were observed. In Figures 3–5, a series of micrographs of titanium plate after various etching processes and at different etching time, from 10 to 17 min, are observed. From Figures 2 and 5, the thickness of titanium plates linearly decreased with etching time and, however, the nonuniformity in thickness became apparently at a longer etching time. For the polishing-chemical etching specimens, as the etching time was equal to and longer 13 min, the thickness become obviously asymmetrical. As shown in Figure 3, the polished surface of Ti plate was changed to be a porous structure after alkali treatment. Apparently, the porosity of the structure was observed to be increased after chemical etching treatment.

For the electrochemical polishing-chemical etching specimens, as Figure 4 shows, asymmetrical in the thickness was not observed even the etching time was 17 min. Figure 4 also shows that even with the electrolytic publishing and chemical etching the Ti plate had a smooth surface. For the results shown in Figure 5, after annealing treatment, the porous structure seemed to become more compact and rigid than before the annealing treatment, and asymmetrical in the thickness was not observed even the etching time was 17 min. A cross-section of the etched Ti plates was observed by using scanning electron microscope (SEM) and the thickness of the Ti plates was measured by SEM. The variation in the thickness of Ti plates under different etching processes as a function of etching time is displayed in Figure 6.

When the etching time was increased from 10 to 17 min, the thicknesses of the polishing-chemical etching, electrochemical polishing-chemical etching, and chemical etching Ti plates decreased from 75.3 μm to 33 μm, from 60.4 μm

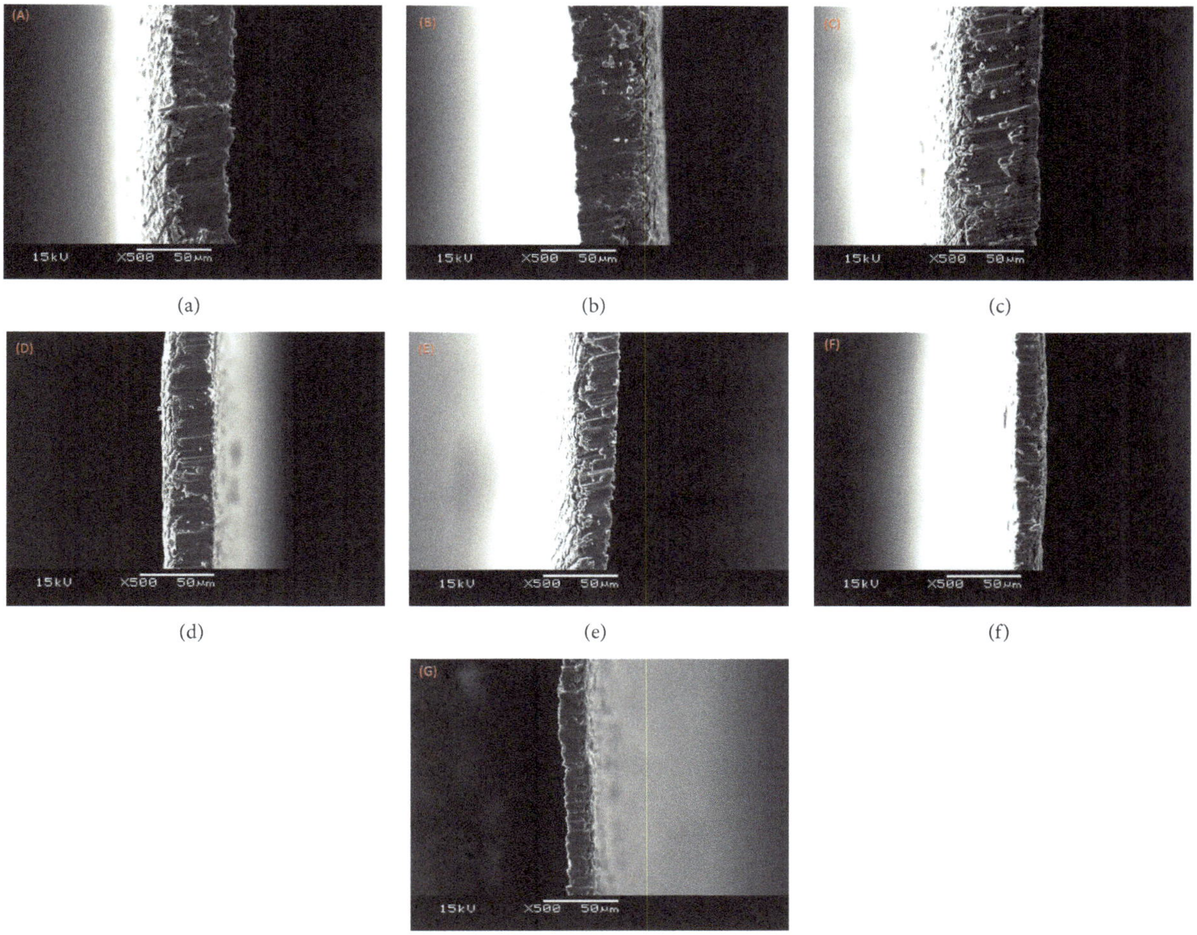

(a) (b) (c) (d) (e) (f) (g)

FIGURE 5: Cross-section morphologies of chemical etching under different times (a) 10 min, (b) 11 min, (c) 12 min, (d) 13 min, (e) 14 min, (f) 15 min, and (g) 16 min.

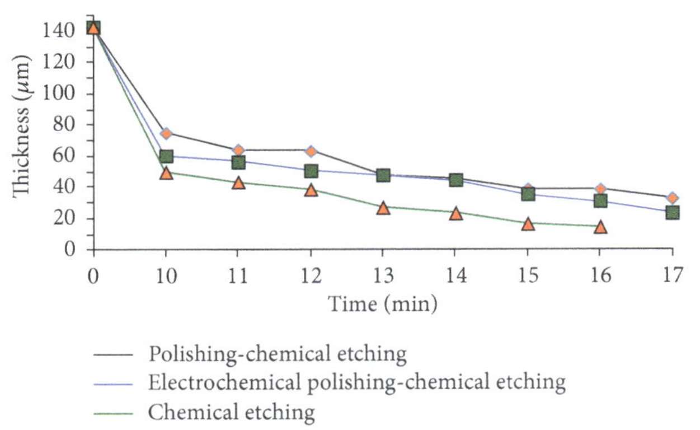

FIGURE 6: Titanium plates thickness curve chart of different methods and etching times.

to 23.6 μm, and from 50 μm to 14.8 μm, respectively. As Figure 6 shows, as the same etching time is used, the thicknesses of the chemical-etching Ti plates are thinner than those of the other two processes-treated Ti plates, and the abrading process is the reason to cause this result. From the results shown in Figures 3–5, the porous structure on the surfaces of Ti plates only existed in polishing-chemical etching Ti plates but not existing in electrochemical polishing-chemical etching and chemical etching Ti plates. A similar porous structure was reported to be produced by the alkali (KOH) treatment of a commercial Ti plate; the porous structure was composed of nanowires with a diameter of less than 30 nm, and the thickness of nanowire layer was estimated to be approximately 500 nm on the Ti substrate [16]. For that, the different etching processes would have large effect on the formation of the TiO_2 nanotube arrays. If the thickness of Ti plates is too thin, the photoelectrodes of the TiO_2 nanotube arrays are not easy to be grown; therefore, the 11 min treated Ti plates are chosen for producing the TiO_2 nanotube arrays.

3.2. Morphology of TiO_2 Nanotube Arrays. From the X-ray diffraction patterns (not shown here), the main crystalline phase of the as-prepared TiO_2 nanotube arrays was amorphous, the diffraction peaks of Ti and rutile TiO_2 phases

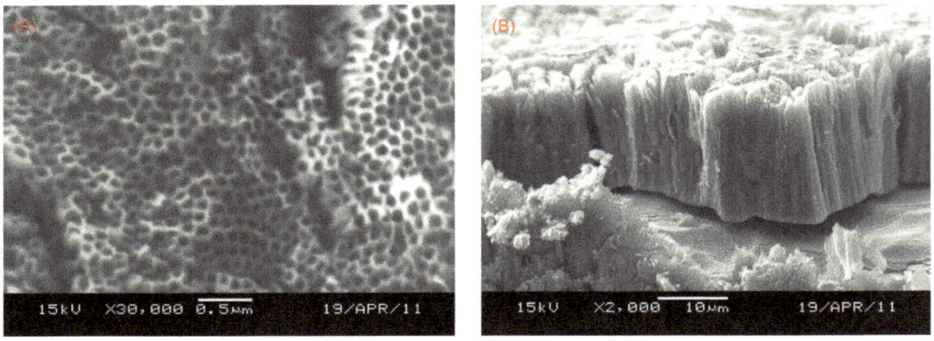

(a) (b)

FIGURE 7: Morphologies of the TiO_2 nanotube arrays grown on polishing-chemical etching Ti plate (a) top view and (b) side view.

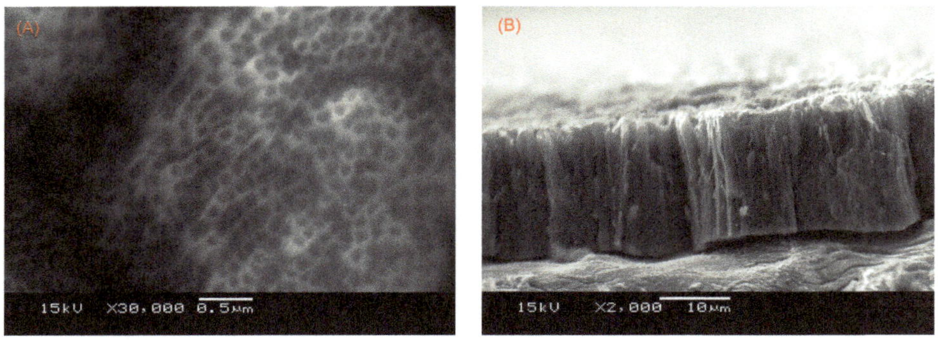

(a) (b)

FIGURE 8: Morphologies of the TiO_2 nanotube arrays grown on electrochemical polishing-chemical etching Ti plates (a) top view and (b) side view.

could be found, and these diffraction peaks were not enhanced as the TiO_2 nanotube array length was changed. It is noted that the anatase TiO_2 peaks are much worse than the Ti and rutile TiO_2 peaks, so it is reasonable to neglect the influences of such trace anatase TiO_2 content in the as-prepared TiO_2 nanotube arrays. As TiO_2 nanotube arrays were annealed at 450°C for 1 h, the diffraction peaks of anatase TiO_2 were clearly observed. Figures 7–9 show the top view and side view of TiO_2 nanotube arrays. The top views in Figures 7(a), 8(a), and 9(a) show that all of the TiO_2 nanotube arrays had the diameters between 100 nm and 140 nm. The surfaces of TiO_2 nanotube arrays with electrochemical polishing-chemical etching and chemical etching processes were smoother than that of TiO_2 nanotube arrays with polishing-chemical etching process. The side views in Figures 7(b), 8(b), and 9(b) show that the lengths of TiO_2 nanotube arrays with the polishing-chemical etching, electrochemical polishing-chemical etching, and chemical etching processes are 19.6 μm, 18.2 μm, and 22.6 μm, respectively. From the side views in Figures 7–9 TiO_2 nanotube arrays with the polishing-chemical etching process are not good as the photoelectrodes of DSSCs because TiO_2 nanotube arrays are easily peeled off. Figures 8(b) and 9(b) show that the TiO_2 nanotube arrays with polishing-chemical etching and chemical etching processes are not easily to be peeled off. Figure 7(b) also shows that the TiO_2 nanotube arrays reveal an unsmooth surface, which will influence the transmission ratio and will be proven in Figure 10.

3.3. Transmission of Photoelectrodes Nanotubes. The transmission ratios of TiO_2 nanotube arrays are shown in Figure 10 as a function of different polishing process and optical wavelength. The measured structure includes the layers of TiO_2 nanotube arrays (photoelectrode), dye, electrolyte, and counter electrode. When the sunlight passes through the photoelectrode layer to dye layer, the transmission ratio has no apparent change. The results in Figure 10 show that the structures with the polishing-chemical etching (70–75%) and electrochemical polishing-chemical etching processes (80–85%) have high transmission ratio in the range of visible light. The structure with the polishing-chemical etching process has the lower transmission ratio which is caused by the being peeled off and unsmooth surfaces. Because TiO_2 nanotube arrays with the chemical etching process have the thicker thickness, the transmission ratio (20–25%) is very low. From those results, the electrochemical polishing-chemical etching and chemical etching processes are the better for formation of TiO_2 nanotube arrays than polishing-chemical etching

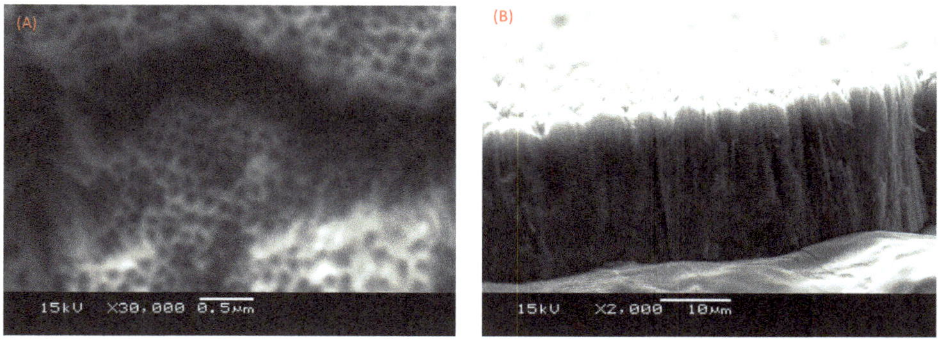

(a) (b)

FIGURE 9: Morphologies of the TiO_2 nanotube arrays grown on chemical etching Ti plates (a) top view and (b) side view.

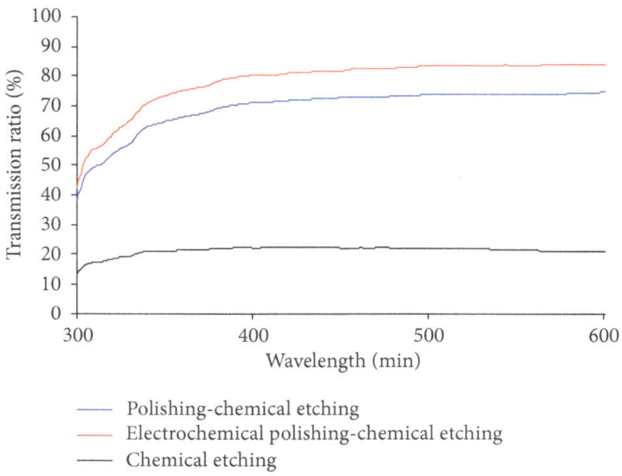

FIGURE 10: Transmission ratio curves of the TiO_2 nanotube arrays grown on etching Ti plates.

process because they have the better structure and morphology factors to get the better characteristics of TiO_2 nanotube arrays as the photoelectrodes of DSSCs.

4. Conclusions

In this study, three different methods were used to etch Ti template from 10 to 17 min, the experimental results were summarized as follows.

(i) The thicknesses of the polishing-chemical etching, electrochemical polishing-chemical etching, and chemical etching Ti plates decreased from 75.3 μm to 33 μm, from 60.4 μm to 23.6 μm, and from 50 μm to 14.8 μm, respectively.

(ii) The TiO_2 nanotube arrays with polishing-chemical etching, electrochemical polishing-chemical etching, and chemical etching processes have the lengths of 19.6 μm, 18.2 μm, and 22.6 μm, respectively, and their diameters were between 100 nm and 140 nm.

(iii) In the range of visible light, the TiO_2 nanotube arrays with polishing-chemical etching (70–75%) and electrochemical polishing-chemical etching (80–85%) processes had the higher transmission ratio than that of the TiO_2 nanotube arrays with chemical etching process (20–25%).

(iv) The electrochemical polishing-chemical etching and chemical etching processes were better for formation of TiO_2 nanotube arrays than the polishing-chemical etching process because they could fabricate TiO_2 nanotube arrays with the better characteristics to be used as the photoelectrodes of DSSCs.

Acknowledgments

The authors acknowledge financial support from NSC 101-2514-S-003-004-, NSC 101-3113-S-262-001-, and NSC 101-2622-E-003-002-CC3.

References

[1] M. Paulose, K. Shankar, S. Yoriya et al., "Anodic growth of highly ordered TiO_2 nanotube arrays to 134 μm in length," *The Journal of Physical Chemistry B*, vol. 110, no. 33, pp. 16179–16184, 2006.

[2] M. Law, L. E. Greene, J. C. Johnson, R. Saykally, and P. D. Yang, "Nanowire dye-sensitized solar cells," *Nature Materials*, vol. 4, no. 6, pp. 455–459, 2005.

[3] G. K. Mor, K. Shankar, M. Paulose, O. K. Varghese, and C. A. Grimes, "Use of highly-ordered TiO_2 nanotube arrays in dye-sensitized solar cells," *Nano Letters*, vol. 6, no. 2, pp. 215–218, 2006.

[4] A. Fujishima and K. Honda, "Electrochemical photolysis of water at a semiconductor electrode," *Nature*, vol. 238, no. 5358, pp. 37–38, 1972.

[5] M. A. Gondal, M. N. Sayeed, and A. Alarfaj, "Activity comparison of Fe_2O_3, NiO, WO_3, TiO_2 semiconductor catalysts in phenol degradation by laser enhanced photo-catalytic process," *Chemical Physics Letters*, vol. 445, no. 4–6, pp. 325–330, 2007.

[6] Y. Li, W. Xie, X. Hu et al., "Comparison of dye photodegradation and its coupling with light-to-electricity conversion over TiO_2 and ZnO," *Langmuir*, vol. 26, no. 1, pp. 591–597, 2010.

[7] N. K. Allam and M. A. El-Sayed, "Photoelectrochemical water oxidation characteristics of anodically fabricated TiO_2 nanotube arrays: structural and optical properties," *Journal of Physical Chemistry C*, vol. 114, pp. 12024–12029, 2010.

[8] M. Gotic, M. Ivanda, A. Sekulic, S. Popovic, A. Turkovic, and K. Furic, "Microstructure of nanosized TiO_2 obtained by sol-gel synthesis," *Materials Letters*, vol. 28, pp. 225–229, 1996.

[9] W. Li, S. I. Shah, C. P. Huang, O. Jung, and C. Ni, "Metallorganic chemical vapor deposition and characterization of TiO_2 nanoparticles," *Materials Science and Engineering B*, vol. 96, no. 3, pp. 247–253, 2002.

[10] J. Choi, R. B. Wehrspohn, J. Lee, and U. Gösele, "Anodization of nanoimprinted titanium: a comparison with formation of porous alumina," *Electrochimica Acta*, vol. 49, no. 16, pp. 2645–2652, 2004.

[11] D. Gong, C. A. Grimes, O. K. Varghese et al., "Titanium oxide nanotube arrays prepared by anodic oxidation," *Journal of Materials Research*, vol. 16, no. 12, pp. 3331–3334, 2001.

[12] V. Zwilling, M. Aucouturier, and E. Darque-Ceretti, "Anodic oxidation of titanium and TA6V alloy in chromic media. An electrochemical approach," *Electrochimica Acta*, vol. 45, no. 6, pp. 921–929, 1999.

[13] R. Beranek, H. Hildebrand, and P. Schmuki, "Self-organized porous titanium oxide prepared in H_2SO_4/HF electrolytes," *Electrochemical and Solid-State Letters*, vol. 6, no. 3, pp. B12–B14, 2003.

[14] S. Sobieszczyk, "Self-organized nanotubular oxide layers on Ti and Ti alloys," *Journal of Materials Science*, vol. 9, no. 20, pp. 25–41, 2009.

[15] K. M. Reddy, S. V. Manorama, and A. R. Reddy, "Bandgap studies on anatase titanium dioxide nanoparticles," *Materials Chemistry and Physics*, vol. 78, no. 1, pp. 239–245, 2003.

[16] J. I. Kim, S. Y. Lee, and J. C. Pyun, "Characterization of photocatalytic activity of TiO_2 nanowire synthesized from Ti-plate by wet corrosion process," *Current Applied Physics*, vol. 9, no. 4, pp. e252–e255, 2009.

Permissions

The contributors of this book come from diverse backgrounds, making this book a truly international effort. This book will bring forth new frontiers with its revolutionizing research information and detailed analysis of the nascent developments around the world.

We would like to thank all the contributing authors for lending their expertise to make the book truly unique. They have played a crucial role in the development of this book. Without their invaluable contributions this book wouldn't have been possible. They have made vital efforts to compile up to date information on the varied aspects of this subject to make this book a valuable addition to the collection of many professionals and students.

This book was conceptualized with the vision of imparting up-to-date information and advanced data in this field. To ensure the same, a matchless editorial board was set up. Every individual on the board went through rigorous rounds of assessment to prove their worth. After which they invested a large part of their time researching and compiling the most relevant data for our readers. Conferences and sessions were held from time to time between the editorial board and the contributing authors to present the data in the most comprehensible form. The editorial team has worked tirelessly to provide valuable and valid information to help people across the globe.

Every chapter published in this book has been scrutinized by our experts. Their significance has been extensively debated. The topics covered herein carry significant findings which will fuel the growth of the discipline. They may even be implemented as practical applications or may be referred to as a beginning point for another development. Chapters in this book were first published by Hindawi Publishing Corporation; hereby published with permission under the Creative Commons Attribution License or equivalent.

The editorial board has been involved in producing this book since its inception. They have spent rigorous hours researching and exploring the diverse topics which have resulted in the successful publishing of this book. They have passed on their knowledge of decades through this book. To expedite this challenging task, the publisher supported the team at every step. A small team of assistant editors was also appointed to further simplify the editing procedure and attain best results for the readers.

Our editorial team has been hand-picked from every corner of the world. Their multi-ethnicity adds dynamic inputs to the discussions which result in innovative outcomes. These outcomes are then further discussed with the researchers and contributors who give their valuable feedback and opinion regarding the same. The feedback is then collaborated with the researches and they are edited in a comprehensive manner to aid the understanding of the subject.

Apart from the editorial board, the designing team has also invested a significant amount of their time in understanding the subject and creating the most relevant covers. They scrutinized every image to scout for the most suitable representation of the subject and create an appropriate cover for the book.

The publishing team has been involved in this book since its early stages. They were actively engaged in every process, be it collecting the data, connecting with the contributors or procuring relevant information. The team has been an ardent support to the editorial, designing and production team. Their endless efforts to recruit the best for this project, has resulted in the accomplishment of this book. They are a veteran in the field of academics and their pool of knowledge is as vast as their experience in printing. Their expertise and guidance has proved useful at every step. Their uncompromising quality standards have made this book an exceptional effort. Their encouragement from time to time has been an inspiration for everyone.

The publisher and the editorial board hope that this book will prove to be a valuable piece of knowledge for researchers, students, practitioners and scholars across the globe.

List of Contributors

Tin-Tai Chow
BEETRU, Division of Building Science and Technology, City University of Hong Kong, Kowloon, Hong Kong

Jie Ji
Department of Thermal Science and Energy Engineering, University of Science and Technology of China, Hefei 230026, China

R. M. Mohamed
Chemistry Department, Faculty of Science, King Abdulaziz University, P.O. Box. 80203, Jeddah 21589, Saudi Arabia
Nanostructured Materials and Nanotechnology Division, Advanced Materials Department, Central Metallurgical Research & Development Institute (CMRDI), P.O. Box 87 Helwan, Cairo 11421, Egypt

M. A. Barakat
Department of Environmental Sciences, Faculty of Meteorology, Environment and Arid Land Agriculture, King Abdulaziz University, Jeddah 21589, Saudi Arabia
Minerals Technology Department, Central Metallurgical Research & Development Institute (CMRDI), Cairo 11421, Egypt

Zhibin Lv, Hongwei Wu, Xin Cai, Yongping Fu, Dan Wang, Zengze Chu and Dechun Zou
Beijing National Laboratory for Molecular Sciences, Key Laboratory of Polymer Chemistry and Physics of Ministry of Education, College of Chemistry and Molecular Engineering, Peking University, Beijing 100871, China

N. Poornima, T. V. Vimalkumar, V. G. Rajeshmon, C. Sudha Kartha and K. P. Vijayakumar
Department of Physics, Cochin University of Science and Technology, Cochin, Kerala 682022, India

Naoyuki Kishimoto
Faculty of Science and Technology, Ryukoku University, 1-5 Yokotani, Setaoe-cho, Otsu 520-2194, Japan

Eri Nakamura
Graduate School of Science and Technology, Ryukoku University, 1-5 Yokotani, Setaoe-cho, Otsu 520-2194, Japan

Wenjie Zhang and Ruyuan Li
School of Environmental and Chemical Engineering, Shenyang Ligong University, Shenyang 110159, China

Hongbo He
Institute of Applied Ecology, The Chinese Academy of Sciences, Shenyang 110016, China

Meng-Hui Wang
Department of Electrical Engineering, National Chin-Yi University of Technology, Taichung 41110, Taiwan

R. Ramaprabha and B. L. Mathur
Department of EEE, SSN College of Engineering, Kalavakkam-603 110, Chennai, India

Yongchun Dong
Division of Textile Chemistry & Ecology, School of Textiles, Tianjin Polytechnic University, 399 Binshui Western Road, Xiqing District, Tianjin 300387, China
State Key Laboratory Breeding Base of Photo catalysis, Fuzhou University, Fuzhou 350002, China

Zhizhong Ding and Bing Li
Division of Textile Chemistry & Ecology, School of Textiles, Tianjin Polytechnic University, 399 Binshui Western Road, Xiqing District, Tianjin 300387, China

Ljubinko Timotijevic and Milos Vujisic
Faculty of Electrical Engineering, University of Belgrade, 11060 Belgrade, Serbia

Irfan Fetahovic
State University of Novi Pazar, 36300 Novi Pazar, Serbia

Djordje Lazarevic
Institute of Nuclear Sciences "Vinca", 11000 Belgrade, Serbia

Chih-Lung Shen and Shih-Hsueh Yang
Department of Electronic Engineering, National Kaohsiung First University of Science and Technology, Nanzih, Kaohsiung 811, Taiwan

Danhong Li, Nengqian Pan, Jianjun Liao, Xiankun Cao and Shiwei Lin
Key Laboratory of Ministry of Education for Advanced Materials in Tropical Island Resources, School of Materials and Chemical Engineering, Hainan University, Haikou 570228, China

Dae-Hwang Yoo and Sung Hong Hahn
Department of Physics, University of Ulsan, Ulsan 680-749, Republic of Korea
Basic Science Research Institute (Energy Harvest-Storage Research Center), University of Ulsan, Ulsan 680-749, Republic of Korea

Woo-Young Kim and Soon-Wook Kim
Department of Physics, University of Ulsan, Ulsan 680-749, Republic of Korea

Eui Jung Kim
Department of Chemical Engineering, University of Ulsan, Ulsan 680-749, Republic of Korea

Chun-Yao Hsu and Peng-Cheng Huang
Department of Mechanical Engineering, Lunghwa University of Science and Technology, Taoyuan 33306, Taiwan

Yu-Yao Chen
Department of Industrial Engineering, Shanghai Dianji University, Shanghai 201306, China

Dong-Cherng Wen
Department of Mechanical Engineering, China University of Science and Technology, Taipei 11581, Taiwan

Jiyong Wei, Baibiao Huang, Peng Wang, Zeyan Wang, Xiaoyan Qin, Xiaoyang Zhang, Xiangyang Jing, Haixia Liu and Jiaoxian Yu
State Key Laboratory of Crystal Materials, Institute of Crystal Materials, Shandong University, Jinan 250100, China

M. Casalino, G. Coppola, M. Iodice, I. Rendina and L. Sirleto
Institute for Microelectronics and Microsystems, National Council of Research, Via P. Castellino 111, 80131 Naples, Italy

Zhang Jinfeng
Department of Physics, Huaibei Normal University, Anhui, Huaibei 235000, China

Yang Yunguang and Liu Wei
Department of Chemistry, Huaibei Normal University, Anhui, Huaibei 235000, China

Zhaoqian Jing and Shiwei Cao
College of Civil Engineering, Nanjing Forestry University, Nanjing 210037, China

Nguyen Tam Nguyen Truong, Thao Phuong Ngoc Nguyen and Chinho Park
School of Chemical Engineering, Yeungnam University, 280 Daehak-Ro, Gyeongsan, Gyeongbuk 712-749, Republic of Korea

Jin-Seok Lee, Bo-Yun Jang and Young-Soo Ahn
Korea Institute of Energy Research, Daejeon 305-343, Republic of Korea

Jiang Liu, Da-Ming Zhuang, Ming-Jie Cao, Chen-YueWang, Min Xie and Xiao-Long Li
Department of Mechanical Engineering, Tsinghua University, Beijing 100084, China

Qingyun Chen, Cheng Suo, Shu Zhang and Yunhai Wang
State Key Laboratory of Multiphase Flow in Power Engineering, School of Energy and Power Engineering, Xian Jiaotong University, Xian 710049, China

Matic Herman, Marko Jankovec and Marko Topic
Laboratory of Photovoltaics and Optoelectronics, Faculty of Electrical Engineering, University of Ljubljana, Trzaska Cesta 25, 1000 Ljubljana, Slovenia

Chin-Guo Kuo
Department of Industrial Education, National Taiwan Normal University, Taipei 10610, Taiwan

Mu-Jung Kao
Department of Vehicle Engineering, National Taipei University of Technology, Taipei 10608, Taiwan

Cheng-Fu Yang
Department of Chemical and Materials Engineering, National University of Kaohsiung, Kaohsiung 81148, Taiwan

Wen-Pin Weng
Department of Chemical and Materials Engineering, Lunghwa University of Science and Technology, Taoyuan 33306, Taiwan

Chi-Cheng Chang
Department of Information Management, Lunghwa University of Science and Technology, Taoyuan 33306, Taiwan

Lih-Ren Hwang
Department of Automation and Control Engineering, Chung Chou University of Science and Technology, Yuanlin 51003, Taiwan

Jian-Lan Nil
Department of Mechatronic Technology, National Taiwan Normal University, Taipei 10610, Taiwan